T0250201

ADVANCES IN NANOTECHNOLOGY AND THE ENVIRONMENTAL SCIENCES

Applications, Innovations, and Visions for the Future

AAP Research Notes on Nanoscience & Nanotechnology

ADVANCES IN NANOTECHNOLOGY AND THE ENVIRONMENTAL SCIENCES

Applications, Innovations, and Visions for the Future

Edited by

Alexander V. Vakhrushev, DSc
Suresh C. Ameta, PhD
Heru Susanto, PhD
A. K. Haghi, PhD

Apple Academic Press Inc.
3333 Mistwell Crescent
Oakville, ON L6L 0A2 Canada

Apple Academic Press Inc.
1265 Goldenrod Circle NE
Palm Bay, Florida 32905 USA

© 2020 by Apple Academic Press, Inc.

First issued in paperback 2021

Exclusive worldwide distribution by CRC Press, a member of Taylor & Francis Group

No claim to original U.S. Government works

ISBN 13: 978-1-77463-446-2 (pbk)
ISBN 13: 978-1-77188-754-0 (hbk)

Library and Archives Canada Cataloguing in Publication

Title: Advances in nanotechnology and the environmental sciences : applications, innovations, and visions for the future / edited by Alexander V. Vakhrushev, DSc, Suresh C. Ameta, PhD, Heru Susanto, PhD, A.K. Haghi, PhD.

Names: Vakhrushev, Alexander V., editor. | Ameta, Suresh C., editor. | Susanto, Heru, 1977- editor. | Haghi, A. K., editor.

Names: AAP research notes on nanoscience & nanotechnology.

Description: Series statement: AAP research notes in nanoscience & nanotechnology | Includes bibliographical references and index.

Identifiers: Canadiana (print) 20190088214 | Canadiana (ebook) 20190088222 | ISBN 9781771887540 (hbk.) | ISBN 9780429425837 (ebook)

Subjects: LCSH: Nanotechnology—Environmental aspects. | LCSH: Green technology.

Classification: LCC T174.7 .A38 2019 | DDC 620/.5—dc23

Library of Congress Cataloging-in-Publication Data

Names: Vakhrushev, Alexander V., editor.

Title: Advances in nanotechnology and the environmental sciences : applications, innovations, and visions for the future / editors, Alexander V. Vakhrushev, DSc, Suresh C. Ameta, PhD, Heru Susanto, PhD, A. K. Haghi, PhD.

Description: Palm Bay, Florida, USA ; Oakville, ON, Canada : Apple Academic Press Toronto, 2019. | Includes bibliographical references and index.

Identifiers: LCCN 2019014398 (print) | LCCN 2019018591 (ebook) | ISBN 9780429425837 (ebook) | ISBN 9781771887540 (hardcover : alk. paper)

Subjects: LCSH: Environmental engineering--Materials. | Nanostructured materials. | Nanotechnology--Environmental aspects. | Green chemistry.

Classification: LCC TD192 (ebook) | LCC TD192 .A38 2019 (print) | DDC 628--dc23

LC record available at https://lccn.loc.gov/2019014398

Apple Academic Press also publishes its books in a variety of electronic formats. Some content that appears in print may not be available in electronic format. For information about Apple Academic Press products, visit our website at **www.appleacademicpress.com** and the CRC Press website at **www.crcpress.com**

ABOUT THE EDITORS

Alexander V. Vakhrushev, DSc

Alexander V. Vakhrushev, DSc, is a Professor at the M. T. Kalashnikov Izhevsk State Technical University in Izhevsk, Russia, where he teaches theory, calculating, and design of nano- and microsystems. He is also the Chief Researcher of the Department of Information-Measuring Systems of the Institute of Mechanics of the Ural Branch of the Russian Academy of Sciences and Head of the Department of Nanotechnology and Microsystems of Kalashnikov Izhevsk State Technical University. He is a corresponding member of the Russian Engineering Academy. He has over 400 publications to his name, including monographs, articles, reports, reviews, and patents. He has received several awards, including an academician A. F. Sidorov Prize from the Ural Division of the Russian Academy of Sciences for significant contribution to the creation of the theoretical fundamentals of physical processes taking place in multilevel nanosystems and honorable scientist of the Udmurt Republic. He is currently a member of editorial board of several journals, including *Computational Continuum Mechanics, Chemical Physics and Mesoscopia,* and *Nanobuild.* His research interests include multiscale mathematical modeling of physical–chemical processes into the nanohetero systems at nano-, micro-, and macro-levels; static and dynamic interaction of nanoelements; and basic laws relating the structure and macro characteristics of nano-hetero structures.

Suresh C. Ameta, PhD

Suresh C. Ameta, PhD, is currently Dean, Faculty of Science at PAHER University, Udaipur, India. He has served as Professor and Head of the Department of Chemistry at North Gujarat University Patan and at M. L. Sukhadia University, Udaipur, and as Head of the Department of Polymer Science. He also served as Dean of Postgraduate Studies. Prof. Ameta has held the position of President of the Indian Chemical Society, Kolkata and is now a life-long Vice President. He was awarded a number of prestigious awards during his career such as national prizes twice for writing chemistry books in Hindi. He also received the Prof. M. N. Desai Award (2004), the

Prof. W. U. Malik Award (2008), the National Teacher Award (2011), the Prof. G. V. Bakore Award (2007), a Life-time Achievement Award by the Indian Chemical Society (2011) as well as the Indian Council of Chemist (2015), etc. He has successfully guided 81 PhD students. Having more than 350 research publications to his credit in journals of national and international repute, he is also the author of many undergraduate- and postgraduate-level books. He has published three books with Apple Academic Press: *Chemical Applications of Symmetry and Group Theory*, *Microwave-Assisted Organic Synthesis*, and *Green Chemistry: Fundamentals and Applications* and two with Taylor and Francis: *Solar Energy Conversion and Storage* and *Photocatalysis*. He has also written chapters in books published by several other international publishers. Prof. Ameta has delivered lectures and chaired sessions at national conferences and is a reviewer of number of international journals. In addition, he has completed five major research projects from different funding agencies, such as DST, UGC, CSIR, and Ministry of Energy, Govt. of India.

Heru Susanto, PhD

Heru Susanto, PhD, is currently Head of the Information Department and researcher at the Indonesian Institute of Sciences, Computational Science & IT Governance Research Group. At present, he is also an Honorary Professor and Visiting Scholar at the Department of Information Management, College of Management, Tunghai University, Taichung, Taiwan. Dr. Susanto has worked as an IT professional in several roles, including Web Division Head of IT Strategic Management at Indomobil Group Corporation and Prince Muqrin Chair for Information Security Technologies at King Saud University. His research interests are in the areas of information security, IT governance, computational sciences, business process re-engineering, and e-marketing.

A. K. Haghi, PhD

A. K. Haghi, PhD, is the author and editor of 165 books, as well as over 1000 published papers in various journals and conference proceedings. Dr. Haghi has received several grants, consulted for a number of major corporations, and is a frequent speaker to national and international audiences. Since 1983, he served as professor at several universities. He is currently Editor-in-Chief of the *International Journal of Chemoinformatics and Chemical Engineering* and *Polymers Research Journal* and on the editorial boards of many International journals. He is also a member of the Canadian Research and Development Center of Sciences and Cultures (CRDCSC), Montreal, Quebec, Canada. He holds a BSc in urban and environmental engineering from the University of North Carolina (USA), an MSc in mechanical engineering from North Carolina A&T State University (USA), a DEA in applied mechanics, acoustics and materials from the Université de Technologie de Compiègne (France), and a PhD in engineering sciences from Université de Franche-Comté (France).

ABOUT THE AAP RESEARCH NOTES ON NANOSCIENCE & NANOTECHNOLOGY BOOK SERIES

AAP Research Notes on Nanoscience & Nanotechnology reports on research development in the field of nanoscience and nanotechnology for academic institutes and industrial sectors interested in advanced research.

EDITOR-IN-CHIEF:

A. K. Haghi, PhD
Associate Member of University of Ottawa, Canada;
Member of Canadian Research and Development Center of Sciences and Cultures
email: akhaghi@yahoo.com

EDITORIAL BOARD:

Georges Geuskens, PhD
Professor Emeritus, Department of Chemistry and Polymers, Universite de Libre de Brussel, Belgium

Vladimir I. Kodolov, DSc
Professor and Head, Department of Chemistry and Chemical Technology, M. I. Kalashnikov Izhevsk State Technical University, Izhevsk, Russia

Victor Manuel de Matos Lobo, PhD
Professor, Coimbra University, Coimbra, Portugal

Richard A. Pethrick, PhD, DSc
Research Professor and Professor Emeritus, Department of Pure and Applied Chemistry, University of Strathclyde, Glasgow, Scotland, UK

Mathew Sebastian, MD
Senior Consultant Surgeon, Elisabethinen Hospital, Klagenfurt, Austria; Austrian Association for Ayurveda

Charles Wilkie, PhD
Professor, Polymer and Organic Chemistry, Marquette University, Milwaukee, Wisconsin, USA

Books in the AAP Research Notes on Nanoscience & Nanotechnology Book Series:

Nanostructure, Nanosystems and Nanostructured Materials:
Theory, Production, and Development
Editors: P. M. Sivakumar, PhD, Vladimir I. Kodolov, DSc,
Gennady E. Zaikov, DSc, A. K. Haghi, PhD

Nanostructures, Nanomaterials, and Nanotechnologies to Nanoindustry
Editors: Vladimir I. Kodolov, DSc, Gennady E. Zaikov, DSc, and A. K. Haghi, PhD

Foundations of Nanotechnology:
Volume 1: Pore Size in Carbon-Based Nano-Adsorbents
A. K. Haghi, PhD, Sabu Thomas, PhD, and Moein MehdiPour MirMahaleh

Foundations of Nanotechnology:
Volume 2: Nanoelements Formation and Interaction
Sabu Thomas, PhD, Saeedeh Rafiei, Shima Maghsoodlou, and Arezo Afzali

Foundations of Nanotechnology:
Volume 3: Mechanics of Carbon Nanotubes
Saeedeh Rafiei

Engineered Carbon Nanotubes and Nanofibrous Material:
Integrating Theory and Technique
Editors: A. K. Haghi, PhD, Praveen K. M., and Sabu Thomas, PhD

Carbon Nanotubes and Nanoparticles: Current and Potential Applications
Editors: Alexander V. Vakhrushev, DSc, V. I. Kodolov, DSc, A. K. Haghi, PhD,
and Suresh C. Ameta, PhD

Advances in Nanotechnology and the Environmental Sciences: Applications,
Innovations, and Visions for the Future
Editors: Alexander V. Vakhrushev, DSc, Suresh C. Ameta, PhD,
Heru Susanto, PhD, and A. K. Haghi, PhD

Chemical Nanoscience and Nanotechnology: New Materials
and Modern Techniques
Editors: Francisco Torrens, PhD, A. K. Haghi, PhD, and Tanmoy Chakraborty, PhD

CONTENTS

Contributors...*xiii*

Abbreviations ....*xv*

Preface ....*xxi*

PART I: Nanotechnology for Environmental Protection....................1

1. **Global Water Crisis, Nanofiltration, the Vast Innovations, and the Vision for the Future**...3
   Sukanchan Palit

2. **Nanofiltration, Nanotechnology, and Chemical Engineering Science: The Scientific Truth and a Vision for the Future** 27
   Sukanchan Palit

3. **Water Purification and Nanotechnology: A Critical Overview and a Vision for the Future** 53
   Sukanchan Palit

4. **Green Nanotechnology: An Approach Towards Environment Safety** 85
   Francisco Torrens and Gloria Castellano

5. **Green Nanotechnology, Green Nanomaterials, and Green Chemistry: A Far-Reaching Review and a Vision for the Future**........93
   Sukanchan Palit

PART II: Advanced Technologies 117

6. **Quantum Dots and Their Applications** 119
   Rakshit Ameta, Kanchan Kumari Jat, Jayesh Bhatt, and Suresh C. Ameta

7. **Polymer Nanocomposites, a Smart Material: Synthesis, Preparation, and Properties**..151
   Sonia Khanna

8. **Theoretical Bases of Simulation of Formation Processes of Nanostructures in a Gas Medium and Results of Numerical Studies of Nanosystems**..165
   A. Yu. Fedotov

9. **Carbon Nanotubes as Chemical Sensors and Biosensors:**
    A Review 203
    Rakshit Ameta, Kanchan Kumari Jat, Jayesh Bhatt, Avinash Rai,
    Tarachand Nargawe, Dipti Soni, and Suresh C. Ameta

10. **Applications of Micellear Phase of Pluronics and Tetronics as**
    Nanoreactors in the Synthesis of Nobel Metal Nanoparticles 241
    Rajpreet Kaur, Navdeep Kaur, Divya Mandial, Lavanya Tandon,
    and Poonam Khullar

11. **Methods for Calculating the Thermoelectric Characterizations**
    of Nanomaterials 281
    A. V. Severyukhin, O. Yu. Severyukhina, and A. V. Vakhrushev

Index.. *297*

CONTRIBUTORS

Rakshit Ameta
Faculty of Science, Department of Chemistry PAHER University, J. R. N. Rajasthan Vidyapeeth (Deemed-to-be University), Udaipur–313003, Rajasthan, India

Suresh C. Ameta
Department of Chemistry PAHER University, Udaipur–313003, Rajasthan, India

Jayesh Bhatt
Department of Chemistry PAHER University, Udaipur–313003, Rajasthan, India

Gloria Castellano
Departamento de Ciencias Experimentales y Matemáticas, Facultad de Veterinaria y Ciencias Experimentales, Universidad Católica de Valencia San Vicente Mártir, Guillem de Castro-94, E-46001 València, Spain

A. Yu. Fedotov
Udmurt Federal Research Center of the Ural Branch of the Russian Academy of Sciences, Institute of Mechanics, Izhevsk, Russia, E-mail: alezfed@gmail.com

Kanchan Kumari Jat
Department of Chemistry, Mohan Lal Sukhadia University, Udaipur-313002, Rajasthan, India

Navdeep Kaur
Department of Chemistry, B.B.K. D.A.V. College for Women, Amritsar–143005, Punjab, India

Rajpreet Kaur
Department of Chemistry, B.B.K. D.A.V. College for Women, Amritsar–143005, Punjab, India

Sonia Khanna
Department of Chemistry, School of Basic Sciences and Research, Sharda University, Greater Noida, India, E-mail: sonia.khanna@sharda.ac.in

Poonam Khullar
Department of Chemistry, B.B.K. D.A.V. College for Women, Amritsar–143005, Punjab, India, E-mail: virgo16sep2005@gmail.com

Divya Mandial
Department of Chemistry, B.B.K. D.A.V. College for Women, Amritsar–143005, Punjab, India

Tarachand Nargawe
Department of Chemistry, PAHER University, Udaipur–313003, Rajasthan, India

Sukanchan Palit
Assistant Professor (Senior Scale), Department of Chemical Engineering, University of Petroleum and Energy Studies, Energy Acres, Post-Office- Bidholi via Premnagar, Dehradun-248007, Uttarakhand / 43, Judges Bagan, Post-Office-Haridevpur, Kolkata–700082, India, Tel.: 0091-8958728093, E-mail: sukanchan68@gmail.com, sukanchan92@gmail.com

Avinash Rai
Department of Chemistry, PAHER University, Udaipur–313003, Rajasthan, India

A.V. Severyukhin
Udmurt Federal Research Center of the Ural Branch of the Russian Academy of Sciences, Institute of Mechanics, Izhevsk, Russia

O. Yu. Severyukhina
Udmurt Federal Research Center of the Ural Branch of the Russian Academy of Sciences, Institute of Mechanics, Izhevsk, Russia

Dipti Soni
Department of Chemistry, PAHER University, Udaipur–313003, Rajasthan, India

Lavanya Tandon
Department of Chemistry, B.B.K. D.A.V. College for Women, Amritsar–143005, Punjab, India

Francisco Torrens
Institut Universitari de Ciència Molecular, Universitat de València, Edifici d'Instituts de Paterna, P. O. Box 22085, E-46071 València, Spain, E-mail: torrens@uv.es

A. V. Vakhrushev
Udmurt Federal Research Center of the Ural Branch of the Russian Academy of Sciences, Institute of Mechanics, Izhevsk, Russia, E-mail: vakhrushev-a@yandex.ru

ABBREVIATIONS

AA	ascorbic acid
AChE	acetylcholinesterase
ACOP	acetaminophen
ACV	antiviral drug acyclovir
ADMET	absorption, distribution, metabolism, excretion, and toxicity
AIBN	azobisisobutyronitrile
AIS QDs	AgInS2 quantum dots
AOPs	advanced oxidation processes
APDs	avalanche photodiodes
API	active pharmaceutical ingredient
ARFF	attribute relation file format
Au NPs	gold nanoparticles
BCPs	block copolymers
BPLC	blue phase liquid crystal
BSA	bovine serum albumin
CADD	computer-aided drug design
CADD	computer analysis drug design
CASD	computer-assisted synthesis design
CEF	cell efficiency factor
CFE	carbon film electrodes
CHI	chitosan
CIP	diethyl phthalate
CNOT	controlled-NOT
CNT	carbon nanotube
CNTA	CNT arrays
CNT-ECIS	carbon nanotube-based electrical cell impedance sensing biosensor
CNT-FET	carbon nanotube field effect transistor
CNT-TF	carbon nanotube thin film
CQD	colloidal quantum dot
CQD/Ag	carbon quantum dot/silver
CQDSCs	colloidal quantum dot solar cells

CV	coefficient of variation
CVD	chemical vapor deposition
CWA	chemical warfare agents
DA	domoic acid
DA	dopamine
DCM	dichloromethane
DDS	drug delivery system
DEP	ciprofloxacin
DET	direct electron transfer
DFT	density functional theory
DLBCL	diffuse substantial B-cell lymphoma
DMMP	dimethoxy methylphosphonate
DNA	deoxyribonucleic acid
DNT	2,4-dinitrotoluene
DPV	differential pulse voltammetry
DTP	developmental therapeutics program
DWNTs	double-walled CNTs
ECB	electrochemical bio-sensor
ECL	electrochemiluminescence
ECL	electrogenerated chemiluminescence
EGDMA	ethylene glycol dimethacrylate
EGF	epidermal growth factor
EGFR	EGF receptors
ELISA	linked immunosorbent assay
EPU	epoxy-polyurethane
ETL	electron transport layer
FET	field effect transistor
FFS	fringing field switching
FITC	fluorescein isothiocyanate
FLIM-FRET	fluorescence lifetime imaging microscopy
FOX	ferrous oxidation-xylenol orange
FRET	Förster resonance energy transfer
GBHC	Gaussian Bayesian various leveled grouping
GC	glassy carbon
GCE	glassy carbon electrode
GMA	glycidyl methacrylate
GNT	green nanotechnology
GO	graphene oxide

GOD	glucose oxidase
GQDPEIs	graphene quantum dots-polyethyleneimines
GQDs	graphene quantum dots
HLB	hydrophilic-lipophilic balance
HNSCC	head and neck squamous carcinoma cells
HRP	horseradish peroxidase
HTS	high throughput screening
IBM	International Business Machines
ICs	integrated circuits
ICT	information communication technology
IDE	interdigitated electrode
IL–6	interleukin–6
ISEs	ion selective electrodes
ISM	ion selective membrane
IT	information system
LB	Langmuir–Blodgett
LCDs	liquid crystal displays
LEDs	light emitting diodes
LMICs	low- and middle- income countries
LOD	limit of detection
LPE	liquid-phase exfoliation
LSCs	luminescent solar concentrators
MAA	methacrylic acid
MDR	multidrug resistant
MIAS	mammography image analysis society
MICP	molecularly imprinted conducting polymer
MIP	molecularly imprinted polymer
MMDB	molecular modeling databases
MR	molar ratio
MSP	metallo-supramolecular polymer
MVA	multi-domain vertical alignment
MWCNT-PAH/SPE	multiwall carbon nanotube-polyallylamine modified screen-printed electrode
MWCNTs	multiwalled carbon nanotubes
MZO-NC	MgZnO nanocrystal
NCI	National Cancer Institute
NCs	nanocomposites
NCs	nanocrystals

NFCB	NiOOH/FeOOH/CQD/BiVO4
NIR	near-infrared
NMPEA	n-methylphenethylamine
NMR	nuclear magnetic resonance
NMs	nanomaterials
NNI	National Nanotechnology Initiative
NOM	natural organic matter
NP	nanoparticle
NW	nanowire
OLEDs	organic light-emitting diodes
OP	organophosphorus
OPH	organophosphorus hydrolase
PANI	polyaniline
PAY/MWCNT	polymer–MWCNTs nanocomposite film
PBASE	1-pyrenebutanoic acid succinimidyl ester
PC	polycarbonate
PCA	principal component analysis
PCE	photoelectric conversion efficiency
PCE	power conversion efficiency
PCL	polycaprolactone
PDDA	poly(diallyldimethylammonium chloride)
PEDOT	poly(3,4-ethylene dioxythiophene)
PEI	poly(ethyleneimine)
PEO	polyethylene oxide
PET	polyethylene terephthalate
PF–4	platelet factor–4
PL	photoluminescence
PMMA	polymethylmethacrylate
PP	polypropylene
PPO	polypropylene oxide
PS	polystyrene
PSA	prostate-specific antigen
PSMA	prostate specific membrane antigen
Pt/f-MWCNT	Pt-decorated functionalized multi-walled carbon nanotubes
PV	photovoltaic
PVC	poly(vinyl chloride)
PVP	poly(vinylpyrrolidone)

QCM	quartz crystal microbalance
QD-LEDs	quantum dot-based light-emitting diodes
QDs	quantum dots
QDs/GR	quantum dots/graphene
QDSSCs	quantum dot-sensitized solar cells
QN	quercetin
QSAR	quantitative structured activity relationship
QY	quantum yield
REs	reference electrodes
RGO	reduced graphene oxide
RH	relative humidity
SAN	styrene–acrylonitrile copolymer
SDS	sodium dodecyl sulfate
SILAR	successive ionic layer adsorption and reaction
SMILES	simplified molecular input line entry system
SOAP	simple object access protocol
SOF	silica optical fiber
SPE	screen-printed electrode
SPES	sulfonated poly(ether sulfone)
SPR	surface plasmon resonance
SRL	strain reducing layer
SSCs	silicon solar cells
SWASV	square wave anodic stripping voltammetry
SWCNTs	single-walled carbon nanotubes
TEF	time efficiency factor
TFT	thin-film transistor
TNP	1,3,6-trinitropyrene
UA	uric acid
UV	ultraviolet
VOCs	volatile chemical compounds
XPS	x-ray photoelectron spectroscopy
XRD	x-ray diffraction

PREFACE

Nanoscience and nanotechnology are at the forefront of modern research. This book is devoted to nanochemistry, a branch of the actively developing interdisciplinary field of nanoscience. It presents some of the latest achievements in nanochemistry and nanomaterials from the leading researchers, offering readers a new and unique approach to the subject. Nanochemistry and nanomaterials are a fast developing field of research, and this book serves as a reference work for both researchers and graduate students.

- This book combines the elements of review and research book, which allows for information on current and prospective directions in nanochemistry.
- This book, on the other hand, presents unified research methodologies to the multidisciplinary world of nanochemistry based on a single paradigm of concepts, terminology, and ideas.
- This book is aimed at all those who are interested in understanding the current research going on in nanomaterial science from the perspectives of engineering applications, including physical chemists, chemical engineers, and material scientists.
- This book provides detailed information on the use of nanomaterials and methodologies in chemistry with possible applications and regulatory barriers to commercialization.
- It presents an interdisciplinary approach that brings together material science, chemistry, and nanotechnology.
- It helps those undertaking research in chemistry and material science to gain a cogent understanding of how nanotechnology is leading to the emergence of new engineering technologies.

PART I

Nanotechnology for Environmental Protection

CHAPTER 1

GLOBAL WATER CRISIS, NANOFILTRATION, THE VAST INNOVATIONS, AND THE VISION FOR THE FUTURE

SUKANCHAN PALIT[1,2]

[1]*Assistant Professor (Senior Scale), Department of Chemical Engineering, University of Petroleum and Energy Studies, Energy Acres, Bidholi via Premnagar (P.O.), Dehradun–248007, India*

[2]*43, Judges Bagan, Haridevpur (P.O.), Kolkata–700082, India, Tel.: 0091-8958728093, E-mail: sukanchan68@gmail.com, sukanchan92@gmail.com*

ABSTRACT

Human mankind and human scientific endeavor are today in the path immense scientific regeneration and vast scientific vision. In a similar manner, environmental engineering science and water science and technology are surpassing vast and versatile scientific frontiers. Today, global water status stands in the midst of deep scientific comprehension and vast scientific imagination. In this chapter, the author deeply comprehends the scientific progress in water technology and gives a vast glimpse in the recent innovations and the visionary technologies in water science and water purification. With assuring cadence and deep introspection, the author delves deep into the next-generation scientific innovations and scientific vision with the prime objective of furtherance of science and engineering. Global water crisis today is in the critical juncture of deep scientific catastrophe and vast scientific vision. The world today witnesses the burning issues of climate change, loss of ecological biodiversity, environmental disasters and depletion of fossil fuel resources. In such a situation, engineering science and technology need to be envisioned and readdressed with the passage of scientific history and visionary timeframe. In this chapter, the author pointedly focuses on traditional and non-traditional environmental engineering techniques such as advanced oxidation processes (AOPs) and membrane separation processes. Scientific innovation, deep scientific instinct, and the vast scientific needs of the human society are the torchbearers towards a newer era in the field of environmental engineering science and water technology. Global water shortage and research and development initiatives in environmental engineering science are today leading a long and visionary way in the true emancipation of global environmental sustainability. This chapter gives a wide glimpse on the vast scientific potential and the scientific success of environmental engineering tools mainly non-traditional techniques with a sole objective of furtherance of science and engineering. Nanofiltration is an area of scientific endeavor which needs to be re-envisioned and reframed as global water crisis surges forward to a deep catastrophe. In this well-researched chapter, the author also elucidates on the need of membrane separation processes such as nanofiltration and reverse osmosis in mitigating the global water shortage. The vast vision and the challenge of membrane science are elucidated in minute details.

1.1   INTRODUCTION

Science and technology are witnessing drastic and dramatic challenges in our present day human civilization. Environmental engineering science is moving from one visionary paradigm towards another. Environmental catastrophes, global climate change, loss of ecological biodiversity, and global water shortage has urged scientists and engineers to move towards the newer vision and newer innovation. The vision and the challenge of technology and engineering science are immense and path-breaking today. In this chapter, the author pointedly focuses on the immense importance of innovations and new technologies in water science and technology and the needs of water purification, drinking water treatment, and industrial wastewater treatment. Technological advancements, scientific vision and the needs of human society will all lead a long and effective way in the true emancipation of environmental engineering science today. This chapter is an eye-opener to the world of challenges in environmental engineering science and water purification. Modern science today stands in the midst of deep scientific introspection and vast technological vision. The needs of modern science are the visionary world of the provision of basic human needs such as food, water, shelter, and electricity. Environmental engineering paradigm is in the midst of deep vision and scientific fortitude. As human civilization moves forward towards a newer visionary era, environmental engineering science and water technology assume immense importance. In this chapter, the author pointedly and instinctively touches upon the grave concerns of global water shortage and global water hiatus and the immediate need for innovations and newer technologies. This chapter will open new avenues and new windows of vision and fortitude in the field of water science and technology in decades to come. The vision of science in global water research and development are immense and groundbreaking. Technology and engineering science have few answers to the scientific intricacies of arsenic and heavy metal groundwater remediation in developing and developed economies. Here needs the importance of environmental engineering tools such as membrane science and advanced oxidation processes (AOPs).

1.2   THE AIM AND THE OBJECTIVE OF THIS STUDY

Science and technology are huge colossus today with a vast vision of its own. Environmental engineering science and environmental protection in a similar manner stand in between vast scientific vision and scientific ingenuity. The aim and objective of this study is to articulate the visionary world of innovations in water science and technology and raise the awareness of global water catastrophe. In this chapter, the author deeply elucidates the scientific success, the scientific potential and the vast scientific imagination in the path towards innovations in environmental protection and water purification. The author in this chapter deeply comprehends the need of technology and engineering science in tackling global water shortage and global water hiatus. The author in this chapter pointedly focuses on human scientific fortitude and deep human scientific far-sightedness in grappling the worldwide problem of drinking water shortage, water purification, and industrial wastewater treatment. Engineering science and technology are today retrogressive with the passage of scientific history and visionary timeframe. Environmental engineering is in a state of immense scientific crisis and scientific acuity. This chapter opens up new thoughts and new ideas in the field of chemical process engineering, environmental engineering, biological sciences and applied geology in the tackling with the enigma of arsenic and heavy metal groundwater remediation. Human mankind's immense scientific divination and prowess, the vast technological profundity and the immediate need of pure drinking water will all lead a long and visionary way in the true emancipation of water purification and drinking water treatment today. Industrial wastewater treatment is another cornerstone of this well-researched chapter. Industrial wastewater treatment is a burning issue of human civilization and human scientific endeavor today. Man's immense scientific journey, mankind's vast scientific vision and the research forays in environmental engineering are the veritable torchbearers towards a newer era in environmental protection. The purpose and the aim of this study are to proclaim the importance of water purification technologies in the sole objective of scientific advancement. Some of the other areas covered in this research treatise are nanofiltration and reverse osmosis and its immense importance in solving global water problems.

1.3 GLOBAL WATER SHORTAGE AND THE SCIENTIFIC DOCTRINE OF ENVIRONMENTAL SCIENCE

Global water shortage and global water hiatus are the scientific blunders of human civilization today. South Asia today stands in the midst of deep crisis and immense scientific fortitude. The scientific doctrine of environmental engineering science today needs to be envisioned and re-organized. Arsenic and heavy metal groundwater contamination is a burning issue in present-day human civilization. Human mankind stands in the midst of deep scientific devastation and in the similar manner vast scientific vision. Environmental engineering science thus needs to be streamlined with newer vision, and newer innovations as science and technology surge forward. Global water shortage and groundwater heavy metal contamination are the burning issues and are a veritable scientific enigma. Technology, engineering, and science have practically no answers to the marauding and monstrous issue of arsenic and heavy metal groundwater remediation. Scientific and academic rigor are today at the helm in every research and development initiatives in water purification and water technology. The vision and challenge of science are immensely retrogressive at this crucial juncture of scientific history and time.

1.4 THE VISION OF ENVIRONMENTAL ENGINEERING TECHNIQUES

The vision of environmental engineering tools is today's surpassing vast and versatile scientific frontiers. Human scientific imagination and scientific candor are in a state of immense devastation. Environmental engineering and chemical process engineering tools such as membrane science and AOPs are the utmost need of the hour as water purification, and drinking water treatment stands in the midst of scientific upheaval. This vision of water purification, drinking water treatment, and industrial wastewater treatment needs to be readdressed and revamped if global water crisis needs to be eliminated. The author in this chapter reiterates the success of the application of membrane science in water purification with the sole objective of furtherance of science and engineering. Desalination and water treatment are today linked by an unsevered umbilical cord. Human scientific progress in membrane science and water technology needs to be equally addressed and envisioned as science and engineering moves from one visionary paradigm to another.

1.5 THE STATUS OF GLOBAL WATER CRISIS

The status of the global water crisis is grave and needs immediate attention. South Asia particularly Bangladesh and India are in the threshold of major environmental crisis. Arsenic groundwater contamination is challenging the scientific firmament of Bangladesh and India. Drinking water contamination is at its helm in South Asia today. Water purification paradigm needs to be envisioned and restructured with the passage of scientific history and time. The state of global water crisis today stands in the midst of deep hope and scientific optimism. Scientific grit, vast scientific vision and the world of scientific challenges will all lead a long and effective way in the true emancipation of water science and water technology. Newer technology, newer innovation, and newer vision are the pivots of scientific endeavor in water purification today. The author pointedly focuses on human scientific progress in membrane science, AOPs, and non-traditional environmental engineering tools. Technology and engineering science are highly advanced today as human civilization moves from one environmental crisis to another. The answers to these environmental catastrophes are the new technologies and newer innovations. In this chapter, the author repeatedly proclaims the importance of membrane science, AOPs, and desalination in the scientific emancipation of water purification technologies. Human mankind needs to be alert to the disasters of climate change today. Climate change and loss of ecological biodiversity are challenging the scientific firmament today. These areas of scientific endeavor need to be streamlined as human civilization moves forward.

1.6 TRADITIONAL ENVIRONMENTAL ENGINEERING TOOLS

Traditional environmental engineering tool was the cornerstones of scientific research pursuit previously. Sedimentation, flocculation, and activated sludge treatment are the conventional areas of industrial wastewater treatment. Industrial wastewater treatment and water purification thus stand in the midst of vast scientific vision and vast scientific cognizance. The immediate need of the hour is the scientific forays into non-traditional environmental engineering techniques such as membrane science and advanced oxidation techniques. Human scientific progress and strides of human civilization today needs to be envisioned and restructured as environmental engineering science moves towards a newer visionary realm. Today, water science and technology and water purification tools are vastly latent and need to be streamlined as human scientific progress surges forward. Science and engineering are two huge colossus today with a definite and purposeful vision of its own. In this chapter, the author reiterates with immense cadence and scientific candor the success of innovation and the need of technological profundity in the furtherance of environmental protection and environmental engineering. Human mankind's immense scientific prowess, the needs of human society and the world of challenges in environmental protection and environmental engineering will all lead a long and visionary way in the true emancipation of environmental sustainability today. Sustainable development and infrastructural prowess are the pillars of human progress today. Environmental and energy sustainability are the need of the hour.

1.7 NON-TRADITIONAL ENVIRONMENTAL ENGINEERING TOOLS

Non-traditional environmental engineering tools are today the cornerstones of environmental engineering scientific endeavor. Membrane science is revolutionizing the scientific fabric of environmental engineering and chemical process engineering. Human mankind's immense scientific vision is in a state of immense catastrophe as environmental disasters redefine the vast history of science and technology. An advanced oxidation process is another branch of non-traditional environmental engineering scientific endeavor. Human scientific pursuit's immense prowess, the vast technological profundity and the world of scientific validation will all lead a long, effective, and visionary way in the true realization of environmental sustainability and environmental protection. Today is the world of space science, nuclear technology, and renewable energy technology. Man's scientific vision are witnessing immense challenges as the energy crisis, and environmental engineering disasters devastate the vast and wide scientific firmament. Ozonation or ozone-oxidation is a branch of advanced oxidation process which today is surpassing vast scientific frontiers. In this chapter, the author deeply ponders on the immense scientific and technological acuity and far-sightedness in the application of AOPs, integrated oxidation processes, and ozonation. The challenge and vision of non-traditional environmental engineering techniques are slowly unfolding today. Scientific progress and emancipation of technology and engineering science in today's world are surpassing vast and versatile frontiers. Today, the world is faced with the issues of groundwater heavy metal and arsenic contamination. Science and engineering have practically no answers to the enigmatic and monstrous issue of groundwater remediation. Here comes the necessity of innovation and deep scientific vision.

1.7.1 SCIENTIFIC DOCTRINE AND DEEP SCIENTIFIC PROFUNDITY IN NANOFILTRATION AND REVERSE OSMOSIS

The science of nanofiltration and reverse osmosis are today surpassing vast and versatile scientific boundaries. Nanofiltration is a relatively recent membrane filtration process used most often with low dissolved solids in water such as surface water and fresh groundwater with the purpose of softening (polyvalent cation removal) and removal of disinfection by-product precursors such as natural organic matter and synthetic organic matter. Nanofiltration is also used in food processing applications such as dairy, for simultaneous concentration and partial demineralization. The world of science today stands in the midst of scientific failures, and scientific difficulties as heavy groundwater metal and arsenic groundwater contamination challenges the scientific firmament. Technology and engineering science have practically no answers to the marauding impact of arsenic drinking water contamination in developing as well as developed countries around the world. Thus, the need of membrane science and nanofiltration.

The reverse is a water purification technology that uses a semipermeable membrane to remove ions, molecules, and larger particles from drinking water. In reverse osmosis, an applied pressure is used to overcome the osmotic pressure, a colligative property, that is driven by potential chemical differences of the solvent, a thermodynamic parameter. Scientific vision and deep scientific fortitude are the pillars of membrane science and novel separation processes today. The author reiterates with vast scientific conscience the need for innovations in nanofiltration and reverses osmosis in solving global water issues.

1.8 THE SCIENTIFIC DOCTRINE OF ADVANCED OXIDATION PROCESSES (AOPs) AND THE VISION FOR THE FUTURE

Human scientific vision and the vast scientific doctrine in AOPs are today surpassing scientific frontiers. Today, AOPs are a visionary scientific endeavor in the field of water purification and industrial wastewater treatment. Non-conventional Advanced Oxidation techniques are challenging the scientific landscape today. Ozone-oxidation, sonochemical industrial wastewater treatment, and electro-Fenton treatment are the branches of environmental engineering tools which have immense scientific potential and replete with scientific profundity. The world of science and technology are faced with immense scientific travails and deep scientific barriers. In this chapter, the author deeply pronounces the need of scientific innovation, the need of scientific ingenuity and the vast scientific cognizance in the field of both traditional and non-traditional environmental engineering applications. Today, the world of environmental engineering and chemical process engineering are huge colossus with a definite and purposeful vision of its own. Human civilization's immense scientific prowess is at a state of immense disaster as frequent environmental disasters destroy the scientific fabric of deep intellect and precision. Zero-discharge norms and global environmental regulations are scarcely followed by developing and developed countries around the world. Here comes the importance of innovation and vision. AOPs and integrated oxidation processes will with immense lucidity and clarity open a new avenue of research pursuit in years to come.

1.9 SIGNIFICANT SCIENTIFIC ENDEAVOR IN ADVANCED OXIDATION PROCESSES

Technology and engineering science are advancing at a rapid pace moving from one visionary paradigm to another. Today, the challenge and the vision of non-traditional environmental engineering techniques such as AOPs are immense and path-breaking. Human scientific regeneration, human scientific cognizance and the vast scientific fervor of environmental engineering will go a long and effective way in the true realization of environmental protection and water purification. Research pursuit in AOPs and integrated AOPs are changing the entire scientific landscape and the vast vision of global scientific regeneration. Ozonation or ozone-oxidation needs to be restructured and envisioned as human scientific progress, science, and engineering moves towards visionary directions. Today is the world of energy and environmental sustainability. Sustainable development, infrastructural development, and water purification are all linked by visionary cords. The fetters and scientific travails of environmental sustainability and environmental protection need to be ameliorated.

Munter [1] discussed with deep and cogent insight current status and prospects in AOPs. The chapter provides a deep overview of theoretical basis, efficiency, economics, laboratory, and pilot plant testing, design, and modeling of different AOPs (combinations of ozone and hydrogen peroxide with UV radiation and catalysts). Hazardous organic wastes from industrial, military, and commercial operations represent one of the greatest challenges in environmental engineering scientific endeavor [1]. Technological vision, vast scientific motivation, and profundity are the veritable pillars of this chapter. Conventional incineration has immense disadvantages as environmental engineering science surges forward. The AOPs have proceeded along the two visionary routes: (1) oxidation with oxygen in temperature ranges intermediate between ambient conditions and those found in incinerators; (2) the use of high energy oxidants such as ozone and hydrogen peroxide and/or photons that are able to generate highly reactive intermediates – the hydroxyl radicals [1]. In 1987, scientists from the United States of America defined AOPs as "near ambient temperature and pressure water treatment processes which involve the generation of hydroxyl radicals in sufficient quantity to affect water purification" [1]. Scientific assertiveness, deep scientific vision, and vast scientific far-sightedness are the pillars of this well-researched paper. The hydroxyl radical is a powerful, non-selective chemical oxidant, which acts very rapidly with most organic compounds [1]. The technology of AOP is deeply addressed in this chapter. The attack by the OH radical, in the presence of oxygen, initiates a complex cascade of oxidative reactions leading to mineralization of the organic compound. The exact routes of these reactions are still unclear. Scientific clarity and deep profundity are the hallmarks of this research endeavor. The routes are- chlorinated organic compounds are oxidized first to intermediates, such as aldehydes and carboxylic acids, and finally to carbon dioxide, water, and chloride ion [1]. In this chapter, the author deeply elucidates with cogent far-sightedness the scientific success of non-photochemical and photochemical methods. These are ozonation, ozone + hydrogen peroxide, ozone + catalyst, Fenton system, ozone/ultraviolet, Hydrogen peroxide/ultraviolet, ozone/hydrogen peroxide/ultraviolet, photo-Fenton, and photocatalytic oxidation [1]. The author in this chapter addressed and thoroughly reviewed various AOPs with the sole objective of furtherance of science and engineering.

Sharma et al. [2] deeply elucidate with deep and cogent insight AOPs for wastewater treatment. Today, water purification paradigm stands in the midst of deep scientific vision and vast scientific cognizance. AOPs consti- tute promising technologies for the treatment of wastewaters containing non-easily removable organic compounds. All AOPs are designed to produce hydroxyl radicals [2]. Technology and engineering science of AOPs are today surpassing vast and versatile scientific boundaries. It is the hydroxyl radicals that act with high efficiency to destroy organic compounds. Scientific ingenuity and vast scientific clarity and lucidity are the utmost need of the hour. This chapter presents a general review of efficient AOPs developed to decolorize and degrade organic pollutants for environmental protection purposes [2]. The deep scientific ardor in the application of AOP in industrial wastewater treatment is thoroughly addressed in this chapter [2].

Gilmour [3] deeply discussed in his doctoral thesis application perspectives in water treatment using AOPs. The challenge, the vision and the targets of AOP are deeply discussed in this thesis. AOPs using hydroxyl radicals and other oxidative radical species are being studied extensively for removing of organic pollutants from industrial waste streams. Scientific validation, the vast scientific fervor and the barriers in the application of AOP are discussed in deep details here. This study focuses on the evaluation of upstream processing and downstream post-treatment analysis of AOP techniques [3]. The function of the pilot-scale immobilized photocatalytic reactor is described in details in this chapter. Technology and engineering science of non-conventional environmental engineering tools are gaining immense importance as human scientific research pursuit treads forward. This is a vast and versatile eye-opener towards scientific emancipation of environmental science. Trace concentrations of numerous organic compounds (emerging contaminants and endocrine disrupting compounds) such as pharmaceuticals and personal care products including prescription drugs and biologics, nutraceuticals, fragrances, sunscreen agents, and numerous others are reported in industrial wastewater [3]. There is a growing health and environmental concern for pharmaceuticals and personal care products (PPCP) [3]. Thus, this research work addresses the techniques to eliminate this organic pollutant in wastewater [3].

Oller et al. [4] with vast and versatile vision discussed the combination of AOPs and biological treatments for wastewater decontamination. Technological vision, vast scientific ardor, and scientific prowess are the pillars of scientific advancements today. Vast and wide scientific ingenuity is the pivot of this entire chapter. Nowadays, there is a continuously ever-growing worldwide concern for the development of alternative water reuse technologies, basically focusing on agriculture and industry. In this vast context, AOPs are considered a highly competitive water treatment technology for the removal of those organic and inorganic pollutants not treatable by conventional techniques due to their high chemical stability and low biodegradability [4]. Scientific vision and vast scientific profundity are the immediate need of the hour as environmental disasters devastate the scientific firmament. Although chemical oxidation for complete mineralization is usually expensive, its combination with a biological treatment is vastly and effectively reported to reduce operating costs. The hallmarks of this treatise are the deep introspection in degradation kinetics and the reactor modeling of the combined process. Human civilization's immense scientific prowess, the needs of human society and the futuristic vision will all lead a long and effective way in the true emancipation of environmental engineering and industrial wastewater treatment [4].

1.10   THE VAST SCIENTIFIC VISION OF MEMBRANE SCIENCE

Membrane science is a branch of the novel separation process in chemical process engineering. The vision of modern science and modern scientific endeavor is vast and versatile. Today, membrane science and water technology are the two opposite sides of the visionary coin. Human civilization stands today in the midst of deep scientific introspection and vast comprehension. Water purification and desalination in the similar vein are the visionary aim and objectives of scientific endeavor today. Technology and engineering science need to be readdressed and re-envisioned as civilization surges forward. Membrane science scientific doctrine involves reverse osmosis, nanofiltration, ultrafiltration, microfiltration, electrodialysis, and some other effective water treatment processes. Human scientific vision and human scientific regeneration in the field of membrane science and desalination are of the highest order, and global research and development initiatives surge forward. Technology needs to be rebuilt, and water technology needs to be given utmost importance as human civilization treads forward towards a newer paradigm and a newer scientific era. Food processing technology in developing nations is at a state of immense scientific distress. This area of engineering science needs to be overhauled with the application of membrane science.

1.11   SIGNIFICANT SCIENTIFIC RESEARCH PURSUIT IN MEMBRANE SCIENCE

Membrane science today is in the path of immense vision and scientific rejuvenation. Global water shortage and lack of clean drinking water are challenging the scientific vision of environmental engineering science and water technology. Human mankind's immense scientific determination, scientific girth, scientific prowess and the utmost need of environmental protection will all lead a long and effective way in the true emancipation of environmental sustainability. Energy and environmental sustainability are the pillars of human civilization and human scientific endeavor today. Water technology and environmental sustainability are the two opposite sides of the visionary coin. Membrane science and environmental engineering techniques are thus the utmost needs of human society today.

Palit [5] discussed with vast foresight in a far-reaching review advanced environmental engineering separation processes, environmental analysis, and application of nanotechnology. The challenge and vision of nanotechnology applications and nanofiltration applications are deeply addressed in this chapter [5]. Challenges, barriers, and difficulties are the focal points in environmental engineering applications today. This treatise, with immense insight, delineates the success of environmental separation processes and the efficiency of advanced environmental analysis. Technological advancements are today in the path of rejuvenation and vision. This treatise delineated with deep and cogent insight into the success of environmental separation processes, mainly membrane separation processes and tertiary treatment tools such as AOPs and integrated AOPs [5].

Palit [6] discussed lucidly with deep and cogent insight application of nanotechnology, nanofiltration, and drinking and wastewater treatment. The world of environmental engineering science and water technology are moving at a faster pace towards newer visionary paradigm. Water process engineering and chemical process engineering are the hallmarks of effective scientific endeavor in water purification technology today [6]. In this treatise, the author delineates with cogent insight and deep introspection the application and importance of nanotechnology in drinking water treatment and industrial wastewater treatment. The author with immense vision targets the recent scientific endeavor in nanofiltration in drinking water treatment and industrial wastewater treatment. The global drinking water crisis is at a devastating peril. The success of membrane science and the vast futuristic vision will all lead a long and visionary way in the true realization of both water purification and environmental sustainability. This chapter redrafts and revisits the entire gamut of water purification technologies and opens new windows of innovation and scientific instinct in decades to come [6].

1.12   OTHER ENVIRONMENTAL ENGINEERING TOOLS

Environmental engineering science and chemical process engineering are today in the path of immense validation and scientific profundity. Industrial wastewater treatment, drinking water treatment, and desalination are the innovations of tomorrow. Primary, secondary, and tertiary wastewater treatments are changing the face of scientific endeavor. Desalination is the technology of utmost need in many developing and developed nations throughout the world. The world of water purification and environmental engineering today stands in the midst of a deep crisis with the growing concern for arsenic and heavy metal contamination of groundwater and drinking water. This scientific paradigm needs to be veritably streamlined and re-envisioned. The science of groundwater remediation in a similar manner needs to be re-envisioned and reframed if the human civilization needs to be survived. Environmental engineering tools are the pillars of human civilization and human scientific genre. This chapter veritably proclaims the success and scientific revelation of traditional and non-traditional environmental engineering tools. The world of science needs to be overhauled and re-structured if the human planet needs to be saved.

1.13　MODERN SCIENCE: DIFFICULTIES, CHALLENGES, AND OPPORTUNITIES

Modern science and engineering endeavor are today veritably the necessities of human civilization. The difficulties, challenges, and opportunities in the application of engineering science in water technology are immense and path-breaking. Human scientific research pursuit in environmental protection, the needs of human survival and the wide vision of environmental engineering will lead a long and effective way in the true scientific understanding of environmental sustainability. Sustainable development and infrastructural development as regards to energy and the environment are the prime objectives of any scientific endeavor today. Modern science today stands in the midst of deep scientific vision and unending scientific introspection. The opportunities of modern science and its immense visionary research pursuit are vast and versatile. Today, the world of science and engineering are in the process of newer regeneration as space technology, and nuclear science research are changing the vast scientific firmament of vision and determination. The technology of nanoscience needs to be revived and restructured with the passage of scientific history and time. Nanoscience and nanotechnology are the opposite sides of the visionary coin of environmental sustainability. Sustainable development whether it is energy or environment is the engineering marvels of modern science. The author deeply comprehends the utmost need of modern science in the refurbishment of environmental engineering science and environmental protection today. Human scientific endeavor and the needs of human society and human advancement are today linked by an unsevered umbilical cord. The difficulties, challenges, and opportunities in modern science are immense and needs to be readdressed and re-emphasized with the passage of scientific history. Human scientific judgment, the vast scientific truth, and the difficult scientific barriers are the torchbearers towards a newer era of scientific emancipation and deep scientific validation. Desalination, water purification, industrial wastewater treatment, and drinking water treatment are today replete with scientific vision and vast scientific overhauling. The grit and determination of modern science will all lead a long and visionary way in the true emancipation of environmental protection and environmental sustainability. The science of environmental protection today is equally ensconced with vision and scientific fortitude. The targets of modern science should be the needs of the provision of pure drinking water. Today, human mankind's scientific prowess and vast stature are in a dismal state. Today's scientific endeavor should lead human civilization towards a newer era in the field of water technology.

1.14   ENVIRONMENTAL SUSTAINABILITY AND THE PURSUIT OF SCIENCE

Environmental sustainability and water purification are two visionary avenues of scientific research pursuit today. The visionary words of Dr. Gro Harlem Brundtland, former Prime Minister of Norway on the science of "sustainability" needs to readdressed and re-emphasized with the progress of human civilization. Global water research and development initiatives, the needs of human scientific advancements and the futuristic vision of environmental protection will all lead a long and effective way in the true emancipation of environmental sustainability today. Drinking water treatment and industrial wastewater treatment are the veritable needs of human society today. Arsenic and heavy metal poisoning of groundwater in developing and developed nations throughout the world is a burning and deeply enigmatic issue. Technology and engineering science has few answers to the scientific enigma of arsenic groundwater remediation. Scientists and engineers around the world are today faced with tremendous academic rigor and vast scientific vision. Human scientific progress thus needs to be streamlined and re-organized as water science and water technology surges forward towards a newer era. Today, the science of "sustainability" is a huge colossus with a vast vision of its own. Energy sustainability and energy security are the other sides of the visionary coin. The pursuit of modern science in water purification will today lead a long and visionary path in the true emancipation of environmental sustainability. Technology and engineering science thus need to be redrafted, and re-envisaged as human civilization and human scientific research pursuit surges forward. The burning issue of arsenic and heavy metal contamination needs to be vehemently addressed as science and engineering moves towards a newer visionary eon. Human scientific stature, scientific genre, and the vast scientific ingenuity are the necessities of scientific research pursuit in water technology today. The challenge and the vision of water technology and industrial wastewater treatment should be the targets of science in both developed and developing nations around the world. In this chapter, the author repeatedly stresses on the scientific success, the vast scientific fortitude and the imminent scientific needs of water purification technology and industrial wastewater treatment technologies in the path towards the emancipation of science and engineering. Thus, the scientific vision of environmental sustainability will be veritably realized.

1.15   WATER PURIFICATION, DRINKING WATER TREATMENT, AND INDUSTRIAL WASTEWATER TREATMENT

Technological and scientific advancements in drinking water treatment and industrial wastewater treatment are the visionary avenues of scientific emancipation today. Human scientific progress and the human scientific genre are the pillars of human civilization and human mankind today. The needs of water purification are immense and ground-breaking today. Every nation around the world are veritably struggling in dealing with the burning issues of groundwater remediation and drinking water treatment. Arsenic and heavy metal groundwater contamination are a bane to human civilization today. South Asian countries such as Bangladesh and India are in the throes of an immense environmental disaster of untold proportions. Science, engineering, and technology have veritably no answers to global research forays in groundwater remediation. Zero discharge norms and environmental regulations are hardly followed in developing and developed countries around the world. The success of human civilization and human scientific endeavor is thus at deep stake. Human scientific forbearance and human scientific ingenuity need to be restructured and overhauled as science and engineering of environmental protection surges forward. The whole world is today in the threshold of a newer era of challenges and immense vision. In a similar vein, environmental engineering science and water purification technologies need to be restructured and revamped with the passage of scientific history and time. In this chapter, the author repeatedly pronounces the needs of innovations in water purification technologies in the path towards scientific emancipation and engineering innovations.

1.16 ARSENIC AND HEAVY METAL CONTAMINATION OF GROUNDWATER AND THE VISION FOR THE FUTURE

Arsenic and heavy metal contamination of groundwater is a deep curse to the human scientific vision and human scientific research pursuit today. Bangladesh and India are today in a state of immense distress as regards successful implementation of environmental sustainability. Answers to groundwater remediation are few, but the scientific vision is immensely resounding. Technology and engineering science need to be redrafted and re-envisioned if arsenic groundwater contamination needs to be ameliorated. Human scientific progress and human scientific paradigm are in a state of immense distress and turmoil. Provision of clean drinking water and the huge domain of industrial wastewater treatment in developed and developing economies need to be vehemently addressed and vastly re-envisioned. Arsenic groundwater contamination in Bangladesh is the world's largest environmental disaster. Human mankind and human scientific profundity need to be re-envisioned if arsenic or heavy metal contamination is to be totally erased. Groundwater exploration should be the science and engineering of tomorrow. In the similar vein, environmental sustainability should be the pillar of all environmental engineering research and development initiatives. Chemical process engineering and basic chemical engineering operations need to be overhauled and re-envisioned in a similar manner if arsenic groundwater remediation needs to be enhanced and water science and technology emancipated. Scientific acuity, deep vision and the vast needs of the human society are the torchbearers towards a newer era in cross-boundary research in groundwater remediation. In this chapter, the author repeatedly proclaims the success, the needs and the profundity in research innovations in groundwater exploration and groundwater remediation. This chapter will surely open out windows of scientific invention and deep scientific instincts in environmental engineering in decades to come.

1.17   THE SCIENCE OF GROUNDWATER REMEDIATION

Groundwater remediation and groundwater exploration are today areas of interdisciplinary research. Applied geology, biological sciences, chemical process engineering, and environmental engineering science, are today molded into one science in the research pursuit in arsenic and heavy metal groundwater remediation. Today, the challenge and vision of water technology and environmental chemistry are far-reaching. The technology of arsenic groundwater remediation in South Asia has few positive answers. The challenge lies in the hands of environmental engineering science and biological sciences. Human scientific pursuit is dismal as science and engineering of groundwater decontamination moves towards a newer paradigm. Global water research and development initiatives should be in the path of a new beginning. Civil society challenges and governmental policies are the torchbearers towards a newer eon in the field of drinking water treatment, wastewater treatment, and water purification.

1.18   SIGNIFICANT SCIENTIFIC RESEARCH PURSUIT IN GROUNDWATER REMEDIATION

Research pursuit in heavy metal groundwater remediation are witnessing immense challenges today. The whole world is awaiting towards newer innovation and a complete scientific revival in arsenic groundwater decontamination. Interdisciplinary research endeavor is the pillar and mainstay of arsenic groundwater remediation. Human scientific challenges are veritably in the state of distress and vision.

Hashim et al. [7] discussed with deep and cogent insight remediation technologies for heavy metal contaminated groundwater. Scientific validation in environmental management, deep scientific fervor and the needs of human society will all lead a long and visionary way in the true emancipation of environmental engineering science and sustainable development. In this research work, 35 approaches for groundwater treatment are reviewed and classified under three categories which are chemical, biological, and physicochemical treatment processes. The authors with deep conscience discussed sources, chemical property and speciation of heavy metals in groundwater [7]. The challenges of science and engineering in this chapter are the technologies for the treatment of heavy metal contaminated groundwater which are chemical treatment technologies, in-situ treatment by using reductants, reduction by dithionite, reduction by using iron-based technologies, soil washing, in-situ soil flushing, in-situ chelate flushing, biosorption treatment technologies, and enhanced biorestoration [7]. Human scientific regeneration and deep scientific acuity are the pillars of this well-researched paper.

Global water crisis and significant technological innovations are the two opposite sides of the visionary coin. Human mankind today stands in the midst of deep scientific introspection and vast vision. The author in this treatise deeply comprehends the vast and versatile technological innovations and its imminent needs to human scientific research pursuit.

1.19 FUTURE RECOMMENDATIONS AND FUTURE FLOW OF SCIENTIFIC THOUGHTS

Science, technology, and engineering science are today moving at a fast pace surpassing one visionary frontier over another. Today, the world of environmental engineering science stands in the midst of deep scientific introspection and vast scientific divination [8, 9]. Climate change, loss of ecological biodiversity and frequent environmental disasters are veritably challenging the scientific firmament in today's present-day human civilization. The challenges and the vision of science are immense and far-reaching. Future of human civilization and human scientific endeavor today lies in the hands of environmental engineers and environmental scientists. In this chapter, the author deeply comprehends and poignantly depicts the scientific success, the scientific adroitness and the vision behind traditional and non-traditional environmental engineering tools. The success and the vision of this chapter goes beyond scientific imagination and scientific excellence [10, 11]. Human scientific progress in environmental protection is at a state of immense scientific distress and in a similar manner at a state of scientific regeneration. Future research trends and future recommendations should be targeted towards newer innovation and a newer visionary scientific era. The science of groundwater remediation and heavy metal groundwater decontamination are the enigmatic and puzzling scientific issues today.

Technology and engineering science has few answers to the burning issue of arsenic and heavy metal groundwater contamination in developing and developed economies throughout the world. Human scientific progress and human scientific paradigm are in the state of immense scientific distress and replete with immense failures [12, 13]. Future recommendations and future flow of thoughts should be targeted towards newer innovations and newer techniques. Environmental engineering science today needs to be veritably overhauled and reorganized as science treads forward in a newer century. Novel separation processes, advanced oxidation techniques, integrated AOPs, desalination, and sonochemistry are the visionary endeavor of tomorrow. The challenge and the vision of science are immense and far-reaching. Water purification and industrial wastewater treatment should involve the above-mentioned processes. Membrane science and other novel separation processes are the utmost needs of water treatment today. In this chapter, the author reiterates the success of novel separation processes and desalination in the provision of pure drinking water to the teeming millions in developing and developed economies around the world. Scientific vision, deep scientific candor, and vast scientific ingenuity stands as the futuristic targets of water purification and industrial wastewater treatment. The challenges in water treatment tools such as nanofiltration and reverse osmosis are immense and far-reaching. Future recommendations for research should be targeted towards the immense scientific understanding and scientific prudence in the field of membrane separation processes and its link with water purification.

1.20　FUTURE RESEARCH TRENDS IN WATER PURIFICATION

Water purification and the vision of science and technology are the needs of human society. Future research trends in water purification need to be streamlined towards novel separation processes and non-traditional environmental engineering techniques. Human scientific paradigm today stands in the midst of deep scientific fortitude and vision. Scientific clarity, scientific acuity, and deep scientific vision are the utmost need of research and development initiatives today. Future trends in water purification should be directed towards more innovations in novel separation processes, traditional, and non-traditional environmental engineering techniques. Science has no answers to the environmental disasters and global climate change today. Human scientific ingenuity is in a state of immense distress. In such critical juncture of human scientific history and time, validation of science is of immense necessity. Scientific motivation in water science, the vast technological vision and the wide world of science will surely open up new vistas of candor and scientific vigor in decades to come. Today, membrane science and AOPs have immense potential in solving environmental engineering problems. The state of the environment is immensely dismal. Future research trends in global water research and development forays should focus towards technologies and innovations in tackling climate change, loss of ecological biodiversity and frequent environmental catastrophes. In this chapter, the author deeply elucidates the success of innovations in traditional and non-traditional environmental engineering paradigm and the vast necessities of science and engineering in confronting environmental engineering disasters. The other facets of this chapter are to open up new avenues of challenges and scientific opportunities in the field of nanofiltration and reverse osmosis.

1.21   CONCLUSION AND SCIENTIFIC PERSPECTIVES

Science and technology are two huge colossus with a definite and purposeful vision of its own. Academic and scientific rigor in global water research and development initiatives needs to be targeted with immense scientific might and vision today. The future of science and engineering in today's global scientific order is bright and path-breaking. Innovations, scientific vision, and vast scientific candor are the pillars of research pursuit today. Developed and developing nations around the world are giving immense and widespread importance to research forays in water technology and water purification technologies. In this chapter, the author deeply reiterates the scientific success of traditional and non-traditional environmental engineering tools with the sole objective of furtherance of science and engineering of environmental protection. Today, environmental engineering is of the utmost need to human society and human scientific endeavor. Technology and engineering science today needs to be re-emphasized and re-organized as the human planet witnesses immense scientific travails and unending environmental disasters. Climate change and loss of ecological biodiversity are the pivots of scientific endeavor and scientific genre today. Future scientific perspectives in global water initiatives and environmental science are facing uphill tasks today as human mankind trudges forward towards newer knowledge dimensions. Membrane science, desalination, and AOPs are the necessities of human science, human acuity, and human innovation. The author in this chapter repeatedly pronounces the success of human scientific research pursuit in both traditional and non-traditional environmental engineering techniques and gives a vast glimpse on the difficulties, challenges, and opportunities in furtherance of science and engineering today. The future perspectives in environmental engineering, chemical process engineering and nanotechnology are wide, vast, and versatile. Today, every branch of scientific endeavor are linked with research pursuit in nanotechnology. Nanotechnology in today's human planet are the engineering marvels of today. This chapter will surely go a long and visionary way in the true emancipation of nanotechnology applications in environmental engineering science and chemical process engineering. The challenges and the vision of science are far-reaching and need to be reorganized. Nanotechnology and its applications in environmental engineering and chemical process engineering are the needs of human civilization today. The author in this chapter repeatedly urges the scientific community and the vast technological domain to reorganize themselves towards a newer era in water technology and environmental engineering. Human scientific vision will then be truly revisited. Water purification deliberations will be incomplete without elucidating the vast importance of nanofiltration, reverse osmosis, and membrane science. Technological validation and vast scientific motivation will all be the forerunners towards a newer era in membrane science and water purification.

KEYWORDS

- advanced oxidation processes
- global water crisis
- non-traditional environmental engineering tools
- pharmaceuticals and personal care products

REFERENCES

1. Munter, R., (2001). Advanced oxidation processes: current status and prospects. *Proceedings of Estonian Academy of Sci. Chem., 50*(2), pp. 59–80.

2. Sharma, S., Ruparelia, J. P., & Patel, M. L., (2011). A general review on advanced oxidation processes for wastewater treatment. *International Conference on Current Trends in Technology* (pp. 1–7). NUICONE, Ahmadabad, India.

3. Gilmour, C. R., (2012). "Water treatment using Advanced Oxidation Processes: Application perspectives." The University of Western Ontario, Canada.

4. Oller, I., Malato, S., & Sanchez-Perez, J. A., (2011). Combination of advanced oxidation processes and biological treatments for wastewater decontamination: A review. *Science of the Total Environment, 409,* pp. 4141–4166.

5. Palit, S., (2017). Advanced environmental engineering separation processes, environmental analysis, and application of nanotechnology-A far-reaching review. In: Mustansar, H., & Boris, K., (eds.), *Chapter–14, Advanced Environmental Analysis: Application of Nanomaterials* (Vol. 1, pp. 377–416). Royal Society of Chemistry, United Kingdom.

6. Palit, S., (2017). Application of nanotechnology, nanofiltration and drinking and wastewater treatment- a vision for the future. In: Alexandru, M. G., (ed.), *Chapter–17, Water Purification* (pp. 587–620). Academic Press, USA.

7. Hashim, M. A., Mukhopadhyay, S., Sahu, J. N., & Sengupta, B., (2011). Remediation technologies for heavy metal contaminated groundwater. *Journal of Environmental Management, 92,* 2355–2388.

8. Palit, S., (2016). Nanofiltration and ultrafiltration- the next generation environmental engineering tool and a vision for the future, *International Journal of Chem. Tech. Research, 9*(5), 848–856.

9. Palit, S., (2016). Filtration: Frontiers of the engineering and science of nanofiltration-a far-reaching review. In: Ubaldo, O. M., Kharissova, O. V., & Kharisov, B.I., (eds.), *CRC Concise Encyclopedia of Nanotechnology* (pp. 205–214). Taylor and Francis.

10. Palit, S., (2015). Advanced oxidation processes, nanofiltration, and application of bubble column reactor. In: Boris, I., Kharisov, O. V., & Kharissova, R. D. H. V., (eds.), *Nanomaterials for Environmental Protection* (pp. 207–215). Wiley, USA.

11. Shannon, M. A., Bohn, P. W., Elimelech, M., Georgiadis, J. A., & Marinas, B. J., (2008). *Science and Technology for Water Purification in the Coming Decades* (pp. 301–310). Nature Publishing Group.

12. Cheryan, M., (1998). *Ultrafiltration and Microfiltration Handbook.* Technomic Publishing Company Inc, Lancaster, Pennsylvania, USA.

13. Mathioulakis, E., Bellesiotis, V., & Delyannis, E., (2007). Desalination by using alternative energy: Review and state-of-the-art. *Desalination, 203,* 346–365.

CHAPTER 2

NANOFILTRATION, NANOTECHNOLOGY, AND CHEMICAL ENGINEERING SCIENCE: THE SCIENTIFIC TRUTH AND A VISION FOR THE FUTURE

SUKANCHAN PALIT[1,2]

[1]*Assistant Professor (Senior Scale), Department of Chemical Engineering, University of Petroleum and Energy Studies, Bidholi via Premnagar (P.O.), Dehradun–248007, Uttarakhand, India*

[2]*43, Judges Bagan, Haridevpur (P.O.), Kolkata–700082, India, Tel.: 0091-8958728093, E-mail: sukanchan68@gmail.com, sukanchan92@gmail.com*

ABSTRACT

Technology and engineering science today are witnessing immense challenges and drastic changes. Human civilization and scientific rigor today stands in the midst of deep scientific ingenuity and scientific provenance. Global climate change, global warming, and depletion of fossil fuel resources are today challenging the vast scientific fabric of might, scientific grit and scientific determination. The world of science today stands in the midst of vast difficulties and deep scientific vision. Frequent environmental catastrophes and the research and development forays in nanoscience and nanotechnology have veritably urged the scientific domain to go forward towards newer innovations and newer scientific instincts. The author in this well-researched treatise deeply elucidates the immense scientific success of nanofiltration applications in environmental protection and chemical engineering science. The recent scientific research pursuit in the field of nanofiltration, nanotechnology, and chemical engineering science are deeply enumerated in this treatise. Human mankind's immense scientific grit and determination, man's vast scientific revelation and the futuristic vision of technology and engineering science surely lead a long and visionary way in the true emancipation of environmental engineering and chemical engineering science. The author in this treatise pointedly focuses on the immense research and development initiatives in nanofiltration and nanotechnology with the sole purpose of the emancipation of science and technology globally. In a separate section, the author elucidates on the present status of environmental engineering science and chemical process engineering globally. The global scenario of environmental protection is rigorously pondered with deep scientific profundity in this chapter.

2.1   INTRODUCTION

The world of environmental protection, chemical process engineering, and environmental engineering science are today in the path of newer scientific regeneration. Today, arsenic groundwater and drinking water contamination are challenging the vast and versatile scientific fabric globally. Frequent environmental disasters and global climate change has urged the scientists and engineers to devise newer tools and newer innovations such as membrane science and advanced oxidation processes (AOPs). Technology and engineering science today has practically few answers to the ever-growing concerns and scientific ingenuity of heavy metal and arsenic groundwater contamination. Depletion of fossil fuel resources is the other major challenge of petroleum engineering science today. Human civilization's immense scientific ingenuity, man's vast scientific grit and determination and the futuristic vision of engineering science will all today lead a long and effective way in the true realization of environmental engineering science and environmental sustainability. Energy and environmental sustainability are the cornerstones of scientific research pursuit today. The science of "sustainability" as defined by Dr. Gro Harlem Brundtland, former Prime Minister of Norway needs to be re-envisioned and redefined with the passage of human scientific history and time. In a similar vein, industrial wastewater treatment and drinking water treatment are today in the path of newer scientific rejuvenation. Global water research and development initiatives today stands in the midst of deep scientific vision and vast scientific forbearance. Today, the needs of human civilization are water, energy, food, and shelter. So the needs of human civilization and human scientific progress are enormous and groundbreaking. Scientific and technological validation are also the needs of human science and human society today. This treatise opens up newer thoughts and newer vision in the field of membrane science and nanofiltration in decades to come. Novel separation processes such as membrane science and a brief deliberation on non-conventional environmental engineering tools are the other visionary avenues of this chapter. The ever-growing concerns of human civilization and human scientific endeavors such as energy and environmental sustainability are vastly enumerated in this chapter. Sustainable development whether it is social, economic, energy or environmental are the hallmarks of scientific progress today. Infrastructural development in energy and environment are equally the needs of human civilization's progress today.

2.2   THE AIM AND OBJECTIVE OF THIS STUDY

Technology and engineering science are today in the path of a newer visionary eon and an avenue of scientific regeneration. The world of science and engineering today stands in the midst of vast scientific vision and scientific fortitude. The vision of this study is to delineate the importance of nanofiltration and nanotechnology in environmental protection. Today, practically, arsenic, and heavy metal groundwater contamination has no solutions. Technology and engineering science has no answers towards the immense monstrous issue of arsenic drinking water problem in South Asia. Here comes the importance of nanofiltration, membrane science, and other novel separation processes. Technology and engineering science are today bereft of scientific vision and vast scientific forbearance as regards environmental protection and global climate change. The global status of environmental engineering science is quite grave and alarming. The aim and objective of this study is to target newer innovations and newer scientific ingenuity in the field of nanotechnology and chemical process engineering. Human scientific genesis and scientific divination stand veritably in the crossroads of disaster and deep comprehension. This chapter will surely open up new future thoughts and newer scientific ideas in the field of environmental pollution control, drinking water treatment and industrial wastewater treatment. In developing countries around the world, implementation of scientific skills from the classroom to the industry is immensely poor and backward. Here comes the importance of research and development initiatives in the field of implementation and emancipation of industry-related problems. Water treatment and air and soil pollution are the other avenues of scientific divination and deep scientific genesis today. The primary objective of this vast treatise is to elucidate on the scientific success, the scientific travails and the scientific difficulties in the field of environmental engineering science, chemical process engineering and nanotechnology. The barriers and the scientific truth behind academic rigor of environmental engineering issues are detailed in this chapter.

2.3   THE NEED AND THE RATIONALE OF THIS STUDY

The need and the rationale of this study are immense and groundbreaking as regards environmental engineering applications and chemical process engineering emancipation. Fuel science and petroleum engineering science today stands in the veritable crossroads of scientific introspection and deep scientific emancipation. Technology and engineering science need to be revitalized as global concerns of environment, and fossil fuel resources are gaining immense momentum. Thus, the need of nanotechnology applications in environmental protection and chemical process engineering. Human scientific regeneration and deep scientific rejuvenation should be the targets of science and engineering research pursuit today. This study pointedly focuses on the immense need of membrane science, other areas of novel separation processes and environmental engineering tools in greater emancipation and greater realization of science and technology today. The challenges and the vision of chemical process engineering and unit operations of chemical process engineering are immense and ground-breaking today. This chapter opens up newer vision and newer innovations in the field of nanotechnology and nanofiltration also with the sole aim of revamping the global environmental engineering order. The world of chemical engineering science is moving towards one scientific frontier towards another. Scientific verve and motivation, technological validation and the vast world of scientific prowess are all the forerunners towards a newer visionary eon in the field of nanotechnology and chemical process engineering. Drinking water treatment, industrial wastewater treatment and the vast world of water purification need to be re-envisioned and re-envisaged with the passage of human scientific history and time. This well-researched treatise opens up and unravels the scientific success, the scientific profundity and the world of scientific validation in the field of both nanotechnology and chemical process engineering.

2.4 WHAT DO YOU MEAN BY NANOFILTRATION AND MEMBRANE SCIENCE?

Nanofiltration is a recently a membrane filtration procedure used most often with low total dissolved solids in water such as surface water and fresh groundwater, with the purpose of softening (polyvalent cation removal) and removal of disinfection by-product precursors such as natural organic matter (NOM) and synthetic organic matter. Nanofiltration is also becoming more widely used in food processing industries such as dairy, concentration, and partial demineralization. Nanofiltration is a membrane filtration based method that uses nanometer-sized through-pores that pass through the membrane. Scientific and academic rigor in the field of membrane science, the utmost needs of the human society and the world of scientific challenges will all lead a long and visionary way in the true emancipation of environmental engineering science, nanotechnology, and chemical process engineering.

Membrane technology covers all engineering approaches for the transportation of substances between two fractions with the help of permeable membranes. In general, mechanical separation processes for separating gaseous and liquid streams use the visionary tool of membrane technology. Membrane separation techniques operate without heating and thus use less energy than conventional thermal separation processes such as distillation, sublimation, and crystallization. Today, in the world of modern science, membrane separation processes and environmental engineering science are two opposite sides of the visionary coin. Membrane science and novel separation techniques are today in the path of newer scientific regeneration. In the present-day human civilization, the challenges, as well as the vision of science, are immense and groundbreaking. The author in this treatise rigorously points out towards the scientific success, the scientific vision and the vast scientific profundity in membrane science applications in environmental protection. The technology of membrane science is not new but immensely path-breaking and visionary. This chapter will surely open up newer thoughts and newer avenues in membrane science intricacies and nanofiltration challenges in decades to come.

2.5   THE SCIENTIFIC DOCTRINE OF NANOTECHNOLOGY AND CHEMICAL ENGINEERING SCIENCE

In modern science and present-day human civilization, nanotechnology, and chemical engineering science are linked by an unsevered umbilical cord. Nanofiltration, reverse osmosis and other branches of membrane science are the utmost needs of scientific imagination and vast scientific evolution today. Scientific and technological validation are of immense importance in today's scientific world. Human environment is today in a state of immense distress. Arsenic and heavy metal groundwater and drinking water contamination are today challenging the vast scientific fabric of scientific might and vision. Chemical engineering science and unit operations of chemical engineering are in a similar vein in a state of immense regeneration. Technology and engineering science today has few answers to the immense environmental catastrophe of arsenic drinking water contamination in South Asia mainly Bangladesh and India. The needs of nanotechnology and chemical process engineering are witnessing immense scientific revelation and deep scientific divination. Chemical process engineering science and the immense vision behind this domain needs to be readdressed and revamped with the passage of scientific history, scientific vision and time. Today, nanotechnology is integrated with every branch of science and engineering. Water purification, drinking water treatment, and industrial wastewater treatment are the scientific imperatives and the deep scientific verve of modern science today. In this chapter, the author rigorously points towards the scientific success behind chemical engineering and nanotechnology applications in human society.

2.6   ENVIRONMENTAL SUSTAINABILITY AND THE VISION FOR THE FUTURE

The science of "sustainability" today stands in the midst of immense difficulties and vast scientific grit and determination. Energy and environmental sustainability are witnessing immense challenges and vast scientific ingenuity. Drinking water treatment and industrial wastewater treatment are the utmost necessities of human civilization and human scientific progress today. The visionary words of Dr. Gro Harlem Brundtland, former Prime Minister of Norway on the science of "sustainability" needs to be redefined and re-envisioned with every step of scientific success and scientific ingenuity. Sustainability in every form such as social, economic, energy, and environmental are the forerunners towards a newer era in engineering science and technology today. Today, the implementation of scientific research in the laboratory to industry and human society are of vast importance and immense vision. The deep scientific genesis, the scientific progeny and vast scientific revelation in the field of sustainability will surely lead a long and visionary way in the true realization of environmental sustainability and environmental protection. Today, science and engineering of environmental protection are immensely visionary and surpassing vast and versatile scientific frontiers. The world of science needs to be re-envisioned as regards the application of traditional and non-traditional environmental engineering techniques. Novel separation processes such as membrane science are today in the path of newer scientific regeneration. Water, energy, food, and shelter are the utmost needs of human civilization today. The science of "sustainability" is linked with environmental protection by an unsevered umbilical cord.

2.7 RECENT SCIENTIFIC ENDEAVOR IN THE FIELD OF NANOFILTRATION AND NANOTECHNOLOGY

Nanofiltration, nanotechnology, and water purification are the challenges of human scientific endeavor today. Drinking water treatment and industrial wastewater treatment are the needs of human civilization and human scientific progress today. Nanofiltration and other membrane separation processes are challenging the vast scientific firmament of vision and scientific forbearance. Human scientific endeavor and academic rigor in the field of nanotechnology and its various branches are today in the process of newer vision and newer scientific regeneration. Membrane science and environmental protection are two opposite sides of the visionary coin. Technological profundity, scientific validation and the futuristic vision of nanotechnology will surely and veritably lead an effective way in the true emancipation of science and engineering today. The scientific truth behind nanofiltration and nanotechnology are vast and versatile as human scientific endeavor surges forward. There are visionary difficulties in the application of nanofiltration and membrane science in the field of environmental protection. In such a situation, membrane separation processes and AOPs assume vast importance. Nanotechnology thus is in the path of newer scientific rejuvenation and vision.

The Royal Society Report [1] discussed with lucid and cogent insight opportunities and uncertainties in the field of nanoscience and nanotechnologies. In the developed and the developing world nanotechnology, today is surpassing one visionary paradigm over another. Technology and engineering science are highly challenged as civilization trudges forward [1]. The authors of this report deeply comprehended nanomanufacturing and the industrial application of nanotechnologies, possible adverse health effects, environmental, and safety impacts, social, and ethical issues, regulatory issues and the greater emancipation of nanoengineering and nano vision [1]. Nanotechnology is veritably changing the scientific fabric of human civilization today. This report opens up and unravels the scientific success, the vast scientific profundity and the immense scientific ingenuity in the field of nanotechnology applications in human society. Nanoscience and nanotechnologies are widely seen as having massive potential to bring benefits to many areas of fundamental research and applications and are attracting huge investments from Governments and from businesses around the globe. In a similar vein, it is recognized that their applications may raise new challenges, newer vision and newer future thoughts in the safety, regulatory, or ethical domains that will require and re-envision societal debate. In June 2003, the UK Government, therefore, commissioned the Royal Society and the Royal Academy of Engineering to carry out this independent study into current and future developments in nanoscience and nanotechnologies and their vast impacts. Technology and engineering science has practically no answers towards the vast concerns of environmental protection and environmental engineering. Nanotechnology is one of the many answers towards the visionary emancipation of science and engineering today [1]. The upshot of this study is:

- Define what is meant by nanoscience and nanotechnologies [1].
- Summarize the current state of scientific knowledge and scientific discernment in nanotechnologies.
- Identify the specific applications of the new technologies [1].
- Carry out a forward outlook in nanotechnology applications.
- Identify what health, safety, environmental, ethical, and societal implications or uncertainties may arise from the use of nanotechnologies.
- Identify areas in which additional scientific vision needs to be re-envisaged [1].

Technology revamping is the need of the hour as human civilization surges forward. In a similar vein, nanoscience, and nanotechnology needs to be re-organized and restructured as global research and development initiatives move forward. A nanometer (nm) is one thousand millionth of a meter. For comparison, a single human hair is about 80,000 nm wide, a red blood cell is approximately 7000 nm wide, and a water molecule is almost about 0.3 nm across [1]. Scientific discernment, the vast world of scientific and technological validation and the futuristic vision of engineering science will surely lead a long and visionary way in the true emancipation of nanoscience and nanotechnology today. People are interested in the nanoscale (which we define to be from 100nm down to the size of atoms (approximately 0.2 nm)) because at this scale that the properties of materials can be very different from those at a larger scale [1]. Nanoscience can be defined as the study of phenomena and manipulation of materials at atomic, molecular, and macromolecular scales, where properties differ significantly from those at a larger scale; and nanotechnologies as the design, characterization, production, and application of structures, devices, and systems by controlling shape and size at the nanometer scale. In some research understanding, nanoscience, and nanotechnologies are not new yet vastly mature. The science of nanotechnology thus needs to be re-envisioned and restructured with the progress of scientific history and the visionary timeframe. The properties of nanomaterials can be different at the nanoscale for two different reasons. First, nanomaterials have a relatively large surface area when compared to the same mass of material produced in a larger form [1]. This can make materials more chemically reactive and affect their strength and electrical properties. Second, quantum effects can begin to dominate the behavior of matter at the nanoscale–particularly at the lower end–affecting the optical, electrical, and magnetic behavior of materials [1]. Materials can be harnessed that are nanoscale in one dimension, in two dimensions or in all three dimensions [1]. Human civilization today stands in the midst of deep scientific vision and a wider scientific forbearance. Nanoscience and nanotechnology in a similar manner are in the path of newer scientific rejuvenation. The authors of this report aim to provide an overview of current and potential future developments in nanoscience and nanotechnologies against which the health, safety, environmental, social, and ethical implications can be considered and justified. Engineered nanomaterials are the other avenue of scientific research pursuit today. This treatise deeply comprehends the immense need of nanotechnology, nanomaterials, and engineered nanomaterials in the path towards newer scientific and engineering emancipation. This report also discusses the application of nanomaterials in metrology, electronics, optoelectronics, ICT, bionanotechnology, and nanomedicine [1]. Concerns have been veritably aired that the very properties of nanoscale particles being exploited in certain applications might also have negative health and environmental impacts [1]. This report vastly elucidates on the scientific success, the vast scientific imagination and the scientific needs of human society as regards nanotechnology applications in human society. This vast vision of nanomaterials and engineered nanomaterials are enunciated in deep details in this report [1].

Hilal et al. [2] discussed with cogent and lucid insight in a comprehensive review of nanofiltration membranes, treatment, pretreatment, modeling, and atomic force microscopy. Nanofiltration membranes have applications in vast and versatile areas. One of the main applications has been in water treatment for drinking water production as well as industrial wastewater treatment [2]. Nanofiltration can either be used to treat all kinds of water including surface, ground, and wastewater or used as a pretreatment for desalination. Technological and scientific prowess, the veritable needs of human society and the vast challenges of engineering science are the veritable forerunners towards a newer visionary era in nanofiltration, membrane science and membrane separation techniques [2]. Membrane science and water purification are the challenges of human civilization and human scientific progress today. Nanofiltration membranes and its applications in water purification are the hallmarks of this well-researched treatise [2]. Scientific vision, deep scientific profundity, and vast scientific fortitude are the veritable needs of human scientific research pursuit and scientific forays in environmental protection today. One of the main applications has been in water treatment in drinking water production and the vast domain of environmental protection. Nanofiltration can either be used to treat all kinds of water including ground, surface, and wastewater or used as a pretreatment for desalination which is the veritable need of water science and technology today. The introduction of nanofiltration as an effective treatment is considered as a significant and challenging breakthrough for the desalination technique. Nanofiltration membranes have been shown to be able to remove effectively turbidity, microorganisms, and hardness as well as a fraction of the dissolved salts. This results in a significantly lower operating pressure and thus provides a more energy efficient environmental engineering technique. Similar to other traditional and non-traditional environmental engineering tools, a major problem in the membrane separation process is fouling [2]. Human scientific research pursuit's immense scientific prowess and scientific grit, the validation of science and the futuristic vision of nanofiltration and nanotechnology will all lead a long and effective way in the true emancipation of scientific truth. Several research forays have investigated the mechanisms of fouling in nanofiltration membranes and with immense scientific conscience suggested methods to minimize and control the fouling of nanofiltration membranes [2]. This review also vastly investigates the application of atomic force microscopy in studying the morphology of membrane surfaces as an avenue of nanofiltration membrane characterization. Technological revamping and scientific build-up are the necessities of nanofiltration and membrane science research forays today. This well-researched treatise opens up newer thoughts and newer vision in the field of nanomembrane characterization [2].

Hong et al. [3] discussed with immense lucidity chemical and physical aspects of NOM fouling of nanofiltration membranes. Fouling of membranes is a crucial issue in the success of the membrane separation process [3]. This chapter pointedly focuses on scientific success, the vast scientific needs and the futuristic vision of membrane science and its branches. The role of chemical and physical interactions in NOM fouling of nanofiltration membranes is deeply investigated in minute details [3]. Results of fouling experiments with three humic acids demonstrate that membrane fouling increases with increasing electrolyte concentration, decreasing solution pH and the addition of divalent cations (Ca^{2+}). In recent years, membrane filtration has emerged as a remarkable and viable tool to comply with the existing environmental regulations [3]. Of the various tools, for the removal of NOM, a precursor of disinfection by-products, nanofiltration has the immense scientific potential [3].

Technological and scientific validation in the field of nanotechnology are the utmost needs of scientific regeneration globally. Human civilization's immense scientific grit and determination and the futuristic vision of nanofiltration and membrane science are the forerunners towards a newer visionary era in the field of environmental protection, environmental engineering science, and nanotechnology. The need for nanotechnology and membrane science to human society is immense and versatile today. The vision and the challenges of environmental protection are similarly immense. The author rigorously points towards the intricacies of the science of nanofiltration in water treatment in deep details with the sole aim to upholding scientific temperament.

2.8   RECENT SCIENTIFIC ENDEAVOR IN THE FIELD OF CHEMICAL ENGINEERING SCIENCE

Humanity and human scientific research pursuit are today in the crossroads of scientific fortitude and vast scientific introspection. Technological challenges, the scientific verve and determination and the futuristic vision of chemical engineering science and nanotechnology will all lead a long and visionary way in the true realization of environmental protection and environmental engineering science today. Chemical engineering is a vast area of scientific research pursuit in modern science and present-day human civilization. Environmental protection and environmental engineering science two visionary areas of chemical engineering science today. Global climate change and global warming are today veritably challenging the scientific firmament of might and vision today. Technology and engineering science has practically few answers towards the heavy metal and arsenic groundwater contamination in developing and developed countries around the world such as India and Bangladesh. The immediate need of research and development initiatives globally is to target and harness the vast domain of environmental engineering tools such as novel separation processes, AOPs and membrane science. Stankiewicz et al. [4] discussed with immense scientific vision and ingenuity process intensification which is transforming chemical engineering science veritably. Emerging equipment, processing techniques, and operational methods envision spectacular and remarkable improvements in a process plant, markedly shrinking their process size and dramatically boosting their efficiency. Chemical process design is the hallmark of scientific research pursuit in chemical engineering science today [4]. The scientific developments associated with chemical process design may veritably result in the extinction of some traditional types of equipment, if not the unit operations. Unit operations in chemical engineering need to be revamped and restructured as science and engineering surge forward. In 1995, while opening the 1st International Conference on Process Intensification in the Chemical Industry, Ramshaw, one of the pioneers in this field defined process intensification as a definitive strategy for making a dramatic reduction in the size of a chemical plant to achieve the final objective of effective process design [4]. These reductions can come from shrinking the size of the individual pieces of the chemical process equipment and also for cutting the number of unit operations or apparatuses involved. In veritably any case, the degree of reduction must be absolutely significant, how significant remains a matter of discussion [4]. Ramshaw speaks about volume reduction on the order of 100 or more; which is quite a challenging number. Process intensification is a heart of chemical process engineering and the vast world of chemical process technology. The human civilization's immense scientific prowess, girth, and determination, the futuristic vision and scientific discernment, and the needs of human society will all today lead a long and effective way in the true emancipation of chemical engineering science [4].

The vast domain of chemical engineering is undergoing a major transformation and moving towards visionary challenges. A new pattern of 'borderless chemical engineering science' is slowly evolving. The demands from the society on 'cleaner' technologies rather than 'clean-up technologies,' the emergence of new and smart materials is today driving the profession towards a newer visionary era. Today, the areas of focus in chemical engineering science are:

1. New separation processes;
2. Novel reactors;
3. Manufacturing science and engineering;
4. Interfacial science and engineering;
5. Engineering of small systems;
6. Microscopic analysis of solids handling;
7. Alternative energy resources;
8. Water resources engineering and management;
9. New modeling techniques and simulation tools.

The vision and the challenges of chemical engineering science are immense and path-breaking. Today, the world of science stands in the midst of scientific vision and deep scientific forbearance. In this entire treatise, the author rigorously points towards the need of innovation and validation of chemical engineering processes and environmental engineering tools. Nanotechnology is the other side of the visionary coin. Human mankind's immense scientific grit and determination will surely open newer thoughts and newer scientific instincts in the field of engineering science and nano-technology in decades to come.

2.9 THE VAST WORLD OF WATER PURIFICATION AND ENVIRONMENTAL PROTECTION

The vast and wide world of water purification and environmental protection are undergoing immense scientific revamping and immense scientific introspection and contemplation. Water science and technology stands in the midst of deep scientific disaster and vast scientific comprehension. Global climate change and frequent environmental catastrophes are changing the face of human civilization today. Water purification and environmental protection are the utmost needs of human civilization and human scientific progress today.

Shannon et al. [5] with immense scientific conscience and lucidity discussed science and technology for water purification in the coming decades. One of the biggest and pervasive problems afflicting people throughout the world is inadequate access to clean water and proper sanitation. Problems with drinking water are expected to grow immensely worse in the coming decades, with water scarcity occurring globally, even in regions currently considered water-rich [5]. Human civilization's immense scientific prowess, the vast world of scientific validation and futuristic vision of water science will all lead a long and visionary way in the true emancipation of environmental protection and environmental engineering science today [5]. The environmental and water situation is immensely grave and out of control. Arsenic drinking water contamination in developed and developing economies around the globe are veritably challenging the vast scientific fabric of scientific might and vision [5]. In Bangladesh and India, environmental protection and environmental engi-neering stand in the midst of a vast scientific disaster and deep compre-hension. Technology and engineering science have practically no answers to the vicious domain of heavy metal drinking water and groundwater contamination. The basic needs of humanity need to be addressed if civi-lization wants to move forward. Here comes the importance of traditional and non-traditional environmental engineering tools and novel separation processes. In this chapter, the author rigorously stresses on the importance of environmental engineering tools such as traditional and non-traditional in the further emancipation of science and engineering globally. The many problems worldwide associated with the lack of clean, fresh water are vastly well known: 1.2 billion people lack access to safe drinking water, 2.6 billion have little or no sanitation, millions of people die annually–3900 children a day–from diseases transmitted through unsafe water or human excreta [5]. Human civilization, as well as human scientific progress, thus stands in the crossroads of vision and scientific grit and determination [5]. Countless more people are sickened from disease and contamination. Intestinal parasitic infections and diarrheal diseases caused by waterborne bacteria and enteric viruses have become a pivotal cause of malnutrition owing to poor digestion of the food eaten by people sickened by impure drinking water [5]. In both developing and developed nations around the world, a growing number of contaminants are entering water supplies from human activity: from traditional compounds such as heavy metals and distillates to emerging micropollutants such as endocrine disrupters and nitrosamines [5]. Today, increasingly public health and environmental engineering concerns drive vast efforts to decontaminate waters previously considered clean. This is the case for arsenic and heavy metals decontamination also. More effective, low-cost, robust environmental engineering tools to disinfect and decontaminate waters from source to point-of-use are veritably needed without further stressing the human environment. The world of challenges and the vision behind environmental engineering science are vast, versatile, and groundbreaking. Water also affects energy and food production, industrial output, energy, and environmental sustainability, affecting the economies of both developed and developing nations around the globe. Human civilization's immense scientific prowess is at risk as water science and technology challenges the vast scientific firmament. With agriculture, livestock, and energy consuming more than 80% of all water for human civilization's use, the demand for fresh water is expected to increase with population increase further stressing the traditional sources of water in both developing as well as developed nations [5]. Fortunately, a recent flurry of activity in water treatment research endeavor offers immense hope in mitigating the impact of impaired waters around the world, and the more scientific surge is re-envisioned as human civilization moves forward [5]. In this chapter, the authors deeply discussed disinfection, decontamination, re-use, and reclamation and desalination of industrial wastewater and contaminated drinking water [5].

2.10 RECENT SCIENTIFIC RESEARCH PURSUIT IN THE FIELD OF ENVIRONMENTAL PROTECTION

Environmental protection and environmental engineering science today stands in the midst of deep scientific vision and scientific forbearance. Arsenic and heavy metal groundwater and drinking water contamination are veritably challenging the vast scientific firmament today. Human scientific research pursuit in drinking water treatment stands in the midst of vision and scientific fortitude. AOPs, novel separation processes and membrane separation processes are the needs of human civilization and research and development initiatives globally. In this section, the author lucidly reviews the scientific needs, the technological ingenuity and the vast scientific profundity in the field of non-conventional environmental engineering tools such as AOPs.

Andreozzi et al. [6] discussed with immense scientific conscience and ingenuity AOPs for water purification and recovery. All AOPs are veritably characterized by a common chemical discernment and knowledge: the vast potential of exploiting the high reactivity of HO radicals in driving oxidation processes which are absolutely suitable for complete mitigation and through mineralization of even less reactive pollutants [6]. Environmental engineering science and the vast domain of pollution today stands in the midst of vision and scientific fortitude. The different AOPs are investigated for carrying out industrial wastewater treatment [6]. The varied experimental apparatus and working procedures which can be adopted for carrying out wastewater treatments by AOP applications are thoroughly explained in minute details. In the last ten years, a rather fast evolution of research and development forays devoted to environmental protection has been recorded as the result of the special attention paid to the environment by social, political, and legislative regulations of human society [6]. The need of the human society and human scientific progress today is the innovation of low-cost environmental engineering techniques and much larger emancipation of science. Chemical oxidation aims at the mineralization of the contaminants to carbon dioxide, water, and inorganics, or at least, at their metamorphosis into less harmful and benign products. In this chapter, the authors pointedly focus on scientific success, the scientific needs and the research domain of AOPs [6]. The authors did a general survey of AOPs, the Fenton processes, photo-assisted Fenton processes, photocatalysis, ozone-water system, and comparative comments in the vision towards greater scientific understanding. The authors target the vast scientific vision in research and development forays in the field of AOPs [6].

Esplugas et al. [7] deeply with immense scientific understanding compared different AOPs for phenol degradation. In this chapter, a comparison of these techniques is undertaken: pH influence, kinetic constants, stoichiometric coefficient, and optimum oxidant/pollutant ratio. The success of different AOPs and its limitations and future opportunities are elaborated in details in this chapter [7]. The authors touched upon AOP chemistry, an experimental investigation of an annular photoreactor, and the futuristic vision of non-conventional environmental engineering techniques [7].

The success of engineering science and technology are today surpassing one visionary frontier over another. In this treatise, the author deeply contemplates the need of novel separation processes and non-traditional techniques such as AOPs and membrane science in the futuristic vision of science and engineering. Another future vision of this entire treatise is the success of laboratory to industrial applications of AOPs. This is the veritable need of science today.

2.11   THE ROAD AHEAD IN THE SCIENCE OF SUSTAINABILITY

The science of sustainability today is in the path of newer scientific regeneration and are moving from one visionary paradigm towards another. Human civilization and human scientific endeavor today stands in the crossroads of vision and scientific grit and determination. Developed and developing nations around the world are today surging ahead with the sole purpose of scientific emancipation of energy and environmental sustainability. At this crucial juncture of human history and time, a brief deliberation of sustainability policies in developed and developing economies around the world is veritably needed. Sustainable development goals as defined by the United Nations are today in the process of deep scientific regeneration. Human civilization and human scientific progress are in the similar vein in the path towards a newer scientific rejuvenation and newer scientific vision. Sustainable development goals as envisaged by the United Nations needs to be re-envisaged and re-envisioned with the passage of human scientific history and the visionary timeframe. The world of human equality, the eradication of poverty, the provision of basic human needs such as water, electricity, food, shelter, and education and the greater emancipation of energy and environmental sustainability stands as a major pivot of the growth of a nation and the growth of human civilization today.

In the year 2015, leaders of 193 countries around the world came together and with immense scientific conscience and vision faced the future of human mankind. A daunting task ensued with immense vision and scientific fortitude. The eradication of famines, droughts, wars, plagues, and poverty stands as the major deliberation of this United Nations conference. The sustainable development goals are:

1. No poverty;
2. Zero hunger;
3. Good health and well being;
4. Quality education;
5. Gender equality;
6. Clean water and sanitation;
7. Affordable and clean energy;
8. Decent work and economic growth;
9. Industry, innovation, and infrastructure;
10. Reduced inequalities;
11. Sustainable cities and communities;
12. Responsible consumption and production;
13. Climate action;
14. Life below water;
15. Life on land;
16. Peace, justice, and strong institutions;
17. Partnerships for the goals.

Human scientific progress and the vast scientific and academic rigor behind the science of 'sustainability' needs to be re-envisioned and re-envisaged with the progress of human civilization today. This vision and scientific fortitude of sustainable development goals need to be vehemently addressed as mankind moves forward.

McKinsey and Company Report [8] deeply elucidated on environmental and energy sustainability with a visionary approach for India. The vast challenges and the vision of the science of sustainability in Indian context needs to be readdressed and reorganized as India moves towards a visionary scientific paradigm. India is already one of the largest economies in the world and will definitely continue its rapid urbanization in the coming decades [8]. The challenges of economic growth in India are: (1)

the rising consumption and demand for energy, (2) increasing greenhouse gas emissions, and (3) the vast constraints on critical natural resources such as land, water, and oil. India today is in the path of newer scientific rejuvenation [8]. Technology and engineering science has few answers towards the science of energy and environmental sustainability and its applications in the Indian scenario today. In 2008–2009, McKinsey, and Company undertook a visionary study to identify and address opportunities for India to meet the closely linked challenges of energy security and environmental sustainability that goes parallel with a nation's economic growth [8]. The core purpose of the report is to investigate which measures have the greatest potential to reduce emissions, and correspondingly energy use, and which are the most feasible given the vast challenges in funding, regulation, technology, capacity, and market inequalities. The science and the vision of sustainable development in India are vast, versatile, and groundbreaking [8]. The study methodology is constructed with an immense scientific conscience on McKinsey's research into greenhouse gas emissions and its effective abatement over the last three years from the report in 18 countries. The economic growth areas targeted are: (1) power; (2) emission-intensive industries such as steel, cement, and chemicals; (3) transportation; (4) habitats; and (5) agriculture and forestry [8]. This report highlights that there are vast benefits from reducing emissions–they include improving energy security, promoting inclusive growth, improving quality of life and the vast emancipation of leadership drive in high-growth business areas. This report also highlights the pivotal challenges of investment, supply, and skill concerns, technology verve and uncertainty, regulation, and market imperfections [8]. The twin objectives of sustainable development and inclusive growth are the areas which the nation needs to envision with the passage of scientific history and time. The challenges of rising emissions, the abatement case, the opportunities, and the abatement potential are vastly elucidated in this well-researched report [8]. As a developing country case study, this report opens up and unravels the scientific imagination and the scientific genesis in the areas of energy and environmental sustainability in decades to come [8].

Environmental Protection Agency Report, Ireland [9] deeply elucidated with scientific vision and conscience science and sustainability and a research-based knowledge for environmental protection. Technological profundity, scientific verve, and vast scientific might and determination are the pillars of economic growth in Ireland today [9]. Targeted and reliable environmental research envisions a positive scientific basis for environmental regulations and decision making. Policy-related research plays a vital role in ensuring that the European Union and national policies are implemented most effectively, thus minimizing the burden to the nation and to business as a whole. Ireland has taken large strides in environmental protection and implementation of environmental restrictions [9]. Ireland today stands in the midst of deep scientific regeneration and vast scientific vision [9]. The overall report of the most recent Environmental Protection Agency State of the Environment Report (Ireland's Environment, 2008) is targeted towards the quality of the environment and the challenges and vision which lay forward [9]. As investigated in the Report, the four main environmental challenges facing Ireland in the coming years are:

1. Limiting, adapting, and deep scientific emancipation of the global climate change [9].
2. Reversing environmental degradation particularly with respect to water pollution.
3. Mainstreaming environmental considerations in various sectors of the nation's economy.
4. Complying with environmental regulations and revitalizing the economy [9].

The report also deeply underlined the role that science, research, and vast innovation play in responding to the immense environmental protection challenges. It also stressed on the factor that high-quality research provides an effective foundation for effective decision making. Human mankind's immense scientific prowess, the vast scientific grit and determination and the futuristic aisles of scientific decision making will all today lead a long and visionary way in the true realization of environmental sustainability globally. Ireland thus in the similar vein is in the path of newer scientific regeneration, and vast scientific frontiers surpassed as regards environmental protection and environmental engineering [9].

The United States Environmental Protection Agency Report [10] discussed with immense scientific conscience and scientific foresight the incorporation of sustainability in the Agency. Recognizing the importance of sustainability at work and operation, the United States Environmental Protection Agency has been deeply examining applications in a variety of areas and visionary avenues in order to better incorporate sustainability in the decision making of the agency [10]. The agency has taken a wide range of initiatives in the successful realization and effective emancipation of energy and environmental sustainability. The areas covered in this report are: (1) the mission and the role of Environmental Protection Agency; (2) history of sustainability; (3) a wide sustainability framework for Environmental Protection Agency; (4) sustainability assessment and management; (5) the visionary tools of risk assessment and risk management; (6) the changing of the culture in Environmental Protection Agency; and (7) the vast and wide benefits of sustainability approach of Environmental Protection Agency, USA [10]. Today, sustainability is based on a rigorous and long-recognized factual scientific understanding and deep scientific fortitude. Everything that humans require for their survival and well being of human civilization depends directly or indirectly on the natural environment. The environment provides the air the humans breathe, the water we drink, and the food we consume. It veritably defines in fundamental quantities the communities in which we live and sustain, and is the source for renewable and non-renewable resources on which the civilization depends. Technological profundity, scientific verve, and immense scientific imagination today stand in the midst of disaster and contemplation as environmental sustainability witnesses immense challenges. This report vastly depicts the need of sustainability efforts in the future path of human civilization. The abyss of scientific and engineering and the immense scientific forbearance of the science of sustainability are deliberated in deep details in this report [10].

The United Nations Environment Programme Report [11] is detailed with vast scientific conscience the global status of environmental protection. UNEP (United Nations Environment Programme) is the principal United Nations body in the field of environment, assisting governments to assist globally regional and national environmental challenges [11]. The state of the global environment is immensely dismal today as science and engineering are highly challenged with the passage of human scientific history and time. UNEP organized Expert meetings in December, 2014 on the cause and concerns of environmental protection and environmental sustainability [11]. As a part of the United Nations support to member countries in the development of Sustainable Development Goals and following UNEP goals, UNEP has been requested to help establish relevant updateable quality assured environmental data flows and indicators. Technology and engineering science has vastly changed since then. These expert meetings involved experts in the field of law, human rights, environment, economics, and also the participation of global environmental organizations [11]. Forests, rangelands, drylands, and bodies of water worldwide are frequently governed by local communities through the participation of rigorous rules such as community-based tenure rights and institutions [11]. Today, the vision and the challenges of environmental protection are highly relevant and groundbreaking in today's scientific world. Empowering local communities with the means and incentives to manage their eco-systems assumes immense scientific importance and vision with the protection of the environment, eradication of extreme poverty and thereby greater emancipation of sustainable development [11]. This report will open up new vistas of scientific research, scientific, and technological verve and positivity in the mitigation of pollution control in decades to come [11].

Science and technology of sustainability today are the forerunners of a visionary global scientific order. The challenges and the vision of environmental protection and environmental sustainability are today immense, vast, and versatile as human civilization surpasses one visionary paradigm over another. In this entire treatise, the author deeply elucidates the scientific success, the scientific ingenuity and the utmost needs of environmental engineering tools such as membrane science, novel separa-tion processes, traditional, and non-traditional environmental engineering techniques. The needs of human scientific imagination and provenance are immense and groundbreaking as civilization moves forward. This treatise moves in the path towards newer revelation and vast scientific vision.

2.12   SCIENTIFIC TRUTH, THE VAST SCIENTIFIC VISION, AND NANOTECHNOLOGY

Scientific and academic rigor in the field of nanotechnology are in the path of newer regeneration and newer vision. Environmental engineering science are today in the midst of vast scientific revamping. Nanotech-nology is veritably advancing at a tremendous pace in today's modern day human civilization. The scientific truth and the futuristic vision of nanotechnology needs to re-envisaged and re-organized with the passage of scientific history and time. Nanotechnology also today has vast limi-tations. This treatise will surely address the difficulties, the barriers and the vision of science behind nanotechnology applications. The targets of science and engineering globally are the vast world of environmental protection, chemical process engineering and the world of petroleum engineering science. In developed and developing economies around the world, provision of pure drinking water is a vexing and a deeply enigmatic issue. Thus, the need of nanotechnology in drinking water treatment and in the industrial wastewater scenario. Human scientific progress will surely be envisioned and enshrined as civilization surges forward.

2.13 FUTURE RECOMMENDATIONS OF THIS STUDY AND THE FUTURISTIC VISION

Technology and engineering science today stands in the midst of vision and scientific fortitude. Future recommendations of this study are to target the vast scientific emancipation of chemical process engineering and nanotechnology. The area of scientific research pursuit which needs to be targeted is the implementation of laboratory scale research to industry. The world today stands enigmatic with the growing concerns of environmental protection and environmental engineering science. Human civilization's immense scientific grit and determination, the technological prowess of environmental protection and the futuristic vision of modern science will all lead a long and effective way in the true emancipation of science and engineering globally. Human race stands in the midst of a certain disaster with the growing concerns of global warming. Human environment stands in the crossroads of vision and vast scientific contemplation. Future research trends in nanotechnology and chemical process engineering should be targeted towards newer avenues of novel separation processes, traditional, and non-traditional environmental engineering tools and the vast world of nanomaterials and engineered nanomaterials. The technology of nanomaterials is not new yet needs to be re-envisioned and restructured with the passage of scientific history and time. Engineered nanomaterials and its applications in chemical process engineering and environmental protection are the other avenues of scientific research pursuit today. Carbon nanotubes and its applications in diverse arenas of scientific pursuit are the other hallmarks of research and development initiatives today.

2.14 FUTURE OF ENVIRONMENTAL ENGINEERING SCIENCE AND NANOTECHNOLOGY

Future of environmental engineering science and nanotechnology are vast and versatile. Nanoscience and nanotechnology are the cornerstones of present-day modern human civilization. Technology and engineering

science are today in the path of deep regeneration and vast scientific rejuvenation. Human society today stands in the midst of immense distress due
to ever-growing concerns of global warming and global climate change.
Nanotechnology and nanovision are the challenges of human scientific
endeavor in modern human civilization today. Nanotechnology is linked
with every diverse branch of science and engineering today. Future of nanoscience and nanotechnology needs to be re-envisioned and restructured as
human civilization progresses towards a newer paradigm and a newer eon.
Technological advancements in nuclear science and space technology are
changing the face of human civilization today. In the similar vein, environmental engineering, nanotechnology, and chemical process engineering are
veritably in the path of newer regeneration and vast scientific rejuvenation.
Frequent environmental catastrophes, deep scientific divination and the
necessities of nanotechnology are veritably changing the face of research
endeavor today. The author in this chapter reiterates and pointedly focuses
on the vast scientific rejuvenation in the field of nanotechnology today.
The vision and the challenge need to be re-envisioned and restructured as
human civilization surpasses one scientific frontier over another.

2.15   CONCLUSION AND ENVIRONMENTAL PERSPECTIVES

The world of science and engineering today stands mesmerized and at the
same time envisioned with the ever-growing concerns of environmental
pollution, climate change, and frequent environmental disasters. Environmental perspectives and environmental engineering avenues today stand
in the vicious crossroads of vision and scientific fortitude. The vision and
the challenges of science and technology are today immense as civilization progresses. Nanotechnology, nanofiltration, and chemical process
engineering need to be vastly envisioned with the progress of scientific
history and time. In this chapter, the author reiterates on the needs of
nanotechnology in human society. Thus, the need of nanotechnology in
diverse areas of engineering science today. Environmental protection and
groundwater contamination by arsenic and heavy metals are challenging
the vast scientific firmament today. The plausible solution to environmental
protection and drinking water issues is nanotechnology and novel separation process such as membrane science. In this treatise, the author deeply
comprehends the vast necessity of membrane science and nanofiltration
towards the greater scientific emancipation of environmental engineering

science and chemical process engineering. Global warming and global climate changes today stands in the midst of deep scientific provenance and scientific comprehension. The scientific surge of civil society is to target innovations and marvels of engineering and science today. Human civilization and human scientific progress thus need to be re-envisioned and restructured with the passage of scientific history and time. The time of scientific revamping and deep scientific understanding is short as global environmental concerns are ever-growing. This treatise opens up new thoughts and newer scientific understanding in the field of environmental protection and chemical process engineering in decades to come. Today is the age of scientific regeneration in the field of nuclear engineering and space technology. In the similar vein, environmental engineering science and the world of protection of the human environment needs to be re-envisioned and re-contemplated as civilization progresses. In this chapter, the author deeply comprehends with vast scientific conscience and scientific forbearance the utmost needs of human society, the vast technological verve and the world of scientific validation in the field of environmental protection, drinking water treatment and industrial wastewater treatment. The success of science and engineering are immense and groundbreaking in modern human civilization today. This chapter opens up newer thoughts and newer scientific understanding in the field of nanotechnology, chemical process engineering and environmental protection in years to come.

KEYWORDS

- chemical engineering science
- nanofiltration
- nanotechnology
- natural organic matter

REFERENCES

1. The Royal Society Report, (2004). *Nanoscience and Nanotechnologies: Opportunities and Uncertainties.* The Royal Society and The Royal Academy of Engineering, United Kingdom.

2. Hilal, N., Al-Zoubi, H., Darwish, N. A., Mohammad, A. W., & Arabi, M. A., (2004). A comprehensive review of nanofiltration membranes: Treatment, pretreatment, modeling and atomic force microscopy. *Desalination, 170*, pp. 281–308.

3. Hong, S., & Elimelech, M., (1997). Chemical and physical aspects of natural organic matter (NOM) fouling of nanofiltration membranes. *Journal of Membrane Science, 132*, pp. 159–181.

4. Stankiewicz, A. J., & Moulijn, J. A., (2000). Process intensification: Transforming chemical engineering. *Chemical Engineering Progress*, pp. 22–34.

5. Shannon, M. A., Bohn, P. W., Elimelech, M., Georgiadis, J. G., Marinas, B. J., & Mayes, A. M., (2008). *Science and Technology for Water Purification in the Coming Decades* (Vol. 452, pp. 301–310). Nature Publishing Group.

6. Andreozzi, R., Caprio, V., Insola, A., & Marotta, R., (1999). Advanced oxidation processes (AOP) for water purification and recovery. *Catalyst Today, 3*, pp. 51–59.

7. Esplugas, S., Gimenez, J., Contreras, S., Pascual, E., & Rodriguez, M., (2002). Comparison of different oxidation processes for phenol degradation. *Water Research, 36*, pp. 1034–1042.

8. Mckinsey and Company Report, (2009). Environmental and energy sustainability: An approach for India.

9. Environmental Protection Agency Report, Ireland, (2010). Science and sustainability: Research-based knowledge for environmental protection, (Authors- Dr. Shane Colgan, Dr. Brian Donlon).

10. United States Environmental Protection Agency Report, (2011). Sustainability and the U.S. EPA, National Research Council of the National Academies, USA.

11. United Nations Environment Programme Report (UNEP), 2015, (Submitted to the 14th session of the UN- Permanent Forum on indigenous issues, United Nations, New York).

CHAPTER 3

WATER PURIFICATION AND NANOTECHNOLOGY: A CRITICAL OVERVIEW AND A VISION FOR THE FUTURE

SUKANCHAN PALIT[1,2]

[1]*Assistant Professor (Senior Scale), Department of Chemical Engineering, University of Petroleum and Energy Studies, Bidholi via Premnagar (P.O.), Dehradun–248007, Uttarakhand, India*

[2]*43, Judges Bagan, Haridevpur (P.O.), Kolkata–700082, India, Tel.: 0091-8958728093, E-mail: sukanchan68@gmail.com, sukanchan92@gmail.com*

ABSTRACT

Human civilization and human scientific endeavor today are in the midst of deep scientific vision and vast scientific ingenuity. Water science and technology and nanotechnology in the similar vein stand today in the visionary scientific directions of scientific fortitude and vast scientific insight. Nanotechnology is the fountainhead of science and engineering today. Technology and engineering science has few answers towards the scientific intricacies of water purification, drinking water treatment and industrial wastewater treatment. Arsenic and heavy metal groundwater contamination are veritably challenging the vast scientific firmament of modern science and present-day human civilization. Developing as well as developed countries around the world are engulfed by the monstrous and ever-growing concerns of heavy metal and arsenic drinking water contamination. The author in this chapter deeply comprehends the scientific travails, the scientific drive and the deep scientific contemplation in the application of nanotechnology in water purification. The science of water purification and industrial wastewater treatment are a huge colossus with a definite vision of its own. This vision and the challenges are explored in minute details in this chapter. The author reiterates the scientific success of drinking water treatment techniques such as novel separation processes and advanced oxidation processes (AOPs). Human mankind's immense scientific adjudication, the scientific truth, and the vast scientific ingenuity are the forerunners towards greater emancipation of environmental and energy sustainability today. The author in a similar vein depicts in details the success of environmental sustainability today.

3.1 INTRODUCTION

The world of environmental engineering science and nanotechnology are moving drastically towards vast and versatile scientific frontiers. Water purification, drinking water treatment, and industrial wastewater treatment is today challenging the vast scientific fabric veritably. Scientific vision, the vast scientific might and the scientific grits are the torchbearers towards a newer visionary era and an era of scientific emancipation today. The author in this chapter deeply elucidates the scientific ingenuity and the scientific discernment in the field of nanotechnology applications in water purification and water pollution control. The areas of science and engineering which are of immense importance are the areas of nanotechnology, nanomaterials, and engineered nanomaterials. Biomaterials are the other areas of scientific research pursuit which needs to be re-envisioned with the passage of scientific history and time. The world of environmental engineering science today stands in the midst of deep scientific vision and vast scientific forbearance. Human civilization's immense scientific prowess, the deep scientific girth of nanotechnology applications to human society and the needs of human society will all lead a long and visionary way in the true realization of science and engineering today. Depletion of fossil fuel resources, global climate change, frequent environmental catastrophes and the world of nuclear proliferation are challenging the scientific firmament of human mankind today. In such a crucial juncture of scientific history and scientific vision, this chapter opens up newer thoughts, newer questions and sound research initiatives in the field of water purification and environmental protection. The author reiterates the tremendous vision and scientific acuity of water treatment techniques and environmental sustainability.

3.2   THE AIM AND OBJECTIVE OF THIS STUDY

The vision and the challenge of the science of water purification and nanotechnology are immense and ground-breaking. The vision, mission, and the goal of this study is to elucidate upon the scientific difficulties and the scientific barriers in nanotechnology applications in water purification, drinking water treatment and industrial wastewater treatment. Arsenic groundwater and drinking water contamination are challenging the vast scientific firmament of scientific might and forbearance. The utmost need of the hour is the newer innovations and the newer scientific tools. Drinking water treatment and water pollution control today stands in the midst of deep scientific vision and vast scientific fortitude. The author with vast scientific conscience elucidates on the needs of nanotechnology in environmental protection and the holistic domain of environmental engineering science. Novel separation processes such as membrane science and non-traditional environmental engineering tools such as AOPs are changing the face of human civilization and the face of vast scientific rigor. Nanofiltration and reverse osmosis are the utmost needs of scientific endeavor today. The aim and objective of this chapter goes beyond scientific imagination and vast scientific fabric in addressing the issue of global drinking water shortage and the ever-growing and monstrous issue of global climate change. Environmental and energy sustainability are the other pillars of this well-researched treatise.

3.3   SCIENTIFIC DOCTRINE AND DEEP SCIENTIFIC VISION OF WATER PURIFICATION

Scientific vision in the field of water purification are ever-growing and are surpassing vast and versatile scientific frontiers. Technological verve, scientific girth and futuristic vision of nanotechnology will veritably open up new chapters in the field of water purification today. Human scientific progress and the vast scientific rigor in the field of environmental engineering science are changing the face of human civilization today.

The author in this chapter reiterates the needs of advanced environmental engineering tools in drinking water treatment, industrial wastewater treatment and the vast world of water purification. Arsenic and heavy metal groundwater contamination are a veritable bane of human civilization today. Engineering science and technology has practically no answers to the monstrous issue of arsenic drinking water contamination in developing and developed nations around the world. Here come the vast needs of environmental engineering techniques such as novel separation processes and AOPs. Scientific and academic rigor in the field of environmental engineering and chemical process engineering are immense and groundbreaking today. Membrane separation processes are one of the novel separation processes. Nanofiltration and reverse osmosis are challenging the vast scientific firmament today. In this treatise, the author with cogent insight elucidates on the challenges, the opportunities and the scientific travails in the field of nanotechnology applications in environmental engineering science.

3.4 MODERN CIVILIZATION, NANOTECHNOLOGY, AND THE VISION FOR THE FUTURE

Modern science and water purification are today in the path of newer scientific regeneration and deep scientific rejuvenation. The scientific paradigm in the field of drinking water treatment are witnessing immense scientific revival today. Arsenic and heavy metal groundwater remediation are the utmost need of the hour. Today, environmental engineering science and chemical process engineering stand in the midst of deep scientific fortitude and scientific ingenuity. Modern civilization today in a similar vein stands in the crucial juncture of vast scientific introspection and immense travails and challenges. The immediate need of the hour for human mankind is the provision of basic human needs such as water, electricity, food, and shelter. Here comes the need of energy and environmental sustainability. Water purification and environmental sustainability are the two opposite sides of the visionary coin today. Sustainable development whether it is energy, environmental, social or economic is the utmost need of the hour. Science and technology are today huge colossus with a definite vision of its own. Modern civilization and modern science need to be re-envisioned as regards nanotechnology applications in human society. The author deeply elucidates the vast scientific prowess, the deep scientific travails and the introspection behind nanotechnology application in water treatment and to human scientific advancement.

3.5 SIGNIFICANT SCIENTIFIC ENDEAVOR IN THE FIELD OF WATER PURIFICATION

Science and technology of water purification today stands in the midst of deep scientific vision and scientific fortitude. Water purification, drinking water treatment and industrial wastewater treatment today are the utmost need of the hour as human civilization moves forward. The author in this treatise reiterates the need of water pollution control and environmental protection in true realization of environmental sustainability. Sustainability whether it is environmental, energy, social or economic are the utmost need of human scientific progress today. Water purification stands as a major scientific imperative towards the advancement of science and engineering today.

Shannon et al. [1] discussed with lucid and cogent insight science and technology for water purification in the coming decades. One of the most burning issues afflicting people throughout the world is inadequate access to clean water and proper sanitation. Problems with water are expected to be more worse in the coming decades, with water shortage occurring globally, even in countries around the world which are considered water-rich. Addressing these problems calls out for a tremendous amount of research and development initiative to be conducted to identify robust new methods of purifying water at lower cost and with less energy, while at the same time minimizing the use of chemicals and the impact of the surrounding environment. Here the authors poignantly depict some of the science and technology being developed to improve the disinfection and decontamination of water as well as the vast initiatives to increase water supplies through the safe re-use of wastewater and effective desalination of seawater [1]. Throughout the world the shortage of clean drinking water is well known: 1.2 billion people lack proper access to safe drinking water, 2.6 billion have little or no sanitation, millions of people die annually–3900 children a day–from diseases transmitted through impure water or human excreta [1]. Technology and engineering science are today highly challenged and stressed as drinking water treatment gains immense importance [1]. Countless citizens around the world are sickened from disease and water contamination. Intestinal parasitic infections and diarrheal diseases caused by waterborne bacteria and enteric viruses have become a pivotal cause of malnutrition owing to poor digestion of the food eaten by people sickened by water pollution and drinking water issues. In both developing and developed countries around the world, a growing number of contaminants are entering water supplies from human activity: from traditional compounds such as heavy metals and distillates to emerging micropollutants such as endocrine disrupters and nitrosamines [1]. In this chapter, the author deeply comprehends the scientific success, the scientific travails, the scientific ingenuity in the field of water purification tools. The author discusses with immense scientific conscience disinfection, decontamination, re-use, and reclamation, and the vast domain of desalination. Many freshwater aquifers are being highly contaminated and overdrawn in populous regions around the human civilization. Scientific vision, deep scientific foresight, and profundity are the pillars of this chapter. The need of water purification and drinking water treatment in present-day human civilization are immense and groundbreaking [1]. Water also strongly and veritably affects energy and food production, industrial output, and the quality of our environment, deeply affecting the economic growth and advancement of developing and developed nations around the world. In this well-researched paper, the authors deeply comprehend the scientific sagacity and the scientific far-sightedness in the applications of water purification in the holistic development of nations around the world. Environmental sustainability is the need of human civilization today [1]. Sustainable development and water treatment are the domains of science which are connected by an unsevered umbilical cord. A recent activity in water treatment and industrial wastewater treatment research offers immense hope and wide scientific vision in mitigating the impact of impaired waters around the world. Technology and engineering science has today few answers to the immense scientific barriers and scientific travails of heavy metal groundwater and drinking water contamination. Conventional methods of water purification, water disinfection, and desalination can address these immense problems but lack in scientific truth and vision. These treatments are often chemically, energetically, and operationally intensive, focused on large systems and thus veritably requires capital, expertise, and infrastructure which are not feasible. Here comes the need of holistic sustainable development and sound environmental sustainability. Furthermore, these systems can veritably add to more problems, and more issues as human civilization move forward, and scientific progress surges ahead [1]. Disinfection is an overarching goal of human scientific endeavor in water purification. Removal of traditional and emerging pathogens is a goal of many research and development initiatives around the globe. Waterborne pathogens have a devastating effect on public health, especially in the developing countries of sub-Saharan Africa and south-east Asia. Also, arsenic groundwater and drinking water contamination in India and Bangladesh are destroying the scientific firmament of research vision and fortitude. In developed nations around the world, disinfection assumes immense importance [1]. Therefore, the effective control of waterborne pathogens in drinking water calls for immense innovations and scientific vision [1]. The primary goal for the future research trends in decontamination is to detect and remove the toxic substances from water in a robust manner. Widely distributed pollutants, such as arsenic, heavy metals, halogenated aromatics, nitrosamines, nitrates, phosphates, and so on are found to be harmful to humans and the surrounding environment. Human scientific and academic rigor in a similar vein today stands in the midst of catastrophe and deep scientific introspection [1]. Two key problems are that the vast amount of suspected harmful agents is growing rapidly and many of these compounds are highly toxic in nature. The challenge, the vision and the targets of scientific research pursuit need re-envisioning and deep re-organization. The overarching goal for the future of reclamation and re-use of water is to capture water directly from non-conventional sources such as industrial and municipal wastewaters. Here comes the need of scientific discernment and scientific wisdom [1]. The other area of research pursuit in this chapter is the vast world of desalination. The overarching goal for the futuristic vision of desalination is to increase the fresh water supply with the help of desalination of seawater and saline aquifers. These sources account for 97.5% of all water on the Earth, so capturing even a tiny fraction will be beneficial to human mankind's progress [1]. Through continuous research and development initiatives and human scientific progress, desalination techniques can be used reliably to desalinate seawater as well as brackish waters from saline aquifers and rivers. This entire chapter opens up and unravels the scientific intricacies, the vast scientific profundity and the vision behind the application of water purification technologies towards a greater emancipation of the science of environmental protection and environmental sustainability.

Hashim et al. [2] discussed with deep and cogent foresight remediation technologies for heavy metal contaminated groundwater. The high contamination of groundwater by heavy metal, origination either from natural soil resources or from anthropogenic sources is a domain of immense concern to the public health. Technological stewardship, scientific motivation and the vast world of scientific validation stands as important scientific imperatives towards greater emancipation of environmental sustainability today. Here comes the importance of groundwater remediation, drinking water treatment and industrial wastewater treatment [2]. The authors in this well-researched treatise reviewed the importance of environmental engineering tools in groundwater remediation. Scientific ingenuity, deep scientific far-sightedness, and scientific insight are the necessities of human innovation and human instinct today. South Asia particularly India and Bangladesh are in the threshold of a monstrous environmental crisis- the arsenic groundwater and drinking water contamination [2]. This difficult challenge and the scientific profundity behind groundwater remediation are deeply comprehended in this chapter. In this chapter, 35 approaches for groundwater treatment and remediation have been reviewed and classified under three large categories viz chemical, biochemical/biological/biosorption and physicochemical treatment techniques [2]. Keeping environmental sustainability in mind and environmental ethics in vision, the authors, elucidated on the technologies encompassing natural chemistry and bioremediation in clear vision towards scientific emancipation. Heavy metal is a collective term which applies to a group of metals and metalloids with an atomic density greater than 4000 kg m^{-3} or 5 times more than water. In the global environmental engineering scenario, the heavy metals are generally more persistent than organic components such as pesticides or petroleum by-products. Technological vision, scientific drive, and vast motivation stand behind global research and development initiatives in environmental protection and water purification [2]. The authors deeply related sources, chemical property and speciation of heavy metals in groundwater. This treatise also deeply comprehends technologies for the treatment of heavy metal contaminated groundwater, chemical treatment technologies, in-situ treatment by using reductants, reduction by dithionite, reduction by using iron-based technologies, soil washing, in-situ soil flushing, in-situ chelate flushing, in-situ chemical fixation, enhanced bio-restoration and in-situ bioprecipitation processes [2]. Science and engineering of environmental protection are a huge colossus with a definite vision of its own. Technology needs re-envisioning and revamping as regards removal of heavy metal from drinking water. This challenge and vision are enumerated in this treatise.

Choudhury et al. [3] deeply discussed with scientific conscience groundwater arsenic contamination in Bangladesh and West Bengal, India. Technology and engineering science has practically no answers to this monstrous environmental crisis. Nine districts in West Bengal, India, and 42 districts in Bangladesh have arsenic levels in groundwater above the World Health Organization maximum permissible limit of 50 µg/l [3]. The area and population of the 42 districts in Bangladesh and nine districts in West Bengal are 92,106 km^2 and 79.9 million and 38,865 m^2 and 42.7 million, respectively. In a preliminary study, the authors have identified 985 arsenic-affected villages in 69 police stations/blocks of nine arsenic affected districts in West Bengal [3]. Thus, this case of vast environmental catastrophe needs to be revamped and re-envisioned with the passage of scientific history and time. This is the world's largest environmental disaster. Arsenic groundwater remediation is the utmost need of the hour in developing economies around the world. The authors are pioneers in the field of arsenic groundwater remediation. Thus, technology and engineering science need to be re-envisioned and re-organized with the passage of scientific vision and time. According to the chapter, approximately 20 incidents of groundwater arsenic contamination have been reported around the world. Of these, four major incidents were in Asia: Bangladesh, West Bengal, India, Inner Mongolia, and Taiwan [3]. The world's two biggest cases of groundwater arsenic contamination and those that affected the largest number of people are in Bangladesh and West Bengal, India. Thus, human civilization stands in the midst of an unending catastrophe and a deep scientific introspection. The author in this chapter reiterates on the scientific success, the vast scientific vision and the scientific profundity behind scientific research pursuit in arsenic groundwater remediation. Not only the developing economies, but developed countries around the world are also in the threshold of an era of scientific introspection and vast scientific might. This challenge and the vision are related in this chapter.

Chakraborti et al. [4] deeply elucidated with cogent insight the future, monstrous issue and the danger of arsenic groundwater contamination in Middle Ganga Plain, Bihar, India. The ever-growing problem of arsenic poisoning due to contaminated groundwater in West Bengal, India, and vast regions in Bangladesh has been thought to be limited to the Ganges delta (the Lower Ganga Plain). Despite, early survey reports of arsenic contamination of groundwater in the Union Territory of Chandigarh and its surroundings in the northwestern Upper Ganga Plain and very recent findings in the Terai area of Nepal [4]. This monstrous and burning issue of arsenic groundwater contamination is veritably challenging the scientific firmament of fortitude and vision. Groundwater arsenic contamination in the Lower Ganga Plain of West Bengal, India was first identified in July, 1983. After that year, the world of science surged forward towards a newer era of research and development initiatives in the field of environmental engineering science. The authors discussed with scientific conscience and deep scientific vision the methods of arsenic analysis, iron analysis, iron concentration in tube-well water, clinical observations and the vast interconnected world of neurology and medical science. Scientific far-sightedness, scientific stewardship, and girth are the necessities of the vast domain of arsenic groundwater contamination and mitigation. This chapter opens up the immediate needs of arsenic groundwater remediation in the pursuit of scientific emancipation [4]. This research work highlights that more than 6 million people from 9 affected districts (population approximately 50 million) of 18 total districts (total population of approximately 80 million) are drinking water containing greater than equal to 50 µg/l arsenic, and greater than 300,000 people may have visible arsenical skin lesions [4]. Human scientific research pursuit in environmental protection, the challenges of human civilization and the futuristic vision of arsenic groundwater remediation will thus lead a long and effective way in the true emancipation of water purification today. The arsenic content of the biological samples indicates that many more may be subclinically affected. Today, arsenic groundwater contamination is a veritable bane to human civilization. Thus, the need of vision and scientific forbearance. In the combined areas of West Bengal and Bangladesh, additionally affected villages are identified frequently with grave concern. Although West Bengal's arsenic problem reached a wide public concern almost 20 years ago, there are still few concrete plans, much less achievements and a larger emancipation of science and engineering. Villagers in West Bengal and Bangladesh are usually more severely affected than they were 20 years ago. Technological drive, scientific verve and the futuristic vision of environmental engineering will surely lead a long and effective way in the true realization of heavy metal decontamination of drinking water today. The source of arsenic in the deltaic plain of West Bengal is considered to be the arsenic-rich sediments transported from the Chotonagpur Rajmahal Highlands and deposited in the sluggish, meandering streams under reducing conditions [4]. The Upper, Middle, and Lower Ganga plains are the most thickly populated areas of India. Technological drive and vast scientific motivation have veritably no answers to the ravenous issue of arsenic groundwater contamination [4]. This treatise highlights that since 1998, the severe arsenic contamination of groundwater in the Lower Ganga Plain of West Bengal and Bangladesh. The authors also found out that there is severe arsenic groundwater contamination in Bhojpur District, Bihar which is in the Middle Ganga Plain. Science and engineering globally are crossing vast and versatile scientific frontiers. This treatise is a severe warning to the ever-growing concerns of heavy metal drinking water and groundwater contamination [4]. The challenges and the vision of engineering science need to be re-envisioned and readdressed as regards environmental protection globally.

Mukherjee et al. [5] discussed with vast scientific conscience arsenic contamination in groundwater and a global perspective with emphasis on the Asian scenario. South Asia particularly India and Bangladesh are in the threshold of a devastating crisis. Arsenic groundwater contamination is a bane to human scientific progress in modern civilization today [5]. This chapter deeply revisits the current scenario of arsenic contamination in countries across the world with an emphasis on Asia [5]. Along with the present situation in vastly affected countries in Asia, such as Bangladesh, India, and China, recent instances and incidents from Pakistan, Myanmar, Afghanistan, Cambodia, etc. are lucidly reported [5]. Over the past two or three decades, the presence of high concentrations of arsenic in drinking water and groundwater has been widely recognized as a major public health concern in several parts of the world [5]. Mitigation of arsenic drinking water contamination is the ultimate vision of scientists around the world [5]. Technology, engineering, and science around the world need to be re-envisioned, reorganized, and revamped with the passage of scientific history and the visionary timeframe. With the discovery of newer sites of devastation in the recent past, the arsenic contamination scenario around the world, especially in Asian countries have changed remarkably [5]. In this chapter, the author deeply elucidated the natural sources in Afghanistan, Argentina, anthropogenic sources in Australia, Brazil, and Bulgaria, natural sources in Cambodia, Chile, People's Republic of China and natural sources in India [5]. Human scientific vision, the needs of scientific profundity and the futuristic scientific ingenuity are the forerunners towards a newer era of innovation and instinct in decades to come in the field of water purification. Human scientific research pursuit in environmental protection today stands in the midst of deep scientific catastrophe and vast scientific introspection [5]. The authors in this chapter with vast scientific conscience and scientific intellect depicts the need of a global research and development initiative in the field of arsenic groundwater remediation poignantly. This success, the vision, and the scientific travails are enumerated in minute details in this well-researched treatise.

Chakraborti et al. [6] discussed with immense scientific vision and scientific sagacity in a 20-years-old study report status of groundwater arsenic contamination in the state of West Bengal India. According to the author, since 1988, the authors have analyzed 140150 water samples from tube wells in all 19 districts of West Bengal for arsenic; 48.1% had arsenic levels above 10 µg/l (WHO guideline value), 23.8% above 50 µg/l (Indian standard) and 3.3% above 300 µg/l (concentration predicting overt arsenical skin lesions) [6]. The situation in the state of West Bengal, India, and Bangladesh are immensely grave as well as far-reaching. Science and engineering today are in the path of a newer scientific regeneration and immense scientific ingenuity [6]. Water is a veritable need of human civilization today. In a similar vein, environmental, and energy sustainability are the needs of human society today. Arsenic groundwater remediation and the successful realization of environmental sustainability should be the targets of human scientific progress today. In this study, the authors effectively dealt with the ever-growing concerns of heavy metal groundwater contamination in Bangladesh and India. In a preliminary study for the last 20 years in India, some areas of all the states (Uttar Pradesh, Bihar, Jharkhand, West Bengal) in Ganga plain are arsenic affected, and millions of humans are at severe risk to arsenic toxicity [6]. The first report of arsenic groundwater contamination and its health risks in the Ganga plain from West Bengal was published in 1984. Since 1988, the team comprising of the authors has been surveying arsenic-affected villages in West Bengal, India [6]. In 1988, when the first survey was done in West Bengal, there were only 22 villages over 12 blocks in five districts [6]. During the time of writing the chapter, the number of affected villages has increased to an alarming 3417 in 107 blocks in some nine districts [6]. Science and technology has vastly changed since then and have witnessed drastic changes in the field of environmental protection. During the last 20 years before this research endeavor, with every additional survey, the authors found out that an increasing number of contaminated villages and more affected people from nine arsenic affected districts. Here comes the global need of research and development initiatives in water purification and environmental sustainability. The authors in this treatise brought forth the ever-increasing issue of groundwater arsenic contamination in the nine districts of West Bengal, India where some groundwater contained arsenic at concentrations of 300 µg/l and above [6]. There is another avenue of immense concern in West Bengal, India- the fluoride contamination of groundwater and drinking water. Fluoride concentration of groundwater has been known since 1937 [6]. At present in India, 62 million people are severely suffering from fluorosis, a crippling disease [6]. The challenge and the immense vision of science and engineering need to be revitalized as regards true implementation of global sustainability initiatives. Although the groundwater in Bankura, Birbhum, and Purulia in the western part of West Bengal seems veritably arsenic safe, there is a fluoride contamination in these areas, and thousands are suffering from fluorosis. In this entire chapter, the authors bring to the forefront the challenge of heavy metal contamination to human society and its effective mitigation. Science and technology will be the veritable forerunners towards a true emancipation of environmental sustainability.

Qu et al. [7] discussed with immense scientific and technological conscience and profundity applications of nanotechnology in drinking water and wastewater treatment. Scientific vision and vast scientific ingenuity are the pillars of this well-researched treatise. Providing clean and cheap water initiatives to meet human needs is a grand challenge of the 21st century [7]. Water issues are exacerbated by population growth, global warming, and water quality deterioration. The need for technological drive and integrated water resource management assumes immense importance as science and engineering surges forward [7]. Nanotechnology holds immense potential in advancing the field of water and wastewater treatment and the true emancipation of the science of environmental sustainability. This vast and well-researched treatise redefines candidate nanomaterials, properties, and mechanisms that enable the applications, advantages, and limitations as compared to the existing processes and the vast domain of scientific travails and scientific barriers in nanotechnology applications in wastewater treatment [7]. This review highlights the vast opportunities and immense limitations to further capitalize on the domain of sustainable and integrated water management. The current global water supply today faces immense challenges [7]. Worldwide more than 780 million people still lack access to improved drinking water resources. Recent advances in water and wastewater treatment offer immense opportunities to develop next-generation water supply systems [7]. Human scientific regeneration in integrated water management system is at a severe stake and at a state of disaster. The highly efficient, modular, and multifunctional processes enabled by nanotechnology are vastly envisioned to provide high performance, affordable water, and wastewater treatment solutions that less rely on large infrastructures [7]. Here comes the visionary need of environmental sustainability and integrated wastewater management. The authors provided an overview of recent nanotechnologies in water and wastewater treatment. The major applications of nanomaterials are vastly and critically reviewed based on their functions in unit operation processes. The authors in this treatise widely discussed adsorption, membranes, and membrane-based processes, nanocomposite membranes, biologically inspired membranes, photocatalysis, and the vast area of application of nanotechnology in disinfection and microbial control. Retention and reuse of materials are the other pivots of this research endeavor [7].

Technology and engineering science are in the path of newer scientific regeneration. Today, water and energy are the utmost needs of human scientific progress. This chapter will surely open up new vistas of scientific ingenuity and technological verve in the field of both nanotechnology and integrated water resource management in decades to come. Human civilization and human scientific progress today stands in the midst of deep scientific vision and scientific contemplation. Water and wastewater treatment are the utmost need of the progress and advancements of human civilization today. Energy, as well as environmental sustainability, is in the similar vein veritable scientific imperatives towards larger scientific emancipation of science and technology today. The author deeply elucidates in this chapter the scientific success, the vast scientific barriers and the scientific re-organization in the field of water purification technologies with a clear vision towards more innovation and scientific research forays.

3.6 RESEARCH ENDEAVOR IN THE FIELD OF NANOTECHNOLOGY APPLICATIONS IN WATER PURIFICATION

The world of science and engineering are moving ahead at a drastic pace. Nanotechnology is linked with every sphere of scientific research pursuit such as water purification. Human scientific research pursuit in the field of nanotechnology and nanofiltration today stands at the crossroads of introspection and vision. Today is the world of interdisciplinary studies of science and engineering. Water purification, drinking water treatment and industrial wastewater treatment today are veritably linked with nanotechnology.

Mansoori et al. [8] discussed with deep and cogent insight discussed environmental applications of nanotechnology. Nanotechnology is an emerging field that encompasses a wide range of technologies which are presently developed in nanoscale [8]. It plays a pivotal role in the development of innovative methods to produce new products, to substitute existing production equipments, and to reformulate new materials and chemicals with improved and visionary performance. Scientific vision, deep scientific fortitude, and ingenuity of technological advancements are the forerunners towards a newer era in the field of nanotechnology today [8]. This article gives a comprehensive overview on the ongoing research and development forays on environmental remediation by nanotechnology. Various environmental treatments and remediations using different types of nanostructured materials from air, contaminated wastewater, groundwater, surface water, and soil are discussed in minute details [8]. The advantages and limitations in the environmental applications are deeply evaluated and compared with each other and with the existing techniques. This report covers the bulk of published researches during the period from 1997 to 2007. Science and engineering of nanotechnology are in the path of newer scientific regeneration. Technology and engineering science of water purification in a similar vein are surpassing visionary boundaries [8]. The authors discussed in deep details nanomaterials and their environmental applications. The nanoparticles are titanium dioxide based nanoparticles, iron-based nanoparticles, bimetallic nanoparticles, nanoparticulate photocatalysts and catalysts, nanoclays, and nanotubes [8]. Nanomaterials and engineered nanomaterials are the utmost need of scientific emancipation today. Nanomembrane and nanosieves are the other pillars of this scientific endeavor.

Stander et al. [9] with immense scientific conscience elucidated on environmental implications of nanotechnology. Nanotechnology is today highly concerned with dealing in environmental implications, and regulatory compliance encompasses practicing areas for scientists and engineers. Areas of immense concern are current or proposed environmental regulations and procedures for quantifying both health and hazard risks. Technological challenges, scientific motivation, and deep scientific profundity are the pivots of scientific endeavor in nanotechnology today. The authors touched on the environmental concerns, environmental risk management, the health risk evaluation process, and the hazard risk evaluation processes [9].

3.7 RECENT SCIENTIFIC RESEARCH PURSUIT IN THE FIELD OF NANOTECHNOLOGY

Human civilization and human scientific progress today are in the midst of deep scientific comprehension and immense scientific profundity. Technological challenges and scientific motivation are immense in modern day human civilization. Nanotechnology today is the fountainhead of scientific research pursuit in present-day science and engineering. Nanotechnology today is veritably surpassing vast and versatile scientific boundaries. Research forays in the field of nanotechnology and water purification need to be re-envisioned and revitalized with the passage of scientific history, scientific vision, and vast scientific forbearance. The author in this well-researched chapter elucidates on the scientific vision, the scientific barriers and the scientific far-sightedness in the research and development initiatives in the field of nanotechnology applications in modern science and modern civilization. The veritable fountainhead of science and engineering are today nanotechnology and sustainability. Human race today stands in the crossroads of vision and fortitude. This chapter opens newer scientific understanding, newer scientific wisdom, and the vast scientific ingenuity in the field of nanotechnology forays.

Filipponi et al. [10] discussed with cogent and lucid insight principles, applications, implications, and hands-on activities in the vast and versatile domain of nanotechnologies. This publication has been developed within the context of the research developed by a European project, funded by the European Union's seventh framework programme. It has been vastly enriched by the authors with numerous and multifaceted inputs, reflections, and insights on societal issues and scientific emancipation of nanotechnology [10]. This treatise has been specifically developed to present in a single compendium, much of the relevant and invaluable educational material that will help inform and vastly motivate young people about nanotechnologies and its applications. The programme focuses on three application fields–information and communication technologies, energy, and environmental sustainability, and the vast world of medicine [10]. It also addresses the associated ethical, societal, legal, and safety aspects to enhance the scientific dialogue [10]. Young Europeans might become increasingly attracted to enhance their awareness and knowledge in these disciplines and their associated visionary impacts. Nanotechnology is one of the very frontiers of science and engineering today [10]. As a visionary matter of fact, nanotechnology would affect us all, beyond nanoparticles, critical length scales and the vast world of nanotools. The authors in this chapter discusses in minute details about the vast domain of nanomaterials, how nanotechnology can be brought to classrooms, the definition of natural nanomaterials, the relevance of Dr. Richard Feynman's vast visionary definition of nanotechnology, material properties at the nanoscale, electric properties, optical properties, magnetic properties, mechanical properties, nanostructured metals and alloys, self-assembled nanomaterials, polymers, semiconductors, ceramic, and glassy materials, composites, and nanocoatings. Characterization methods discussed are microscopy and spectroscopy methods [10]. Fabrication methods include lithography. The other facets of this treatise involve environmental remediation and mitigation, pollution prevention, environment sensing, and food packaging and monitoring [10]. A deep deliberation of energy engineering's such as solar energy, hydrogen energy, thermoelectricity, rechargeable batteries, energy savings and the vast world of information and communication technologies are done in this well-researched report [10]. Nanotechnology can be defined as the study of phenomena and manipulation of materials at atomic, molecular, and macromolecular scales, where properties differ significantly from those at larger scales. Human technological divination, scientific revelation and the futuristic vision of nanotechnology will all lead a long and visionary way in the true emancipation of science and engineering today. Nanotechnologies are the visionary design, characterization, production, and application of structures, devices, and systems by controlling shape and size at the nanometer scale [10]. When the term was first used in 1959, it was used in the singular, nanotechnology [10]. In the last decade, the field has vastly evolved in terms of science and technology development. Scientists and engineers have also started to address the safety, ethical, and societal impact of 'nanotechnology' [10]. Today the world of science and engineering are moving at a rapid pace and surpassing vast and versatile scientific boundaries. The concept of ambient intelligence, computation, and communication are deeply elucidated in this report. The authors in this report also gave a wide definition of nano-materials and engineered nanomaterials [10].

Kumar [11] deeply discussed with vast far-sightedness nanotechnology development in a developing country like India. Nanotechnology has been heralded as a revolutionary technology by scientists and engineers worldwide [11]. Nanotechnology is today an enabling technology and has the tremendous potential to open up newer avenues in the field of research and development initiatives in various multiple disciplines in a wide range of sectors such as healthcare/medicines, electronics, textiles, agriculture, construction, water treatment and food processing to cosmetics. Human scientific challenges and the vast scientific vision in the domain of nano-technology are immense and far-reaching. Much of these applications are pertinent for a developing country like India. Today, the government has been playing an important role in fostering and enhancing the nanotechnology R&D in India since the early 2000s [11]. Technological drive, scientific verve and the vast challenges in research and development initiatives in nanotechnology are the utmost needs of science and technology forays today. This report opens up newer avenues and newer visionary roads of success in the field of nanotechnology today. Nanotech-nology, being an enabling technology of vast importance and of emerging techno-economic paradigm is still in the nascent stage of research and development [11]. Worldwide governments have vastly launched many nanotechnology-specific initiatives to gain the potentialities and scientific vision of nanotechnology for social and economic gains. In 2005 itself, more than 62 countries launched national nanotechnology-specific activities throughout the world. Human civilization and human scientific research pursuit thus stand in the crossroads of vision and scientific forti-tude. Nanotechnology is the veritable vision of tomorrow. The research and development effort was significantly promoted world over with the announcement of the National Nanotechnology Initiative (NNI) in 2001 by USA [11]. Most advanced countries around the world have based their nanotechnology initiatives on the foundation lain by NNI in the USA. The NNI is the most comprehensive and a detailed R&D programme in nanoscience and technology in the global environment [11]. The focus of NNI is on research and development forays of nanoscale science and technology for economic benefit and national security. Scientific imagina-tion, the futuristic vision of nanotechnology and the world of scientific challenges are the today's forerunners in greater scientific emancipation and technological motivation. NNI programmes are vastly aligned with the goals of the participating organizations. In the case of Europe, most developed countries have government supported major nanotechnology research and development programmes [11]. The 9th Five Year Plan in India (1998–2002) had mentioned with immense scientific vision and might for the first time that national organizations and core groups were set up to promote research facilities in frontier areas of science and technology which redefined superconductivity, robotics, neurosciences, and carbon and nanomaterials. Nanotechnology is today that branch of scientific research pursuit which needs to be revamped and re-envisioned with every step of scientific history and visionary timeframe. In India, the thrust of research and development initiatives came with the launch of "Programme on nanomaterials: Sciences and devices" in 2000 by the Department of Science and Technology [11]. DST launched a special initiative to generate and support some end-to-end projects leading to tangible processes, products, and technologies. In this report, the author deeply elucidates on major actors in nanotechnology research, development, and innovation in India. The government and research and development organizations in India played a pioneering role in promoting and enhancing nanotechnology R&D in India. The status of nanoscience and nanotechnology initiatives in India is bright as well as groundbreaking [11]. Similarly, nanotechnology applications in water purification need to be re-envisioned with government's intervention and vast scientific emancipation. Water and energy are the lifelines and fountainheads of human scientific progress today. This report opens up the challenges and the scientific targets in the field of nanotechnology applications in a developing nation such as India [11].

Roco et al. [12] discussed with vast scientific conscience nanotechnology research directions for the societal needs in 2020 along with an instinctive retrospective and outlook. The accelerating pace of scientific discovery and innovation and its wide interdisciplinary nature leads to the emergence of newer capability and a newer vision in the field of nanotechnology applications and vast scientific emancipation [12]. This pace has been discovered from the confluence of discoveries in interdisciplinary areas such as physics, chemistry, biology, material science and computational science around the year 2000. At that time, a global scientific and societal endeavor emerged. It focused on two factors: (1) an integrative definition of technology; (2) articulation of long-term visions and goals that included a twenty-year vision for the scientific success and vast and wide technological prowess [12]. This report highlights the foundational knowledge and infrastructural development achieved by the domain of nanotechnology in the last few years and explores the potential of the United States and the global nanotechnology enterprise to 2020 and beyond. Only ten years after adopting the definition and long-term vision for nanotechnology, the NNI and other programs around the world have achieved remarkable results in terms of scientific discoveries and vast scientific innovations. Key emphasizes described in the report are:

- Integration of knowledge at the nanoscale and of nanocomponents in nanosystems with complex behavior [12].
- Better control of molecular self-assembly, quantum behavior, the creation of new molecules, and the interaction of nanostructures with external fields [12].
- Understanding of biological processes and of nanobio interfaces with abiotic materials [12].
- Governance to increase innovation and public-private partnership [12].

The success of this report and the research and development initiatives in nanotechnology are immense and groundbreaking today. The authors in this report elucidated on: (1) the long view of nanotechnology development; (2) investigative tools along with theory, modeling, and simulation; (3) the challenges of enabling and investigative tools; (4) synthesis, processing, and manufacture of components and devices; (5) environmental and health issues; (6) the holistic world of science of sustainability; and (7) applications such as nanobiosystems, nanoelectronics, nanophotonics, catalysis by nanostructured materials and high performance engineered materials [12]. The responsible governance of nanotechnology are the other pivots of this well-researched study.

Nanotechnology vision is the cornerstone of scientific progress today. Human civilization and human scientific research pursuit need to be re-organized as environmental engineering science, and chemical process engineering enters a newer era of scientific grit and scientific determination. In this chapter, the author deeply comprehends the vast needs of conventional and non-conventional techniques of environmental engineering science with the sole purpose of scientific emancipation of environmental protection.

3.8 HEAVY METAL AND ARSENIC CONTAMINATED DRINKING WATER TREATMENT: A VISION FOR THE FUTURE

Heavy metal and arsenic groundwater contamination are challenging the vast scientific fabric of might and scientific fortitude today. Developed as well as developing countries around the world are in the threshold of a major environmental engineering catastrophe. Bangladesh and the state of West Bengal, India are the regions of the largest water and environmental engineering calamity. The contamination of groundwater and drinking water by heavy metal results from either natural soil sources or from anthropogenic sources. Hassan [13] deeply elucidated on the crisis of arsenic groundwater contamination in Bangladesh and India. The author in this well-researched treatise discussed with vast scientific conscience the topic of arsenic poisoning in Bangladesh for ages. They dealt with the global scenario of groundwater arsenic disaster [13]. The areas covered in this book are groundwater arsenic discontinuity, spatial mapping, spatial planning, an environmental engineering perspective of chronic arsenic exposure to drinking water, epidemiological, and spatial assessment, experiences from arsenicosis patients, policy response and arsenic mitigation in Bangladesh, legal issues of responsibility, and the futuristic vision of arsenic groundwater remediation science [13]. It is estimated that more than 300 million people in 70 countries worldwide are at a vicious risk of groundwater arsenic poisoning. Apart from Bangladesh and the neighboring Indian state of West Bengal, which have the largest problem, there have been warnings from Argentina, Chile, Taiwan, Vietnam, China, Pakistan, Thailand, and even the southwestern part of the USA [13]. The human struggle has many forms today in developing and least developed countries around the world. Mankind's immense scientific prowess, the scientific travails, and difficulties and the needs of scientific progress will lead a long and visionary way in the true realization of environmental engineering science. In this chapter, the author reiterates the success of science and engineering in solving arsenic drinking water issues globally [13].

3.9 SIGNIFICANT SCIENTIFIC RESEARCH ENDEAVOR IN MEMBRANE SCIENCE

Membrane technology covers all engineering approaches and forays for the transport of substances between two fractions with the help of a permeable membrane. In general mechanical separation processes for separating gaseous or, liquid streams use membrane technology. Scientific vision and vast scientific girth and determination are the pillars of research and development initiatives in membrane science today. Nanofiltration is a branch of scientific endeavor which needs to be pursued with the vision of industrial wastewater treatment, drinking water treatment and environmental sustainability. Sustainable development whether it is environmental or energy are today surpassing vast and versatile scientific frontiers. Novel separation processes such as membrane science are the utmost needs of human civilization and human scientific progress today. In this chapter, the author profoundly depicts the vast scientific success, the scientific ingenuity, and profundity of membrane science applications in water purification. Membrane science, nanofiltration, and reverse osmosis are in the path of newer scientific regeneration and deep scientific rejuvenation. Water treatment and environmental engineering tools are today opposite sides of the visionary coin today. A burning and complex issue in the field of membrane science is membrane fouling and biofouling. Arsenic and groundwater heavy metals contamination are today challenging the very scientific firmament of vision, girth, and scientific determination.

Cheryan [14] dealt with lucid and cogent insight ultrafiltration and microfiltration and its applications and greater emancipation in environmental engineering tools. The development of Sourirajan-Loeb synthetic membrane in 1960 provided and envisioned a valuable separation tool to the process industries, but it considerable difficulties in its initial years. Human civilization's immense scientific prowess, the futuristic vision of environmental engineering science and the needs of human society will surely lead a long and effective way in the true emancipation of novel separation processes and environmental sustainability today [14]. The situation of research forays is different today: membranes are more robust, modules and equipment are better designed, and there are a better scientific understanding and far better scientific wisdom of the fouling phenomenon and how to minimize these effects. Scientific vision and vast scientific profundity are the necessities of innovation and truth in the field of environmental engineering science and environmental protection today. Developments in nanofiltration, gas separations, pervaporation, and bipolar membrane electrodialysis have vastly enhanced the applicability of membranes, thus attracting more immediate attention. The author in this book deeply discussed with vast foresight membrane chemistry, structure, and function, performance, and engineering models fouling, and cleaning, process design and the applications of nanofiltration and ultrafiltration [14].

Alomair [15] discussed with deep and cogent insight a novel approach to fabricate zeolite membranes for pervaporation separation processes. Technological advancements, scientific verve, and vision stand as pillars of this doctoral thesis [15]. The production of zeolite membranes has vastly developed over the last decade, and the membranes have been used extensively in pervaporation separation processes due to their resistance to chemical and thermal operating conditions. However, the conventional methods used in preparing anisotropic zeolite membranes such as the secondary growth and in-situ crystallization methods involve long and complex procedures that surely requires the preparation of zeolite aluminosilicate gel prior to the fabrication process [15]. Therefore, the vision of this study was to develop and test an easier, less expensive and less time-consuming technique to fabricate different types of zeolite anisotropic membranes [15]. Science and engineering of fabrication of membranes are today in the path of newer scientific regeneration. Membrane separation processes and other novel separation processes are in the process of newer scientific rejuvenation and vast scientific forbearance. The fabrication of zeolite membranes using inexpensive kaolin raw materials taken straight out of the ground was taken into account and deeply investigated. Within this visionary framework, a novel tool of converting raw source alumina and silica, to a pure useful material of zeolite A was developed without any form of pretreatment. Technological validation and scientific girth and strong determination are the pillars of this well-researched treatise. Anisotropic membranes of zeolite A, mordenite, and ZSM–5 were successfully fabricated using a simple and economical method. This is breakthrough research in the field of zeolite membranes and will surely open up newer thoughts and newer visionary areas in the field of membrane fabrication in decades to come [15].

Mohammadi et al. [16] discussed with lucid and cogent insight membrane fouling. Fouling of ultrafiltration membranes in milk industries is mostly due to precipitation of microorganisms, proteins, fats, and minerals on the membrane surface. Technological and scientific validation, the futuristic vision of novel separation processes and the vast scientific imagination in the field of research pursuit will all lead a long and effective way in the true scientific emancipation. Chemical cleaning of membranes is highly essential [16]. In this well-researched paper, results obtained from investigations on a polysulfone ultrafiltration membrane fouled by precipitation of milk components have been deeply presented in minute details [16]. Membrane fouling is a major operational issue in the membrane separation process, and today it is replete with vast scientific and academic rigor [16]. Membrane fouling phenomena present important limitations and vital challenges on the technology applied. Fouling is defined as existence and growth of microorganisms and irreversible collection of materials on the membrane surface which result in a decrease in flux [16]. To mitigate this issue, a cleaning process must be used. Cleaning usually performs in three forms: physical, chemical, and biological. Chemical methods are used most often. The first and preliminary step of chemical washing is finding appropriate materials as cleaning methods. The choice of the best materials depended on feed composition and precipitated layers on the membrane surface and in most cases are performed by trial and error [16]. The selected materials should be chemically stable, safe, cheap, and washable with water. Some of the cleaning agents are acids, bases, enzymes, surfactants, disinfectants, and combined cleaning materials. Scientific research pursuit in the field of fouling is latent yet vastly advanced [16]. This chapter with immense vision and scientific conscience unfolds the scientific intricacies and scientific ingenuity behind membrane fouling. Human scientific rigor and academic rigor are today in the path of newer scientific rejuvenation. The status of global environmental protection is immensely dismal and at a state of grave distress. Technology and engineering science need to be re-envisioned and revamped as regards novel separation processes and membrane science.

Science and engineering globally are in the process of newer regeneration and vast revamping. Technology has today no answers to the heavy metal and arsenic groundwater contamination in South Asia. Here comes the importance of novel separation processes, nanofiltration, and reverse osmosis. The author in this entire chapter focuses with deep scientific diligence the needs of water treatment tools such as membrane science and AOPs. Mankind's immense scientific prowess, the vast world of technological validation and the ever-growing global water crisis will surely lead a long and visionary way in the true emancipation of science and engineering today (Figures 3.1 and 3.2).

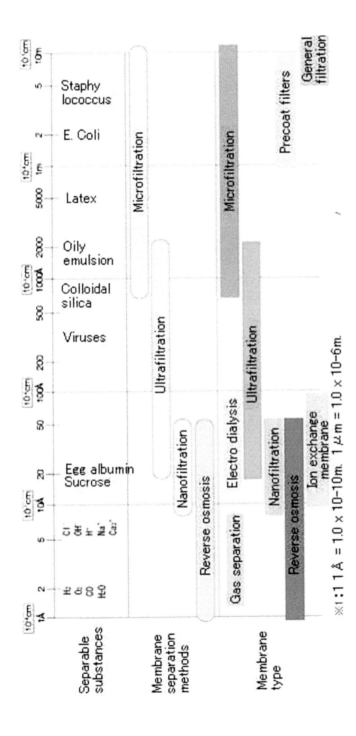

FIGURE 3.1 Classification of various membrane separation processes according to the size of particles.
(Source: Reprinted with permission from Palit, Sukanchan. (2014). Reverse Osmosis and Nanofiltration- Visionary Separation Tools for the Future. i-manager's Journal on Future Engineering and Technology. 10. 1-11. 10.26634/jfet.10.1.2905.)

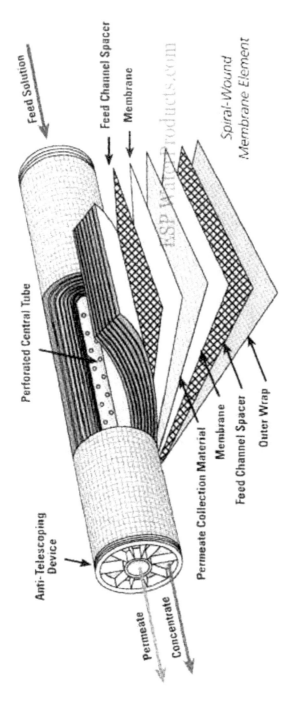

FIGURE 3.2   A sectional view of a spiral wound membrane.

3.10 SIGNIFICANT SCIENTIFIC RESEARCH PURSUIT IN THE FIELD OF ADVANCED OXIDATION PROCESSES (AOPS)

AOPs in a wider sense are a set of chemical treatment techniques designed to remove organic (and sometimes inorganic) materials in water and wastewater by oxidation through reactions with hydroxyl radicals. In real-world applications of wastewater treatment, however, this term usually refers more specifically to branches of such chemical processes that employ ozone (O_3), hydrogen peroxide (H_2O_2) and/or UV light. One such branch of AOP is in-situ chemical oxidation. AOPs rely on in-situ production of highly reactive hydroxyl radicals. These reactive species are the strongest oxidant that can be applied in water and can veritably oxidize any compound present in the water matrix, often at a diffusion controlled reaction speed.

Sharma et al. [17] reviewed with deep scientific vision AOPs for industrial wastewater treatment. Human civilization and human scientific progress today stands in the midst of introspection and immense forbearance. Technological challenges and scientific validation of utmost need in the pursuit of environmental engineering science today. The world of science and engineering needs to be reorganized as regards industrial wastewater treatment and drinking water treatment. AOPs comprise of a remarkable technology for the treatment of wastewaters containing non-easily removable organic compounds. It is the hydroxyl radicals that act with high efficiency to degrade organic compounds [17]. AOP combine ozone, ultraviolet, hydrogen peroxide and catalyst to offer a powerful water treatment solution for the reduction and removal of residual organic compounds as measured by COD, BOD or TOC [17]. This chapter vastly presents a general review of efficient AOPs developed to decolorize and degrade organic pollutants. Human mankind's immense scientific girth and determination, man's scientific ingenuity and the futuristic vision of AOPs will lead a truly greater emancipation of science and engineering today. The authors touched upon advantages and disadvantages of AOPs and thus unearthed the scientific travails and scientific intricacies of this technique.

Gilmour [18] with vast scientific far-sightedness discussed application perspectives of water treatment using AOPs. AOPs using hydroxyl radicals and other oxidative radical species are being studied deeply for removal of organic compounds in industrial wastewater. Human civilization and human scientific progress today stands in the midst of deep

scientific vision and scientific fortitude [18]. Non-conventional environmental engineering tools such as AOPs are the utmost need of the hour.
Large scale applications of these highly effective technologies in water
and wastewater treatment are still very limited due to cost and inadequate
information about the resultant water quality [18]. Mankind's immense
scientific grit and determination, man's vast scientific profundity and the
futuristic vision of environmental protection will all be the forerunners
towards a newer visionary era in the field of environmental engineering
science and chemical process engineering [18]. Today, environmental
engineering science and chemical process engineering are two opposite sides of the visionary scientific coin. This well-researched treatise
focuses on the evaluation of the upstream processing and downstream
post-treatment analysis of selective AOPs. In the first stage of research,
the performance of a proprietary catalyst (VN-TiO$_2$) was compared with
the industry standard TiO$_2$ for the use in a pilot scale immobilized photocatalytic reactor [18]. In the second stage of the study, two bioassays
were used to investigate and compare the toxicity of bisphenol A, and
its degradation intermediates formed in three AOPs namely, ultraviolet/
hydrogen peroxide, ozonation, and photocatalysis [18]. The continuing
process of industrial progress and vast urbanization due to population
growth, deforestation, and pollution are exerting immense pressure on the
depletion of freshwater resources around the world and the case of environmental sustainability. Sustainable development whether it is energy
or environmental is the utmost need of human scientific progress and the
vast world of scientific emancipation today. This thesis opens up newer
thoughts and newer vision in the field of AOPs.

3.11 FUTURE RECOMMENDATIONS OF THE STUDY AND FUTURE RESEARCH THOUGHTS

Drinking water treatment and water purification are the utmost need of the
hour for modern science and present-day human civilization. The challenges and the vision of environmental protection are today immense and
groundbreaking. Future recommendations of the study should be targeted
towards more scientific innovations and path-breaking discoveries. Future
research thoughts in environmental engineering science and chemical
process engineering should be more inclined towards a greater emancipation

of environmental protection and water purification. Scientific progress is today replete with vast scientific and academic rigor. The purpose of this study is to open up new areas of challenges and vision in the areas of nanotechnology applications in environmental engineering and chemical process engineering. Research thoughts and futuristic vision of technology and engineering science will surely lead a long and effective way in the true emancipation of environmental sustainability and water purification. Future recommendations of this study are immense and far-reaching. Research thoughts need to be streamlined as well as envisioned as regards nanofiltration and water purification today. Nanotechnology is today a veritable pillar of scientific progress in modern civilization today. Technological verve, scientific profundity and the vast world of scientific validation are the utmost needs of scientific research pursuit today. This chapter will surely be an eye-opener towards newer innovations and newer ventures in the field of nanofiltration, membrane science, and nanotechnology.

3.12 FUTURE RESEARCH TRENDS AND THE VISION OF SCIENCE

Nanoscience and nanotechnology today needs to be re-envisioned and re-envisaged as global environment, and the holistic domain of environmental engineering science are in the avenues of newer scientific rejuvenation. Arsenic and heavy metal groundwater contamination are changing the face of scientific endeavor today. Global climate change, frequent environmental catastrophes and the depletion of fossil fuel sources are challenging the global research and development initiatives in the field of water and energy. Environmental, as well as energy sustainability, are the torchbearers towards a newer era in the field of science and engineering. The author in this chapter pointedly focuses on the scientific success, the vast scientific imagination and the utmost needs of nanotechnology innovations in environmental protection. Futuristic vision of chemical process engineering and environmental engineering science needs to be re-envisioned and revamped with the passage of scientific history and the visionary timeframe. Human civilization's greatest boon is the basic needs such as water, food, and electricity. In the developing and developing economies around the world, the status of the provision of clean drinking water is dismal and catastrophic. Arsenic and heavy metal groundwater contamination in groundwater and drinking water are troubling and challenging the vast scientific firmament of human civilization today. Thus, the future research trends should be directed towards a greater scientific and technological emancipation of drinking water treatment as well as global climate change mitigation. Science is a colossus and vastly ebullient as human mankind moves forward. This chapter opens up newer thoughts and newer vision in the field of drinking water treatment along with the application of nanotechnology. Nanofiltration and vast avenues of membrane science are today in the midst of scientific introspection and a realm of vision and scientific forbearance. The author in this chapter unravels the scientific difficulties and the scientific barriers in the field of nanofiltration applications in water purification.

3.13   CONCLUSION AND SCIENTIFIC PERSPECTIVES

Technology and engineering science are today in the path of newer scientific regeneration and vast scientific vision. Scientific perspectives in the field of water purification and nanotechnology are highly challenged today with the passage of human scientific history and time. Today, human scientific progress is veritably linked to the science of sustainability whether it is energy, environmental, social or economic. Scientific forbearance, deep scientific intuition, and vast intellect are the utmost needs of research and development in water purification and nanotechnology today. Future perspectives of science and engineering in water treatment should be targeted towards more innovations and scientific instinct. Today, science and engineering are highly challenged with the burning issue of arsenic and heavy metal groundwater contamination. Civil societies and governments around the world need to readdress and surge towards a newer era in the field of water purification, drinking water treatment and industrial wastewater treatment. Human scientific vision and scientific challenges are immensely stressed as global research and development initiatives in water science move forward. Global climate change, frequent environmental disasters, and depletion of fossil fuel resources are changing the face of scientific firmament today. In this chapter, the author with immense vision, scientific girth and scientific determination elucidate on the applications of nanotechnology in environmental protection. The scientific strategy and the scientific vision are to open up newer scientific understanding and scientific wisdom in the field of environmental protection and the vast world of environmental engineering science.

KEYWORDS

- advanced oxidation processes
- hydrogen peroxide
- nanocoatings
- nanomaterials
- nanotechnology
- novel separation processes
- ozone
- UV light

REFERENCES

1. Shannon, M. A., Bohn, P. W., Elimelech, M., Georgiadis, J. G., Marinas, B. J., & Mayes, A. M., (2008). *Science and Technology for Water Purification in the Coming Decades* (Vol. 452, pp. 301–310). Nature Publishing Group, USA.
2. Hashim, M. A., Mukhopadhyay, S., Sahu, J. N., & Sengupta, B., (2011). Remediation technologies for heavy metal contaminated groundwater. *Journal of Environmental Management, 92,* 2355–2388.
3. Choudhury, U. K., Biswas, B. K., Roy, C. T., Samanta, G., Mandal, B. K., Basu, G, C., et al., (2000). Groundwater arsenic contamination in Bangladesh and West Bengal, India. *Environmental Health Perspectives, 108*(5), 393–397.
4. Chakraborti, D., Mukherjee, S. K., Pati, S., Sengupta, M. K., Rahman, M. M., Chowdhury, U. K., et al., (2003). Arsenic groundwater contamination in the Middle Ganga Plain, Bihar, India: A future danger? *Environmental Health Perspectives, 111*(9), 1194–1201.
5. Mukherjee, A., Sengupta, M. K., Hossain, M. A., Ahamed, S., Das, B., Nayak, B., et al., (2006). Arsenic contamination in groundwater: A global perspective with emphasis on the Asian scenario. *Journal of Health Population and Nutrition, 24*(2), 142–163.
6. Chakraborti, D., Das, B., Rahman, M. M., Chowdhury, U. K., Biswas, B., Goswami, A. B., et al., (2009). Status of groundwater arsenic contamination in the state of West Bengal, India: A 20-year study report. *Mol. Nutr. Food Res., 53,* 542–551.
7. Qu, X., Alvarez, P. J. J., & Lin, Q., (2013). Applications of nanotechnology in water and wastewater treatment, *Water Research, 47,* 3931–3946.

8. Mansoori, G. A., Bastami, T. R., Ahmadpour, A., & Eshaghi, Z., (2008). Environmental application of nanotechnology. *Annual Review of Nanoresearch, 2,* Chapter 2, 1–73.

9. Stander, L., & Theodore, L., (2011). Environmental applications of nanotechnology–an update. *International Journal of Environmental Research and Public Health, 8,* 470–479.

10. Filipponi, L., & Sutherland, D., (2013). *Nanotechnologies: Principles, Applications, Implications and Hands-on Activities, (A Compendium for Educators).* Directorate-General for Research and Innovation, Industrial Technologies, European Commission, Luxembourg. ISBN: 9789279214370.

11. Kumar, A., (2014). *Nanotechnology Development in India: An Overview.* Research and Information System for Developing Countries, India. Discussion paper number 193.

12. Roco, M. C., Mirkin, C. A., & Hersam, M. C., (2010). *Nanotechnology Research Directions for Societal Needs in 2020: Retrospective and Outlook.* World Technology Evaluation Centre, Inc, (WTEC), USA Report.

13. Hassan, M. M., (2018). *Arsenic in Groundwater: Poisoning and Risk Assessment.* CRC Press, USA, Taylor and Francis Group LLC, USA (Book).

14. Cheryan, M., (1998). *Ultrafiltration and Microfiltration Handbook.* Technomic Publishing Company, Inc., USA (Book).

15. Alomair, A. A. S., (2013). A novel approach to fabricate zeolite membranes for pervaporation separation processes. Doctor of Philosophy Thesis, School of Chemical Engineering and Analytical Science, University of Manchester, United Kingdom.

16. Mohammadi, T., Madaeni, S. S., & Moghadam, M. K., (2002). Investigation of membrane fouling. *Desalination, 153,* 155–160.

17. Sharma, S., Ruparelia, J. P., & Patel, M. L., (2011). A general review on advanced oxidation processes for wastewater treatment. *International Conference on Current Trends in Technology,* NUICONE, 8–10 December, 2011.

18. Gilmour, C. R., (2012). "Water treatment using advanced oxidation processes: application perspectives." The University of Western Ontario, Canada.

CHAPTER 4

GREEN NANOTECHNOLOGY: AN APPROACH TOWARDS ENVIRONMENT SAFETY

FRANCISCO TORRENS[1] and GLORIA CASTELLANO[2]

[1]Institut Universitari de Ciència Molecular, Universitat de València, Edifici d'Instituts de Paterna, P. O. Box 22085, E-46071 València, Spain, E-mail: torrens@uv.es

[2]Departamento de Ciencias Experimentales y Matemáticas, Facultad de Veterinaria y Ciencias Experimentales, Universidad Católica de Valencia San Vicente Mártir, Guillem de Castro-94, E-46001 València, Spain

ABSTRACT

Sustainable or green nanotechnology (GNT) is commonly used in the development of clean technologies. This work discusses different aspects of GNT in dissimilar fields. It is reasonable to assume that religious root born of material conditions of every society guided certain people by the way of placing growth avidity in front of harmony with the environment. This caused fast development of important philosophical structures that rested on the technical advance, and also set the part of humanity that took the chief role in them in a limit situation in which its own survival capacity is threatened. Concept sustainable development connects directly with West anthropocentric cultural tradition.

4.1 INTRODUCTION

Sustainable or green nanotechnology (GNT) is the use of nanotechnology to conserve and protect the environment. The GNT is commonly used in the development of clean technologies (pollution-free environment). Venturi group reviewed molecular devices and machines, concepts, and perspectives for the nanoworld [1]. They revised the electrochemistry of functional supramolecular systems [2]. They discussed molecular devices and machines from hybrid organic-inorganic structures [3]. They developed sustainability *via* scientific and ethical issues [4]. Hazard screening methods for nanomaterials (NMs) were comparatively studied [5].

Technological revolutions that occurred (e.g., Industrial Revolution of 19th century, technological revolution that was based on the first transfer variator (*transistor*, 1947), integrated circuits (ICs, 1958) [6, 7], solid-state circuits and semiconductors in 20th century] had a transformative impact in society, culture, and scientific community [8]. Scientific and technological transformations that were caused by ICs development were mainly caused by their components miniaturization, based on Moore's law (1965). Miniaturization leads to circuits construction at the molecular or atomic level, which will open the way to molecular-electronics development.

In earlier publications, fractal hybrid-orbital analysis [9, 10], resonance [11], molecular diversity [12], periodic table of the elements [13, 14], law, property, information entropy, molecular classification, simulators [15–23], labor risk prevention, preventive healthcare at work with NMs [24–26], and science and ethics of developing sustainability *via* nanosystems and devices [27] were reviewed. Many researches were performed in different areas in the past many years related to nanotechnology. The present work discusses the dissimilar aspects of GNT in diverse fields. The aim of this work is to initiate debate by suggesting a number of questions (Q), which can arise when addressing subjects of GNT in different fields, and providing, when possible, answers (A) and hypotheses (H).

4.2   GREEN NANOTECHNOLOGY (GNT): AN APPROACH TOWARDS ENVIRONMENT SAFETY

Singh proposed a hypothesis and questions (H/Q) on GNT for environment safety [28].

- H1. GNT motives: products eco-friendly production; developing environment-benefiting products.
- Q1. How is nanotechnology used in biodiversity?
- Q2. How does it help in human health and conservation?

Researches were performed in areas (*cf.* Figure 4.1). He proposed an additional hypothesis.

H2. Photovoltaic (PV)-cell operation needs: light absorption generating electron/hole pairs or *excitons*; separation of opposite-type charge carriers; separate carriers extraction to an external circuit.

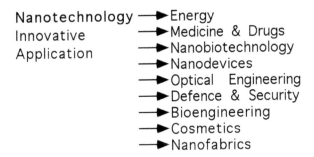

Nanotechnology ➤ Energy
Innovative ➤ Medicine & Drugs
Application ➤ Nanobiotechnology
➤ Nanodevices
➤ Optical Engineering
➤ Defence & Security
➤ Bioengineering
➤ Cosmetics
➤ Nanofabrics

FIGURE 4.1 Areas of nanotechnological research.

4.3 QUANTUM COMPUTING IN INTERNATIONAL BUSINESS MACHINES (IBM)

Gálvez Ramírez proposed Q/H/A on quantum computing in International Business Machines (IBM) [29].

Q1. How are IBM quantum processors (*cf.* Figure 4.2)?

Q2. How do they work?

H1. (Moore, 1965). His law: The number of transistors in a dense integrated circuit doubles approximately every two years.

Q3. (Moore). How much time have people to reach the quantum behavior of nature?

Q4. (Feynman, 1981). Why quantum computing [30]?

A4. (Feynman, 1981). *...nature isn't classical, dammit, and if you want to make a simulation of nature, you'd better make it quantum mechanical, and by golly, it's a wonderful problem because it doesn't look so easy.*

Q5. What does quantum computing solve?

A5. Evaluating functions (Deutsch's algorithm); unordered searches (Grover's algorithm); factoring large numbers (Shor's algorithm).

Computación Cuántica en IBM
IBM Quantum Experience

FIGURE 4.2   Quantum computing in International Business Machines.

4.4   DISCUSSION

As discussed earlier, it is reasonable to assume that religious root born of material conditions of every society has guided certain people by the way of placing growth avidity in front of harmony with the environment. It caused the fast development of important philosophical structures that rested on the technical advancement. Also, it set the part of humanity that took the chief role in them in a limit situation in which its own survival capacity is threatened. Arrived to the point, it is not waited that a dysregulated social evolution produce values that tend to solve the problem before this turn irreversible, and an accorded policy at an international level and strictly measured must be directed to promote the values and produce a pertinent legislation, which is the policy proposed in the report *Our Common Future*.

Concept sustainable development connects directly with West anthropo-centric cultural tradition. One of the greatest achievements is emphasizing, on rising them to the level of international political discussion, synchronic, and diachronic solidarity concepts. Finally, sustainable development passes by understanding that the Earth is a finite system, with a limited external contribution of power, conditions that restrict productive model, and that economy is the tool that human being can handle to orientate development direction (or quality). Although different agents wanted to dispute the concept, this is clear in the terms in which it is defined in the report, being, however, actual policies still to be defined.

4.5 FINAL REMARKS

From the present discussion, the following final remarks can be drawn.

1. To assume that religious root born of material conditions of every society has reasonably guided certain people by the way of placing growth avidity in front of harmony with the environment. This caused fast development of important philosophical structures that rested on the technical advancement. Also, it has placed the part of humanity that took the chief role in them in a limit situation in which its own survival capacity is threatened.
2. Concept sustainable development connects directly with west anthropocentric cultural tradition.

ACKNOWLEDGMENTS

The authors thank the support from Generalitat Valenciana (Project No. PROMETEO/2016/094) and Universidad Católica de Valencia *San Vicente Mártir* (Project No. UCV.PRO.17–18.AIV.03).

KEYWORDS

- **green nanotechnology**
- **human health**
- **nanofabrication**
- **nanomaterials**
- **nanoproduct**
- **nanotechnology**
- **photovoltaic**
- **pollution control**
- **sustainability**

REFERENCES

1. Balzani, V., Credi, A., & Venturi, M., (2008). *Molecular Devices and Machines: Concepts and Perspectives for the Nanoworld.* Wiley-VCH: Weinheim, Germany.
2. Ceroni, P., Credi, A., & Venturi, M., (2010). *Electrochemistry of Functional Supramolecular Systems.* Wiley, New York, NY.
3. Venturi, M., Iorga, M. I., & Putz, M. V., (2017). Molecular devices and machines: Hybrid organic-inorganic structures. *Curr. Org. Chem., 21,* 2731–3759.
4. Venturi, M., (2017). Developing sustainability: Some scientific and ethical issues. In: Putz, M. V., & Mirica, M. C., (eds.), *Sustainable Nanosystems Development, Properties, and Applications* (pp. 657–680). IGI Global: Hershey, PA.
5. Sheehan, B., Murphy, F., Mullins, M., Furxhi, I., Costa, A. L., Simeone, F. C., & Mantecca, P., (2018). Hazard screening methods for nanomaterials: A comparative study. *Int. J. Mol. Sci., 19*(649), 1–22.
6. Uribe, G. M., & López, J. L. R., (2007). Nanoscience and nanotechnology: An ongoing revolution. *Revista Perfiles Latinoamericanos, 14*(29), 161–186.
7. Poole, C. P., Jr., & Owens, F. J., (2007). *Introduction to Nanotechnology,* Reverté: Barcelona, Spain.
8. Ojeda, S. J. H., Cortés, P. J. C., Gómez, C. J. A., & Duque, C. A., (2018). Current's fluctuations through molecular wires composed of thiophene rings, *Molecules, 23*(881), 1–19.
9. Torrens, F., (2001). Fractals for hybrid orbitals in protein models. *Complexity Int., 8*(01), 1–13.
10. Torrens, F. Fractal hybrid-orbital analysis of the protein tertiary structure. *Complexity Int.,* (submitted for publication).
11. Torrens, F., & Castellano, G., (2009). Resonance in interacting induced-dipole polarizing force fields: Application to force-field derivatives. *Algorithms, 2,* 437–447.
12. Torrens, F., & Castellano, G., (2012). Molecular diversity classification *via* information theory: A review. *ICST Trans. Complex Syst., 12*(10–12), e4-1–8.
13. Torrens, F., & Castellano, G., (2015). Reflections on the nature of the periodic table of the elements: Implications in chemical education. In: Seijas, J. A., Vázquez, T. M. P., & Lin, S. K., (eds.), *Synthetic Organic Chemistry* (Vol. 18, pp. 8-1–15). MDPI: Basel, Switzerland.
14. Putz, M. V. *New Frontiers in Nanochemistry: Concepts, Theories, and Trends,* Apple Academic–CRC Press: Waretown, NJ, in press.
15. Torrens, F., & Castellano, G., (2015). Reflections on the cultural history of nanominiaturization and quantum simulators (Computers). In: Laguarda, M. N., Masot, P. R., & Brun, S. E., (eds.), *Sensors and Molecular Recognition* (Vol. 9, pp. 1–7). Universidad Politécnica de Valencia: València, Spain.
16. Torrens, F., & Castellano, G., (2016). Ideas in the history of nano/miniaturization and (quantum) simulators: Feynman, education and research reorientation in translational science. In: Seijas, J. A., Vázquez, T. M. P., & Lin, S. K., (eds.), *Synthetic Organic Chemistry* (Vol. 19, pp. 1–16). MDPI: Basel, Switzerland.
17. Torrens, F., & Castellano, G., (2016). Nanominiaturization and quantum computing. In: Costero, N. A. M., Parra, Á. M., Gaviña, C. P., & Gil, G. S., (eds.), *Sensors and Molecular Recognition* (Vol. 10, pp. 31-1–5). Universitat de València: València, Spain.

18. Torrens, F., & Castellano, G. (2018). Nanominiaturization, Classical/Quantum Computers/ Simulators, Superconductivity and Universe. In: *Methodologies and Applications for Analytical and Physical Chemistry*, (pp. 27–44). Haghi, A. K., Thomas, S., Palit, S., & Main, P., (eds.); Apple Academic Press–CRC Press: Waretown, NJ.

19. Torrens, F., & Castellano, G. (2018). Superconductors, Superconductivity, BCS Theory and Entangled Photons for Quantum Computing. In: *Physical Chemistry for Engineering and Applied Sciences: Theoretical and Methodological Implication* (pp. 379–387), Haghi, A. K., Aguilar, C. N., Thomas, S., Praveen, K. M., (eds.); Apple Academic Press–CRC Press: Waretown, NJ.

20. Torrens, F., & Castellano, G. (2019). EPR Paradox, Quantum Decoherence, Qubits, Goals and Opportunities in Quantum Simulation. In Theoretical Models and Experimental Approaches in Physical Chemistry: Research Methodology and Practical Methods (Vol 5, pp. 319–336), Haghi, A. K., Thomas, S., Praveen, K. M., Pai, A. R., (eds.); Apple Academic Press–CRC Press: Waretown, NJ.

21. Torrens, F., & Castellano, G. (2019). Nanomaterials, Molecular Ion Magnets, Ultrastrong and Spin–Orbit Couplings in Quantum Materials. In: Physical Chemistry for Chemists and Chemical Engineers: Multidisciplinary Research Perspectives (pp. 181–190), Vakhrushev, A. V., Haghi, R., de Julián-Ortiz, J. V., (eds.); Apple Academic Press–CRC Press: Waretown, NJ.

22. Torrens, F., & Castellano, G. (2019). Nanodevices and Organization of Single Ion Magnets and Spin Qubits. In: *Chemical Science and Engineering Technology: Perspectives on Interdisciplinary Research* (pp. 67–74), Balköse, D., Ribeiro, A. C. F., Haghi, A. K., Ameta, S.C., Chakraborty, T., (eds.); Apple Academic Press–CRC Press: Waretown, NJ.

23. Torrens, F., & Castellano, G. (2019). Superconductivity and Quantum Computing via Magnetic Molecules. In New Insights in Chemical Engineering and Computational Chemistry (pp. 201–209), Haghi, A. K., (ed.); Apple Academic Press–CRC Press: Waretown, NJ.

24. Torrens, F., & Castellano, G., (2016). *Book of Abstracts, Certamen Integral de la Prevención y el Bienestar Laboral* (p. 3). València, Spain, Generalitat Valenciana– INVASSAT: València, Spain.

25. Torrens, F., & Castellano, G. (2018). Nanoscience: From a Two-Dimensional to a Three-Dimensional Periodic Table of the Elements. In Methodologies and Applications for Analytical and Physical Chemistry (pp. 3–26), Haghi, A. K., Thomas, S., Palit, S., Main, P., (eds.); Apple Academic Press–CRC Press: Waretown, NJ.

26. Torrens, F., & Castellano, G. (2018). Work with Nanomaterials: Reductionism/Positivist and Ethics Philosophical Considerations. In *Tecnología e Innovación Social: Hacia un Desarrollo Inclusivo y Sostenible* (pp. 11–35), Feltrero, R., (ed.); Desafíos Intelectuales del Siglo XXI No. 1, Global Knowledge Academics: Cantoblanco, Madrid, Spain.

27. Torrens, F., & Castellano, G. (2019). Developing Sustainability via Nanosystems and Devices: Science–Ethics. In *Chemical Science and Engineering Technology: Perspectives on Interdisciplinary Research* (pp. 75–84), Balköse, D., Ribeiro, A. C. F., Haghi, A. K., Ameta, S. C., Chakraborty, T., (eds.); Apple Academic Press–CRC Press: Waretown, NJ.

28. Singh, A., (2018). Green nanotechnology: An approach toward environmental safety. In: Haghi, A. K., Balköse, D., Mukbaniani, O. V., & Mercader, A. G., (eds.), *Applied Chemistry and Chemical Engineering* (Vol. 1, pp. 245–252). Apple Academic–CRC: Waretown, NJ.

29. Gálvez, R. F. J. *Personal Communication.*

30. Feynman, R. P., (1982). Simulating physics with computers. *Int. J. Theor. Phys., 21,* 467–488.

CHAPTER 5

GREEN NANOTECHNOLOGY, GREEN NANOMATERIALS, AND GREEN CHEMISTRY: A FAR-REACHING REVIEW AND A VISION FOR THE FUTURE

SUKANCHAN PALIT[1,2]

[1]Assistant Professor (Senior Scale), Department of Chemical Engineering, University of Petroleum and Energy Studies, Bidholi via Premnagar (P.O.), Dehradun–248007, Uttarakhand, India

[2]43, Judges Bagan, Haridevpur (P.O.), Kolkata–700082, India, Tel.: 0091-8958728093, E-mail: sukanchan68@gmail.com, sukanchan92@gmail.com

ABSTRACT

The world of science and engineering are today witnessing immense scientific challenges. Science and technology are today two huge colossus with a definite and purposeful vision of its own. Similarly, nanotechnology and green nanomaterials are today in the path of newer scientific regeneration. In this chapter, the author deeply elucidates the scientific success, the vast scientific ingenuity and the utmost engineering needs of human civilization today. This well-researched treatise deeply points out towards the recent scientific advances in the field of green nanotechnology, green nanomaterials and green chemistry with a clear vision towards greater emancipation of science and engineering globally. The author also elucidates with lucid and cogent insight the application of the science of sustainability in the future scientific progress and the march of human civilization. Today, scientific and academic rigor in the field of nanotechnology and green nanotechnology stands in the midst of vast vision and deep scientific forbearance. This chapter deeply with immense scientific conscience comprehends the necessity of green nanotechnology and green nanomaterials in the future of human civilization and human scientific progress. The other areas of nanotechnology applications deeply comprehended in this chapter are the areas of arsenic and heavy metal groundwater remediation and the recent scientific advances in the field of sustainability science. This chapter will surely open up newer research thoughts and newer scientific ingenuity in the field of green nanotechnology in decades to come.

5.1   INTRODUCTION

The domain of nanoscience and nanotechnology are today moving at a rapid pace surpassing one visionary frontier over another. Global climate change and global warming are today challenging the vast scientific firmament of vision and scientific determination. Depletion of fossil fuel resources is also devastating the vast scientific landscape. Environmental engineering and petroleum engineering science are the branches of engineering science which are highly challenged and thus needs to be re-envisioned. Heavy metal and arsenic groundwater contamination are changing the face of human civilization and human scientific progress today. In this well-researched compendium, the author deeply elucidates on the scientific success, the scientific travails and the vast scientific profundity in nanotechnology applications in the field of environmental engineering, petroleum engineering, and green engineering. The field of environmental engineering is highly challenged today. Frequent environmental disasters are causing immense concerns to the scientific community. In this entire treatise, the author depicts the scientific needs, the vast scientific profundity and the immense scientific fortitude in green nanotechnology applications in a human society profoundly. Green nanotechnology and green engineering are the utmost needs of human civilization today. Green engineering and environmental protection are the two opposite sides of the visionary coin. This chapter opens up newer thoughts and newer visionary avenues in the field of environmental engineering and green nanotechnology in decades to come.

5.2   THE AIM AND OBJECTIVE OF THIS STUDY

Human civilization and human scientific rigor are today in the path of newer scientific regeneration and deep scientific vision. Academic and scientific rigor in the field of green nanotechnology and environmental engineering science needs to be re-envisioned and readdressed with every step of human scientific advancements. Thus, the aim and objective of this treatise is to widely open up new vistas of scientific research pursuit in nanotechnology, green nanotechnology, and green nanomaterials. Human civilization's immense scientific prowess, the futuristic vision of technological advancements and the utmost needs of the science of nanotechnology will all lead a long and visionary way in the true emancipation of science and engineering. The author in this treatise deeply elucidates the recent advances in the field of green nanotechnology, green nanomaterials, and green chemistry. The sole aim of this chapter is to widely elucidate the technological advances, the scientific motivation and the scientific subtleties in the field of nanoscience and nanotechnology. The world of environmental engineering science today stands deeply devastated with the frequent environmental catastrophes and the vast necessity of engineering and science. The author deeply focuses on the scientific success and the technological ingenuity in the path towards emancipation in environmental engineering and nanotechnology. Green engineering and green chemistry are the hallmarks of scientific research endeavor in science and technology in present-day human civilization. This chapter will veritably open newer thoughts and newer ingenuity in nanotechnology applications in diverse areas of science and engineering today.

5.3   THE NEED AND THE RATIONALE OF THIS STUDY

The challenges and the vision of science and engineering are immense and ground-breaking today. Human civilization and human scientific progress today stands in the crossroads of vision and scientific fortitude. The research and development initiatives of today's science should be directed towards greater emancipation of environmental protection and nanotechnology applications. The need and the rationale of this study go beyond scientific imagination and scientific rationality in the research and development forays in environmental engineering and nanotechnology. Science and engineering of nanotechnology are today huge colossus with a definite vision of its own. Green nanotechnology and green nanomaterials are the marvels of science today. This is the primary aim of this research initiative. Global warming, frequent environmental disasters and the cause of provision of basic human needs will surely be the forerunners towards a newer era in the field of environmental protection today. Technology today is in the path of visionary advancements. Nuclear science and space technology are today integrated with nanotechnology in every knowledge dimensions, and thus, the vital need of this well-researched treatise. The scientific judgment, deep scientific discernment, and the futuristic vision of nanotechnology will all lead a long visionary in the real emancipation of green nanotechnology and green nanomaterials today.

5.4   GREEN NANOTECHNOLOGY AND ITS VAST SCIENTIFIC DOCTRINE

Green nanotechnology refers to the use of nanotechnology to enhance the field of sustainability of processes producing negative externalities. It also refers to the use of products of nanotechnology to enhance sustainability. It includes making green-nanoproducts and using nanoproducts in support of sustainability and the holistic world of sustainable development. The vision, the challenges and the scientific doctrine of green nanotechnology are vastly enumerated in this chapter.

Today environmental engineering science stands in the midst of deep vision and solecism. Nanotechnology applications should be at the forefront of environmental engineering science and green engineering globally today. Scientific profundity, deep scientific subtleties, and the vast scientific ingenuity are the hallmarks of scientific research pursuit in green nanotechnology and green nanomaterials. Nanotechnology should be integrated with green engineering, and green chemistry as human civilization and human scientific progress surges forward towards a new knowledge dimension. The futuristic vision of science and engineering, the needs of human society and the world of scientific validation will all lead a long and visionary way in the true emancipation of the science of nanotechnology applications. Nanotechnology is a vastly interdisciplinary field of scientific endeavor that demands to manage, produce, and develop novel opportunity to use science, engineering, and newer approaches with the nanoscale invention to increasingly support human and environmental health. Nanotechnology encompasses scientific interest that deals with the manipulation of individual molecules at a supramolecular range of 100 nanometers. Vast nanotechnology research and development initiatives reflect the improvement in design and application of nanomaterials, devices, and models that exhibits more visionary and sustainable future. Thus, the recognition of such types of nanotechnology that led to the development of the "green nanoscience." Green nanotechnology has foundations in the field of green nanochemistry that reflects the main aim of nanotechnology to create eco-friendly nanoobjects to reduce environmental hazards by application of green nanoproducts. Technological verve, scientific articulation, and deep scientific validation are the utmost needs of green nanotechnology applications today. Green nanotechnology with precise details refers to the field of nanotechnology that propels environmental sustainability and maintains the eco-friendly environment. Green nanoparticles synthesized from the different green nanotechnological approaches consist of well-defined chemical composition, size, and applications in many technological and scientific arenas. Green nanoparticles developed through eco-friendly tools have considerable and vast importance in areas of medical biology, industrial microbiology, environmental microbiology, bioremediation, environmental nanotechnology, clean technology, and electronics.

5.5 GREEN NANOMATERIALS–THE SCIENTIFIC TRUTH AND THE VISION FOR THE FUTURE

Green nanomaterials along with green nanotechnology are today at the forefront of human scientific endeavor. The scientific vision and the vast scientific truth needs to be re-envisioned and re-envisaged with the passage of human scientific history and time. Nanotechnology uses key methods to generate new products and to enhance the properties of a broad range of market products of electronics, packaging, healthcare, and coatings. This use of nanotechnology is enhanced by green nanotechnology. There are two methods for the synthesis of nanoparticles, one includes chemical synthesis and the other is focused on green synthesis. The importance of nanotechnology in research areas emphasize on the synthesis of nanoparticles with different chemical compositions, sizes, morphologies, and controlled dispersities. The nanoparticles synthesized by through the chemical methods involve chemical reduction, using different metals and chemicals such as sodium citrate, ascorbate, sodium borohydride, etc. The scientific truth and the vast scientific vision behind green nanomaterials applications to human society are path-breaking and surpassing vast and versatile scientific frontiers. Engineering nanomaterials and its vast and versatile applications in green engineering and green chemistry are the utmost need of the hour. Human mankind's immense scientific prowess and scientific determination, the futuristic vision of nanotechnology and the utmost needs of green engineering will all lead a long and visionary way in the true emancipation of science and engineering globally. Thus, the immediate need of green engineering and green nanotechnology. Nano-technology is a relatively a prominent and a new discipline. It is the latest research and development foray of modern technology and has applications in several human disease-related problems. Studies at the nanoscale were enabled and envisioned by the visionary and practical development of instruments in the 1980s by talented visionaries and scientists including and not limited to Binnig, Roher, Gerber, and Quake working at IBM and beyond. Science and engineering of nanotechnology thus ushered in a newer era in the field of scientific validation and technological validation. Thus, scientific innovation resulted in a newer knowledge dimension in the field of nanotechnology. Specifically, green nanotechnology encour-ages collaborations and research and development initiatives that veritably uses scientific research to surge towards sustainability and sustainable development fundamentally.

5.6　GREEN CHEMISTRY AND THE SCIENTIFIC PROFUNDITY BEHIND IT

Green chemistry and green engineering today are at the forefront of scientific research pursuit globally. Scientific profundity, deep scientific ingenuity, and scientific vision are the hallmarks of research endeavor in the field of green chemistry and green engineering today. Environmental protection and environmental engineering science are the utmost needs of human civilization today. Green chemistry also called sustainable chemistry is an area of chemistry, and chemical engineering focused on the designing of products and chemical processes that minimize the use and generation of hazardous substances. Whereas environmental chemistry encompasses the effects of polluting chemicals on nature and its surroundings, green chemistry focuses on the environmental impact of chemistry, including vast technological approaches to mitigating pollution and reducing the consumption of non-renewable resources. Environmental and energy sustainability are the necessities of human civilization and human scientific progress today. The visionary definition of the science of "sustainability" by Dr. Gro Harlem Brundtland, former Prime Minister of Norway needs to be re-envisioned and re-envisaged as human civilization surges forward. The over-arching goals of green chemistry are more resource efficient and inherently safer design of molecules, materials, products, and processes and can be pursued in a wide range of scientific contexts. Green chemistry vastly emerged from a variety of existing scientific ideas and research pursuits (such atom economy and catalysis)in the period leading up to 1990s in the vast and versatile contexts of increasing attention to problems of chemical pollution and loss of resources and also fossil fuel resources. Human civilization's immense scientific girth and determination, the immense scientific and academic rigor behind green nanotechnology and the futuristic vision of nanomaterials will all lead a long and visionary way in the true emancipation of science and engineering.

5.7 ENVIRONMENTAL SUSTAINABILITY AND THE MARCH OF HUMAN CIVILIZATION

Today environmental engineering science and environmental sustainability stand in the midst of deep scientific vision and immense scientific fortitude. There is an immense need of sustainable development in present-day human civilization. Environmental and energy sustainability are today the cornerstones of human scientific rigor. Human civilization today stands amidst deep scientific catastrophe as well as scientific comprehension. Science today is a huge colossus with a definite vision and a definite goal of its own. The march of human civilization today is path-breaking as well as visionary. The vision of Dr. Gro Harlem Brundtland, former Prime Minister of Norway, needs to be readdressed and re-envisaged with the passage of scientific history and the visionary timeframe. Today, the vast domain of environmental protection, drinking water treatment, and industrial wastewater treatment stands in the crossroads of vision, scientific grit, and immense determination. Heavy metal and arsenic groundwater and drinking water contamination in South Asia are veritably challenging the vast scientific firmament of immense might and vision. In India and Bangladesh, arsenic groundwater and drinking water contamination are changing the face of human scientific progress and the scientific rigor. Health issues engulfing the drinking water arsenic contamination are veritably challenging the vast scientific fabric. In this chapter, the author profoundly discusses the utmost needs of environmental sustainability and the innovations of environmental protection science to human society.

5.8   GREEN NANOMATERIALS AND THE SCIENCE OF SUSTAINABILITY

Today human civilization and human scientific rigor stand in the midst of deep scientific vision and immense scientific introspection. Green nanotechnology and environmental sustainability are veritably linked by an unsevered umbilical cord. In this chapter, the author deeply reviews the utmost necessity and the human scientific ingenuity in the application of energy and environmental sustainability in the progress of human civilization. Water purification also is linked with environmental sustainability today. Provision of basic human needs such as water, food, shelter, and energy are the primary objectives of human civilization today. Thus, the immediate needs of environmental and energy sustainability. Green nanotechnology, green nanomaterials and the vast vision of nanotechnology are the frontiers of scientific endeavor today. This chapter opens up newer scientific thoughts and newer scientific vision in the field of nanotechnology and green engineering in decades to come.

5.9   RECENT SCIENTIFIC ADVANCES IN THE FIELD OF GREEN NANOMATERIALS

Green nanomaterials are the smart materials of today. The futuristic vision of green nanotechnology, the needs of human science and society and the scientific prowess of human civilization will surely lead a long and visionary way in the true emancipation of nanoscience and nanotechnology today. Nanomaterials, engineered nanomaterials, and green engineering are today in the path of a newer scientific regeneration. There are today many issues as regards application of nanomaterials and engineered nanomaterials to human society. Thus, the scientific progress and the scientific and academic rigor in the field of nanomaterials needs to be revamped and re-organized. In this chapter, the author pointedly focuses on scientific success, the deep scientific profundity and the needs of the science of green nanotechnology and green engineering in the progress of human civilization. Technological vision and scientific validation need to be deeply revisited and reorganized as human civilization, and human scientific fortitude surpasses one visionary frontier over another. The challenges and the vision of nanoscience and nanotechnology are enormous and groundbreaking. This treatise opens up newer vision and newer thoughts in the scientific research pursuit in the field of engineered nanomaterials.

Lu et al. [1] deeply discussed with immense scientific far-sightedness green nanomaterials and its track for a sustainable future. Human civilization today is in the path of newer scientific regeneration as regards sustainable development [1]. Nanotechnology is one of the most prominent scientific breakthroughs and research and development initiative of the twenty-first century. With vast applications that surpasses scientific frontiers- from electronics to medicine, to advanced manufacturing, to cosmetics-nanoscience and nanotechnology has the visionary potential to dramatically change lifestyles, jobs, and the whole economy of a nation globally [1]. However, in the majority, many of the materials and processes currently used are not only dependent on nonrenewable resources but also create hazardous wastes. Technology and engineering science today globally are in the midst of immense scientific vision and scientific fortitude [1]. The combination of green chemistry techniques and green engineering with nanotechnology applications has thus become a key component of nanotechnology future globally [1]. The science and engineering of nanotechnology are witnessing immense and drastic challenges. The use of natural ingredients to synthesize nanomaterials and design environmentally benign synthetic processes has been extensively investigated globally. While many of the so-called green technologies are now finding their visionary way from the laboratory to commercial applications, green nanotechnology still today faces immense challenges. The march of science and engineering of nanoscience and nanotechnology are today surpassing vast and versatile scientific boundaries [1]. In this chapter, the author deeply discusses the recent advances and challenges in green nanotechnology and suggest improvements for the commercial readiness of these visionary and scientifically inspiring technologies. In the past few decades, the application of nanomaterials has assumed immense importance and nanomaterials have demonstrated superior applications in medicine, energy engineering, material science, and advanced manufacturing. But there are environmental engineering constraints to the research and development forays of nanomaterials and nanotechnology. Thus, the need of a research and development journey in green nanotechnology. Green nanotechnology, the combination of nanotechnology and the incisive principles of green chemistry may hold the key to an environmentally sustainable future of human scientific progress. Academic and scientific rigor in the field of green nanotechnology and green nanomaterials is vastly changing as research endeavor is interspersed with vision and deep scientific determination [1]. Green chemistry and green engineering are a set of basic principles or rather a chemical philosophy that vastly encourages the design of products and sustainable processes that reduce or ameliorate the use and generation of hazardous end products [1]. The authors deeply discussed with scientific conscience and vast scientific jurisprudence natural, renewable sources of reducing agents, natural sources as precursors for carbon nanomaterials, the applications of nanocellulose as a green nanomaterial, green processing, and the vast instinctive research challenges [1]. Natural product chemistry is a visionary avenue of scientific research pursuit today. Nature provides us with numerous chemical substances that serve as reducing agents for the synthesis of nanoparticles, including plant extracts, biopolymers, vitamins, proteins, peptides, and sugars [1]. Plant extracts are the most studied category of natural product chemistry today. Given their immense abundance, plant extracts are regarded as one of the most promising natural reducing agents [1]. The scientific success of the synergy between biotechnology and nanotechnology are deeply discussed in this chapter. One area of immense success in green nanotechnology is the synthesis of metal nanoparticles, useful in electronics and medical applications, using plant extracts as reducing agents [1]. Biomedical applications such as drug and gene delivery using gold and silver nanoparticles has been an active research pursuit [1]. To vastly improve the immense biocompatibility, non-toxic, green reduction agents, plants, algae, bacteria, and fungi were used [1]. Technological vision and vast scientific profundity are the pillars of this research endeavor. Biopolymers are another family of natural sources used as reducing and stabilizing agents for metal nanoparticle synthesis [1]. These polymeric carbohydrate molecules have already been used in various industries and thus are readily available for large scale production of nanoparticles [1]. Examples of biopolymers for nanoparticle synthesis are cellulose, chitosan, and dextran which are isolated from plants, the exoskeleton of crustaceans, and sugarcane respectively [1]. Biotechnology and nanotechnology are today connected veritably by an unsevered umbilical cord [1]. One promising and emerging area of green nanotechnology research is the use of natural resources as precursors for carbon nanomaterials useful in a host of applications in engineering and science [1]. For example, vegetable oil has vastly proven itself to be a viable precursor for high-quality carbon nanotubes using a spray pyrolysis approach [1]. A significant research and development initiative of green nanotechnology is the vigorous development of a green chemistry approach to synthesize nanoparticles [1]. However, recently much attention is given to the use of natural nanomaterials as alternatives to synthetic products. Technological vision, the world of scientific validation and the futuristic vision of green nanotechnology are the veritable forerunners towards a newer era in nanotechnology, green engineering and green chemistry [1].

Nath et al. [2] discussed with deep details green nanomaterial and how green they are as a biotherapeutic material. [2] Technological vision of human mankind, the scientific prowess of nanomaterial and the needs of human society such as water and energy will all lead a long and visionary way in the true realization of science and engineering today. Nanotechnology is a relatively a newer discipline. It is the latest type of modern technology and has vast and varied applications in diverse areas including drug delivery and therapy. [2] Green nanomaterial is showing the path of a newer scientific rejuvenation and a newer vision [2]. These application opportunities are based on the unique properties (e.g., magnetic, optical, mechanical, and electronic) that vary continuously or abruptly with the changes in the size of the materials at the nanoscale. Advances in the vast domain of nanotechnology have significantly affected the field of therapeutics delivery [2]. Although there has been a promising progress the design of disease-targeted NPs allows new treatments with improved specificity, only a few NP-based medicines have reached the global healthcare market [2]. There now needs a new domain–nanotoxicology–that will evaluate the health effects posed by nanoparticles, and would enable safe development of the emerging nanotechnology industry related to biotherapy. Technological innovations and vast scientific validation are the needs of research and development pursuit today. In this chapter, the author deeply comprehends the areas of nanobiotechnology, biotechnology, and the imminent needs of nanotoxicology [2]. Green nanotechnology gives the opportunity in lowering the risk of using nanomaterials, limiting the risk of using nanomaterials, and using nanomaterials to lower the risk of producing unwanted chemical intermediates and end-products [2]. Nanotechnology is a relatively promising and visionary arena of scientific research pursuit. It is the latest hype of modern technology and has tremendous applications in several human disease-related problems [2]. The vast challenges and the instinctive vision in nanotechnology applications in medical science and agriculture need to be addressed and re-envisioned with the passage of scientific history and the visionary global timeframe [2]. Studies at the nanoscale started the practical development of instruments in the 1980s by talented nanotechnologists and physicists but not limited to Binnig, Roher, Gerber, and Quake working at IBM and beyond [2]. Nanotechnology is an understanding and control of matter at the nanoscale, at dimensions between 1 and 100 nanometers, where unique phenomena enable novel applications. Novel technologies today also involve application areas of nanofiltration and other membrane separation processes in environmental engineering science [2]. Encompassing nanoscale science, engineering, and technology, nanotechnology also involves imaging, measuring, modeling, and manipulating matter at this length scale.

European Commission Report [3] discussed with deep scientific conscience and immense lucidity challenges and opportunities to green and nanotechnologies. This is a critical report of opportunities and risks in nanotechnologies on the present day human scientific progress. Nanotechnologies are the science and the business of manipulating matter at atomic scale [3]. Materials veritably produced with the aid of nanotechnologies are starting to be used in many areas of human life (cosmetics, clothing fabrics, sports equipment, paints, packaging, food, etc.). As the vast and varied applications expanded, many nanotechnology proponents are positioning green nanotechnology towards a more environmentally sustainable future [3]. Thus, the review of the scientific difficulties and the scientific success in green nanotechnology applications and the vast forays of green nanotechnology in diverse areas of science and engineering. Today, the scientific barriers of nanotechnology applications are wide and groundbreaking. It is essential for environmental NGOs to gain knowledge on different aspects of the vastly emerging technology development and governance debates, in respect to critically discussing the promotion of nanotechnologies for use in green nanotechnologies [3]. As the applications expand, many proponents of nanovision are positioning nanotechnologies as a part of a greener sustainable future [3]. There is a sound basis to these claims and nanotechnologies will lead to more toxic materials, more production and consumption, and a certain decrease of control over how to create and live our lives [3]. In this important context, it is highly essential for environmental NGOs to gain immense knowledge on different aspects of emerging nanotechnology development and governance debates, primarily in relation to critically discussing the promotion of nanotechnologies for use in green technologies (i.e., for renewable energy production and water filtration) [3]. Technological vision, scientific motivation and the vast world of scientific validation are the challenges and the vision of the science of nanotechnology today. This report opens up newer thoughts and newer visionary aisles in the field of nanotechnology applications in diverse areas of science and engineering. This series of papers is meant to serve as a capacity-building technique empowering environmental NGOs to work actively in the field of sustainable governance and the use of nanotechnologies and nanomaterials. The outlines and the salient features in this issue are:

- Challenges and opportunities to green nanotechnologies [3].
- Environment, health, and safety research and vast emancipation [3].
- Regulatory status and initiatives in Europe and the rest of the world on nanomaterials.
- NGO guidelines on sustainability assessment of nanotechnology and nanomaterials [3].

The series of scientific deliberations will examine a wide number of questions around nanotechnologies development and use,

- How does the use of nanotechnologies and materials have an impact on biodiversity, resource conservation, ecosystems, and human health? [3]
- What are the uncertainties regarding their environmental and health effects? [3]
- Do the risks outweigh the benefits or do the benefits outweigh the risks
- What are the socio-political implications in the application of nano-technology to human society? [3]

The author deeply gleans the challenges and opportunities to green nanotechnologies. The authors of this report questions, the challenges, and the solutions of the science of nanotechnology. Climate change, over-dependence on finite fossil fuels for energy generation, over-exploitation, and depletion of fossil fuel resources as well as the impacts of the Western economies being based on excessive production and consumption are amongst the biggest challenges of the 21st century. Human civilization's immense scientific prowess as regards nanotechnology, the futuristic vision of nanoscience and the needs of sustainable development will all lead a long and visionary way in the true emancipation of science, engineering, and the diverse areas of engineering science [3]. The purpose of this chapter is to review the promises and opportunities that nanotechnological solutions offer in the above areas and to assess if these promises can be fulfilled [3]. Human technological advancements and the vast scientific profundity are today in the path of newer regeneration and vision. Nanotechnology has been positioned as the next technological revolution, but as such it does not occur in isolation [3]. Any technology is not just a set of engineering feats but is centrally positioned within the scientific profundity and scientific ingenuity of cultural boundaries. Science and engineering are a huge colossus with a vast vision of its own today. Nanotechnology is today surpassing every scientific boundary. The report also gives details of green chemistry and green technology. Green chemistry, green technology, and green engineering are a set of design and manufacturing principles trying to address the demands by finding ways to eliminate the use of toxic ingredients, to manufacture at low temperatures, to use less energy and use renewable inputs wherever possible and finally apply life cycle thinking to design and engineering of products. This entire treatise widely elucidates on the vast scientific success and the scientific ingenuity in nanotechnology applications to human society. This deliberation is of utmost necessity in the future emancipation of nanoscience and nanoengineering [3].

5.10 SIGNIFICANT SCIENTIFIC RESEARCH PURSUIT IN THE FIELD OF GREEN NANOTECHNOLOGY

Green nanotechnology and green nanomaterials are the challenges and the vision of engineering science and technology today. Human scientific ingenuity as respects environmental protection and environmental engineering science are today in a state of immense scientific distress. Today, technology and engineering science has practically no answers to the monstrous issues of drinking water treatment and industrial wastewater treatment. Here comes the importance of scientific articulation in green nanotechnology.

Verma et al. [4] discussed with immense vision and scientific foresight the science of green nanotechnology. Nanotechnology covers a unique phenomenon that veritably enables novel applications in different fields. Nanotechnology promises a sustainable future by its growth in green chemistry to develop into green nanoscience and nanotechnology [4]. This review widely reflects how nanotechnology can be advantageous as a green alternative in different aspects of nanoparticle synthesis. As the scientific community knows, nanotechnology is an interdisciplinary field of science that demands to manage, produce, and develop novel opportunity to use science, engineering, and new approaches with nanoscale innovation to support human and environmental health [4]. The march of science and engineering are today surpassing vast and versatile visionary boundaries. In a similar vein, green nanotechnology and green nanomaterials are in the avenues of newer scientific regeneration and vast scientific rejuvenation. Green nanomaterials synthesized from the different green nanotechnological approaches consist of well-defined chemical composition, size, and applications in many scientific and technological arenas [4]. Green nanomaterials developed through eco-friendly techniques have considerable importance in areas of medical biology, industrial microbiology, environmental microbiology, bioremediation, clean technology, and electronics. Scientific evolution, deep scientific ingenuity, and vast scientific imagination are the needs of innovation and deep scientific invention today [4]. Globally the innovative approaches in nanotechnology can be spread through people forming societies to acknowledge the research of green and chemical methods for the successful application of sustainability in the engineering field and medical biology [4]. Today, nanotechnology uses key methods to generate new methods to generate new products and to enhance the properties of a broad range of market products of electronics, packaging, healthcare, and coatings [4]. The vast use of nanotechnology is enhanced by green nanotechnology. There are two methods for the synthesis of nanomaterials, one includes Chemical Synthesis, and the other focused on Green Synthesis [4]. The vast importance of nanotechnology in research field emphasize on the synthesis of nanoparticles with different compositions, sizes, morphologies, and controlled dispersities [4]. The nanoparticle synthesized through chemical methods involves chemical reduction using different metals and chemicals such as sodium citrate, ascorbate, sodium borohydride, etc. [4]. Whereas in Greenway for the synthesis of nanoparticle green reducing agents are employed using phytochemical extracts of different natural products such as leaf extract, juice extract, extract from medicinal plants, etc., to provide unlimited opportunities for new discoveries. Green nanotechnology has been making great forward progress, but the growth of nanoproducts and nanoparticles is challenged by different processes of commercialization as green synthesis requires improvements in specific characterization, development of design, and use of Nanoparticles [4]. There is an increasing need for research in the development of design and use of green nanoparticles [4]. The world of scientific challenges, the futuristic vision of the domain of nanotechnology and the needs of human civilization will all today lead a long and visionary way in the true realization and the true emancipation of the colossus of science that is nanotechnology [4]. In nanomaterials and molecular nanotechnology, authors deeply emphasize on environmental nanoscience research by veritably explaining that there is an immediate need to develop engineered nanomaterials. Today, technological profundity, vast scientific vision, and the scientific travails are the forerunners towards a newer visionary era in the field of nanomaterials and engineered nanomaterials. This chapter elucidates on the vast necessity of engineered nanomaterials in diverse branches of science and engineering [4].

Dhingra et al. [5] deeply discussed sustainable nanotechnology through green methods and life-cycle thinking. Citing the myriad and diverse applications of nanotechnology, this chapter emphasizes the need to conduct "life-cycle"-based assessments as early in the new product development process as possible, for a better understanding of the potential environmental and human health consequences of nanomaterials over the entire life-cycle of a nano-enabled product [5]. Technology and engineering science has today few answers to the scientific travails and the scientific difficulties in nanotechnology applications in human scientific progress and human scientific regeneration. Incorporating life cycle thinking for making improved decisions at the product design stage, combining life cycle and risk analysis using sustainable manufacturing practices and green chemistry tools are seen as viable solutions [5]. Sustainability and futuristic studies are linked to each other, the time domains involved may be different from the individual viewpoints of the stakeholders depending on whether they are futurists or environmentalists [5]. Science, engineering, and technology are today in the path of newer scientific regeneration and vast revamping globally. This chapter is a watershed text in the field of green nanomaterials and evokes a newer scientific regeneration and a deep scientific understanding in the field of green nanotechnology and life-cycle science in decades to come [5].

OECD Policy Papers [6] discussed with deep scientific conscience and vast foresight nanotechnology for green innovation. This chapter brings together information collected through discussions and projects undertaken by the OECD Working Party on Nanotechnology relevant to the development and use of nanotechnology for green innovation and deep scientific instinct [6]. The authors of the report deeply discussed the green innovation through nanotechnology, strategies for green innovation through nanotechnology, and the impact of nanotechnology [6]. Green innovation is that innovation which reduces environmental impact; by increasing energy efficiency, by reducing waste gas emissions, and minimizing the consumption of renewable raw materials. OECD countries and emerging economies around the globe are veritably seeking new ways to use green innovation for increasing competitiveness in business. The need for the development of technologies and affordable and safe ways of addressing global challenges, in areas such as energy, environment, and health has never been more pressing than now [6]. The global demand for energy has been expected to increase by more than 30% between 2010 and 2035 [6]. Green innovation, green chemistry, and green engineering are one such way of addressing global sustainability challenges [6]. Energy and environmental sustainability are the technological and scientific challenges of today's science [6]. This policy paper revisits and re-envisions the need of the science of sustainability in the true emancipation of green nanotechnology and green engineering globally. Green innovation is an innovation which reduces environmental impacts: by increasing energy efficiency, by reducing waste or greenhouse gas emissions and by the consumption of non-renewable raw materials. The authors of this report also elucidated on green nanotechnology in the context of a green innovation transition [6]. Today, technology and engineering science are in the path of immense scientific catastrophe as the global scenario is highly stressed as regards sustainability, energy, water, food, and shelter. Technological revamping, scientific urges and the needs of human society are the veritable forerunners towards a newer visionary era in the field of science and engineering of nanotechnology. The report also trudges upon sustainable manufacturing and the global investment in environmental health and safety [6]. A paper on the impact of green nanotechnology ushers in newer vision and newer scientific ingenuity in the field of nanoscience and nanotechnology in decades to come.

Demirdoven et al. [7] deeply discussed with deep and cogent insight green nanomaterials with examples of applications. During the past few decades, a wide range of advanced materials has drastically transformed the vast landscape of science and engineering. Nanotechnology, in most cases, serves as a general heading for all manners of material investigations at the nanoscale [7]. The continuous pursuit in stronger and lighter materials that can perform multiple functions, self-repairing systems, or materials leads up nanotechnologies enabling sustainable design [7]. The potential contribution to sustainability makes nanotechnology one of the key technologies in green building area. A vast investigation into the use of nanotechnology for green building materials aims to: (1) identify nanotechnology developments, that are relevant to green building design; (2) provide direction for future nanotechnology developments that could be in use green building practices. The study thoroughly classifies the application areas of nanomaterials for green buildings as: (1) safety and security; (2) indoor quality; (3) material surface advancement; (4) energy generation and storage, and (5) environmental impact and control [7]. Technology and engineering are advancing at a rapid pace surpassing one visionary scientific frontier over another [7]. Green nanomaterials today stands in the midst of deep scientific vision and vast scientific fortitude. The necessities of green nanotechnology to human society are immense and groundbreaking. This chapter reiterates the scientific facts and deep scientific ingenuity in the domain of green nanomaterials applications to human scientific endeavor and human scientific progress [7].

5.11 NANOTECHNOLOGY AND THE VAST DOMAIN OF WATER PURIFICATION

Nanotechnology and the vast domain of water purification are today linked by an unsevered umbilical cord. Environmental and energy sustainability are the needs of human civilization and human scientific progress today.

Human mankind's immense scientific prowess, grit, and determination are the forerunners towards a newer visionary era in the domain of water purification. Provision of pure drinking water to the teeming millions around the globe is highly neglected today. Global warming and global climate change are areas which need to be re-envisioned and re-envisaged as human civilization surges forward. Global warming and global scientific breakthroughs in environmental engineering science today stands in the midst of deep vision and vast scientific fortitude. Water, energy, food, and shelter are the areas of human progress which are highly stressed and deeply neglected. Technology and engineering science has practically no answers to the monstrous environmental engineering issue of arsenic groundwater contamination. Thus, the need of energy and environmental sustainability. Today, sustainable development and water pollution control are the two opposite sides of the visionary coin. Millions of people in developing and developed countries around the world are witnessing the agony of immense proportions because of a shortage of drinking water. In this chapter, the author pointedly focuses on the needs of environmental sustainability and water science and technology in the true emancipation of science and engineering. The contribution of green nanotechnology and green nanomaterials in the success of water purification techniques are of immense importance today. The author deeply reiterates the need and the vast necessity of environmental sustainability in the future of science of water purification.

5.12   HEAVY METAL GROUNDWATER CONTAMINATION AND THE APPLICATION OF NANOTECHNOLOGY

Science and technology are today in the path of newer scientific rejuvenation. Heavy metal groundwater contaminations are ravaging the vast scientific landscape of might and vision. Arsenic groundwater and drinking water contamination are veritably challenging the research and development initiatives in water science and technology. One of the prominent problems afflicting people throughout the world is inadequate access to clean water and sanitation. Technology and engineering are highly challenged as regards application of nanotechnology in water pollution control. Problems with water are expected to grow immensely worse in the coming decades with water scarcity growing globally even in regions currently considered water rich. "Heavy metal" is a general collective term which applies to the group of metals and metalloids with an atomic density greater than 4000 kg m^{-3} or 5 times more than water and they are natural components of the earth's crust [8, 9]. In the environment, the heavy metals are generally more persistent than organic components such as pesticides or petroleum byproducts. Application of nanotechnology is a newer area of scientific research pursuit in groundwater remediation. In this chapter, the author deeply depicts the scientific ingenuity in nanotechnology emancipation in water science and technology [8, 9].

5.13   SUSTAINABLE DEVELOPMENT AND NANOTECHNOLOGY

The science of environmental and energy sustainability today stands in the crossroads of vision and deep scientific fortitude. The vast domain of environmental protection and petroleum engineering science are in the midst of scientific rejuvenation and immense technological profundity. The vision and scientific subtleties of nanotechnology are today reaching out to vast and versatile scientific boundaries. Nanotechnology is a giant research and development of tomorrow. Thus, in a similar vision, the science of sustainability needs to be re-envisioned and re-organized with the march of science and engineering. The visionary words of Dr. Gro Harlem Brundtland, former Prime Minister of Norway on the science of "sustainability" needs to have a newer beginning and a newer vision. Developing as well as developed nations around the world are confronting environmental engineering catastrophes of immense proportions such as scarcity of pure drinking water [8, 9]. Here comes the need of environmental sustainability and nanovision. In this chapter, the author deeply depicts with immense profundity the need of nanotechnology in designing environmental engineering processes and techniques. Thus, human civilization will usher in a newer era in the field of scientific adjudication and nanotechnology applications [8, 9].

5.14   FUTURE SCIENTIFIC THOUGHTS AND FUTURE RECOMMENDATIONS OF THIS STUDY

Scientific thoughts and scientific vision are today in the path of regeneration as regards environmental protection and nanotechnology. Future of human civilization and the future of sustainable development are in the midst of a deep crisis as civilization faces global warming, global climate change and depletion of fossil fuel resources [8, 9]. Green nanotechnology and green nanomaterials are the visionary answers towards the scientific travails of human mankind. The difficulties and barriers of human scientific endeavor need to be revamped and re-organized as science and technology surge forward. Future research and development initiatives should be targeted towards the greater application of nanotechnology in environmental pollution control. Future strategies should be targeted toward research in novel materials and novel processes. Novel separation processes and membrane science are one of the many examples of water treatment initiatives. Future recommendations of the study should be directed towards technological innovations in the field conventional and non-conventional environmental engineering techniques. In this chapter, the author deeply pronounces the utmost need and the visionary arenas of nanotechnology-based water treatment procedures. Advanced oxidation processes are non-conventional environmental engineering techniques. The harnessing of these techniques can be a real emancipation towards the advancement of science and engineering globally. Thus, a newer scientific order can be realized if nanotechnology and green nanomaterials can be integrated with the domain of water purification and drinking water treatment [8, 9].

5.15   FUTURE SCIENTIFIC PERSPECTIVES IN THE FIELD OF NANOTECHNOLOGY

Science and technology today are in the midst of deep scientific vision and immense scientific grit and determination. Future scientific perspectives in the field of nanotechnology and green engineering are path-breaking and visionary as human civilization moves forward. In this chapter, the author reiterates with immense vision and scientific conscience the need of nanotechnology and green chemistry and engineering to human progress [8, 9]. The world of scientific validation, the needs of human society and the futuristic vision of green nanotechnology and green chemistry will veritably lead a long and effective way in the true realization of science and technology today. Nanoscience and nanotechnology are today in the path of newer vision and deep scientific provenance. Water science and technology is a visionary area of science and engineering which needs to be re-envisioned and re-organized with the passage of scientific history and the visionary time frame. Green chemistry and green engineering are the necessities of human civilization today. Future scientific thoughts and future recommendations of the study will be directed towards more research initiatives in the field of nanotechnology applications in water purification, drinking water treatment, and industrial wastewater treatment. This chapter opens up newer visionary thoughts, the needs of human scientific progress and the scientific success in nanotechnology applications in almost every domain of science and engineering. The ever-growing concerns of drinking water treatment need to be re-addressed and re-envisaged with immense conscience and far-sightedness in this well-researched chapter.

5.16   CONCLUSION AND SCIENTIFIC PERSPECTIVES

Scientific perspectives in the field of green nanotechnology and green chemistry are vast and versatile. Human civilization and human scientific initiatives are similarly are in the path of a newer era in chemistry, engineering science and nanotechnology. Developed as well as developing economies around the world are in the midst of immense scientific revamping and scientific determination. Water purification problems are destroying and challenging the global scientific firmament. Technological vision and scientific objectives are the necessities of science and engineering today. In this entire treatise, the author delves deep into the scientific success and the deep scientific profundity in green nanotechnology and green engineering applications in human society. Novel separation processes and non-traditional environmental engineering tools are the other salient features of this chapter. Membrane science is thus in the midst of deep vision and scientific introspection. The author in this treatise pointedly focuses on the success of nanotechnology applications in diverse branches of science and engineering today. This chapter opens up newer thoughts and newer visionary avenues in the field of green engineering and green chemistry in decades to come. Mankind's immense scientific prowess and determination, the vast world of scientific validation and the utmost needs of the human society will lead a long and visionary way in the true emancipation of green nanotechnology and green chemistry today. Water science and technology, drinking water issues and the ever-growing concerns for groundwater heavy metal contamination are veritably changing the scientific landscape in present-day human civilization. The author in this chapter reiterates the scientific ingenuity, the deep scientific provenance and the vision behind nanotechnology and green chemistry applications in human scientific progress. Today, the civil society and the world of scientists and engineers need to take a proactive role in the advancement of the domain of water purification and environmental protection. Thus, environmental engineering science and green chemistry will surely open up new doors of innovation and scientific instinct in science in decades to come.

KEYWORDS

- green
- green nanotechnology
- medicine
- nanomaterial
- sustainability
- vision

REFERENCES

1. Lu, Y., & Ozcan, S., (2015). Green nanomaterials: On track for a sustainable future, *Nano Today, 10*, 417–420.
2. Nath, D., Banerjee, P., & Das, B., (2014). Green nanomaterial-how green are they as biotherapeutic tool. *Journal of Nanomedicine and Biotherapeutic Discovery, 4*(2), (Open Access Journal), doi: 10.4172/2155–983X.1000125.
3. European Commission Report, (2009). Challenges and opportunities to green nanotechnologies, *Nanotechnologies in the 21ˢᵗ Century, April 2009, Issue 1.*
4. Verma, A., Sharma, M., & Tyagi, S., (2017). Green nanotechnology: Research and reviews. *Journal of Pharmacy and Pharmaceutical Sciences, 5*(4).
5. Dhingra, R., Naidu, S., Upreti, G., & Sawhney, R., (2010). Sustainable nanotechnology: Through green methods and life-cycle thinking. *Sustainability, 2*, 3323–3338.
6. OECD, (2013). "Nanotechnology for green innovation," OECD science. *Technology and Industry Policy Papers, No. 5*. OECD Publishing.
7. Demirdoven, J. B., & Karacar, P., (2015). *Green Nanomaterials with Examples of Applications*. GreenAge Symposium, Mimar Sinan Fine Arts University Faculty of Architecture, Istanbul, Turkey.

8. Hashim, M. A., Mukhopadhyay, S., Sahu, J. N., & Sengupta, B., (2011). Remediation technologies for heavy metal contaminated groundwater. *Journal of Environmental Management, 92*, pp. 2355–2388.
9. Shannon, M. A., Bohn, P. W., Elimelech, M., Georgiadis, J. G., Marinas, B. J., & Mayes, A. M., (2008). *Science and Technology for Water Purification in the Coming Decades* (pp. 301–310). Nature Publishing Group, USA.

PART II
Advanced Technologies

CHAPTER 6

QUANTUM DOTS AND THEIR APPLICATIONS

RAKSHIT AMETA[1,2], KANCHAN KUMARI JAT[3], JAYESH BHATT[1], and SURESH C. AMETA[1]

[1]Department of Chemistry, PAHER University, Udaipur-313 003, Rajasthan, India

[2]Faculty of Science, Department of Chemistry, J. R. N. Rajasthan Vidyapeeth (Deemed-to-be University) Udaipur–313 002, Rajasthan, India

[3]Department of Chemistry, Mohan Lal Sukhadia University, Udaipur–313 002, Rajasthan, India

ABSTRACT

Quantum dots (QDs) are a semiconductor of a very small size (2–10 nm). The band of energies is changed to the discrete energy level in QDs. The energy will depend on its size. These are considered dimensionless. Use of QDs is versatile that is in solar cells, laser, memory, LEDs, photovoltaics, photodetectors, photocatalyst, quantum computers, etc. QDs will occupy a prominent position in these fields in years to come.

6.1 INTRODUCTION

Quantum dots (QDs) cause the band of energies to change into discrete energy levels, and they are a semiconductor of very tiny size. It is known that band gaps and their related energy largely depend on the relationship between the exciton radius and size of the crystal. The height and the energy between different energy levels depend inversely on the size of QD. Smaller is the size of a quantum dot, higher will be its energy. QDs are defined as very small semiconductor crystals of size ranging from nanometer scale to a few microns. They are so small that they are considered dimensionless and are capable of showing many different chemical properties. These QDs generally house the electrons in the same way as the electrons would have been present in any atom on the application of a voltage. Hence, they are very rightly named as the artificial atoms. The application of voltage may lead to the modification in the chemical nature of the material. QDs are having numerous applications like light-emitting diodes (LEDs), solar cells, memory elements, biosensors, imaging, lasers, quantum computation, photovoltaic devices, photodetectors, etc. (Figure 6.1). It is interesting to note that nanoparticles are able to display any color in the complete ultraviolet-visible spectrum with a small change in their size or composition.

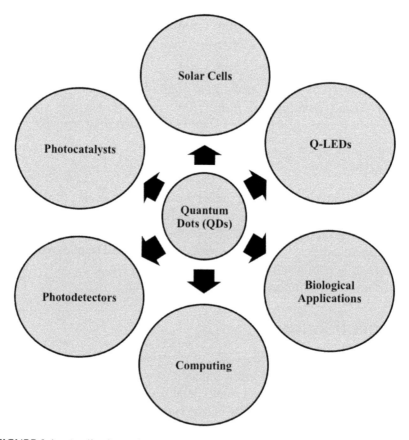

FIGURE 6.1   Applications of quantum dots.

QDs were first characterized by Brus [1, 2]. Almost simultaneously Ekimov and Onushchenko [3] also observed the formation of color in semiconductor-doped glasses. These are small semiconductor spheres in a colloidal suspension. When the radius of a semiconductor sphere becomes very small, i.e., of the order of a few nanometers (2–10 nm), then the Bohr radii of the charge carriers become larger in size as compared to a sphere. Such a confinement to the sphere causes an increase in their energy. Thus, the quantum dot is considered as a finite spherical potential well, which incorporates both; confinement effects and Columbic attraction between charge carriers. A band gap of QDs is easily tunable, which makes them especially suited for optical applications. This behavior makes them useful as qubits, as excited electron spins in individual QDs can be controlled electronically.

6.2   SOLAR CELLS

One of the great challenges for state-of-the-art solar cells is to generate electricity in all weather conditions. The emission of multiple near-infrared (NIR) photons for each photon (ultraviolet/visible) absorbed can be realized by quantum cutting, so as to improve significantly the photoelectric conversion efficiency (PCE) of solar cells The realization of high efficiency and strong bending-durability is a burning issue for possible applications of colloidal quantum dot solar cells (CQDSCs) in flexible devices.

Colloidal quantum dots (CQDs) can be considered promising materials for large-scale and low-cost photovoltaics. The development of quantum dot inks has overcome the prior necessity for solid-state exchanges, which has a high cost, complexity, and morphological disruption to the quantum dot solid, but these inks remain limited because of the photocarrier diffusion length. Yang et al. [4] devised a strategy based on n- and p-type ligand to judiciously shift the quantum dot band alignment. It leads to ink-based materials, which retained the independent surface functionalization of QDs creating distinguishable donor and acceptor domains for bulk heterojunctions. Interdot carrier transfer and exciton dissociation were confirmed by efficient charge separation at the nanoscale interfaces between the two classes of QDs. A first mixed-quantum-dot solar cell was fabricated to achieve a power conversion of 10.4%, surpassing the performance of exciting bulk heterojunction quantum dot devices almost two-fold, indicating that there is a great potential of the mixed-quantum-dot approach.

Vyskočil et al. [5] designed suitable underlying and covering layers of InAs/GaAs QDs (QDs) so as to increase the carrier extraction rate in the QD solar cell structures. When QDs was covered by a GaAsSb strain reducing layer (SRL) with type II band alignment significantly, it improved photogenerated carrier extraction from InAs QDs. An additional thin InGaAs layer below InAs QDs may enhance the extraction of photogenerated carriers. Properties of QD structures are compared:

- without any SRL;
- with GaAsSb covering SRL; and
- with a combination of thin below-QDs InGaAs and GaAsSb covering SRLs.

They were of the opinion that thin below-QDs InGaAs SRL together with the increasing profile of antimony concentration in covering GaAsSb SRL could significantly improve the properties of the solar cell with InAs QDs.

Wang et al. [6] fabricated flexible CQDSCs with a three-dimensional electron transport layer (ETL), which was composed of ZnO nanowire (NW) array for efficient carrier collection. They probed the effect of the NW array on the bonding ability of flexible cells. This ability of cells with ZnO NWs was verified from photovoltaic performance during mechanical bending treatment at various bending angles and cycles. It was observed that the efficient release of stress inside the three-dimensional CQDSCs with the ZnO NW array maintained 97% of the initial power conversion efficiency (PCE) at a bending angle of 160°. Crack formation on the common planar ETL of CQDSCs by ZnO nanoparticles resulted in a decrease of device performance to 77% of the initial value at the same angle.

ETL, extracting electrons from the carbon quantum dots (CQDs) solid layer, needs to be processed at a low-temperature and also suppress interfacial recombination to achieve highly efficient and flexible CQD solar cells. CQD solar cells have a great potential for designing an efficient and lightweight energy supply for flexible or wearable electronic devices. A highly stable MgZnO nanocrystal (MZO-NC) layer has been reported by Zhang et al. [7] for efficient flexible PbS CQD solar cells. It was observed that solar cells fabricated with MZO-NC ETL had a high PCE of 10.4% and 9.4%, on the glass and flexible plastic substrates, respectively. It was revealed that the MZO-NCs significantly enhance charge extraction from CQD solids and diminish the charge accumulation at the ETL/CQD interface; thus, suppressing charge interfacial recombination. Hence, low-temperature processed MZO-NCs are promising for use in efficient flexible solar cells or other flexible optoelectronic devices.

A rapid conversion of CQDs from carbohydrates (including glucose, maltol, or sucrose) has been reported by Tang et al. [8] for an all-weather solar cell. It comprises of a CQD-sensitized mesoscopic titanium dioxide/long-persistence phosphor (m-TiO$_2$/LPP) photoanode, a I$^-$/I$_3^-$ redox electrolyte, and platinum as a counter electrode. In virtue of the light storing and luminescent behaviors of LPP phosphors, this type of generated all-weather solar cells not only convert sunlight into electricity on sunny days but these realize electricity output continuously in all dark–light conditions. The optimum PCE achieved was as high as 15.1% for such all-weather CQD solar cells in dark conditions.

The ETL plays important roles in charge extraction and determining the morphology of the perovskite film in planar n-i-p heterojunction perovskite solar cells. Li et al. [9] reported a solution-processed CQDs/TiO$_2$ composite with a negligible absorption in the visible spectral range, which is a very attractive feature for perovskite solar cells. An efficiency of ~19% has been achieved using this novel CQDs/TiO$_2$ ETL in conjunction with a planar n-i-p heterojunction. They observed that a combination of CQDs/TiO$_2$ increases both; the open circuit voltage (Voc) and short-circuit current density (Jsd) as compared to using TiO$_2$ alone. It was proposed that CQDs increases the electronic coupling between the CH$_3$NH$_3$PbI$_{3-x}$Cl$_x$ and TiO$_2$ ETL interface as well as energy levers that contribute to carulli electron extraction.

Zhou et al. [10] fabricated a novel type of quantum cutting material, CsPbC$_{11.5}$Br$_{1.5}$:Yb^{3+}, Ce^{3+} nanocrystals (NCs). It has larger absorption cross-section, and weaker electron-phonon coupling, with a higher inner luminescent quantum yield (QY) (146%), and therefore, doped perovskite NCs were successfully used as a down converter of commercial silicon solar cells (SSCs). Noticeably, the PCE of these SSCs was found to be improved from 18.1% to 21.5%, with a relative enhancement of 18.8%. The present work provides a low cost, convenient, and effective way to enhance the PCE of SSCs, which may be commercially useful in the future.

Bulk heterojunction solar cells based on blends of QDs and conjugated polymers are considered a promising configuration to achieve high-efficiency and low cost fabricated solution-processed photovoltaic devices. These devices are of significant importance due to their potential to leverage the advantages of both the types of materials:

- high mobility;
- band gap tenability;
- possibility of multiple exciton generations in QDs;
- high mechanical flexibility; and
- large molar extinction coefficient of conjugated polymers.

Although having these advantages, the PCE of these hybrid devices still remained relatively low (about 6%), well below that of all-organic or all-inorganic solar cells. It is all due to major challenges in controlling the film morphology and interfacial structure to ensure efficient charge transfer and charge transport. Kisslinger et al. [11] presented recent development of bulk heterojunctions made from conjugated polymer-quantum dot blends, ongoing strategies attempted to improve the performance of the device and also highlighted the key areas of research that need to extensive research in this direction.

Meinardi et al. [12] used indirect-bandgap semiconductor nanostructures like highly emissive silicon quantum dots. Silicon is a non-toxic, low-cost, and ultra-earth-abundant element, which have limitations for the industrial scaling of QDs as compared to low-abundance elements. Suppressed reabsorption and scattering losses lead to nearly ideal luminescent solar concentrators (LSCs) with an optical efficiency of $\eta=2.85\%$, matching state-of-the-art semi-transparent LSCs. It was indicated that optimized silicon quantum dot LSCs have a clear path to $\eta>5\%$ for 1 m^2 device. Finally, they were able to realize flexible LSCs with good performances comparable to those of flat concentrators, opening a new way.

Muthalif et al. [13] probed the effect of Cu-doping in CdS QDs to improve the photovoltaic performance of the quantum dot-sensitized solar cells (QDSSCs). The Cu-doped CdS photoanodes were prepared by successive ionic layer adsorption, and reaction (SILAR) method and the cell devices were fabricated using CuS as counter electrodes using a polysulfide electrolyte. The photovoltaic performance data demonstrated that 3 mM Cu-doped CdS QDs based QDSSCs exhibited the efficiency (η) of 3% including JSC = 9.40 mA cm^{-2}, VOC = 0.637 V, and FF = 0.501, which were higher than that with bare CdS (η = 2.05%, JSC = 7.12 mA cm^{-2}, VOC = 0.588 V, FF = 0.489). It was indicated that Cu-dopant could inhibit the charge recombination at the photoanode/electrolyte interface; thus, extending the lifetime of electrons.

Graphene quantum dots (GQDs) are quite interesting materials that are attracting the attention of scientists and technologists worldwide because of their unique optical properties. Teymourinia et al. [14] prepared GQDs from corn powder and applied it in DSSCs as down-conversion materials. As-prepared GQDs were found to convert UV light to 450 and 520 nm light, which is favorable for DSSCs. A modified solar cell based on GQDs showed 21% enhancement in short-circuit photocurrent density compared to the reference cell.

A search was on for highly efficient PbS CQDSCs based on an inverted structure, but certain limitations were also there like the construction of an effective p–n heterojunction at the illumination side with smooth band alignment and absence of serious interface carrier recombination. In this context, solution-processed nickel oxide (NiO) was used as the p-type layer and lead sulfide (PbS) QDs with iodide ligand as the n-type layer to build a p–n heterojunction at the illumination side [15]. There was a large depletion region in the QD layer at the illumination side, which leads to a high photocurrent. Interface carrier recombination at the interface was effectively restricted by inserting a layer of slightly doped p-type QDs with 1,2-ethanedithiol as ligands. This resulted in an improved voltage of the device. The efficiency of inverted structural heterojunction PbS QD solar cells was found to be improved using this graded device structure design to 9.7%.

Veerathangam et al. [16] deposited undoped and Ag-doped CdS QDs on spin-coated TiO$_2$ photoanode material by SILAR method. They successfully assembled FTO/TiO$_2$/undoped CdS and FTO/TiO$_2$/Ag-doped CdS QDs to study the photovoltaic performance under standard one sun illumination of AM 1.5,100 mWcm.$^{-2}$ Different photovoltaic parameters, like short-circuit photocurrent density, open circuit voltage, fill factor and conversion efficiency were found to be 1.37 mA cm^{-2}, 0.72 V, 61.5% and 0.61%, respectively, for undoped CdS while Ag-doped CdS QDs showed enhanced photovoltaic parameters as 2.35 mA cm^{-2}, 0.74 V, 71.8% and 1.25%, respectively.

CQDs are low cost, chemically stable, and environmentally friendly photosensitizer, which can dramatically broaden the light absorption range to the complete visible range. Ye et al. [17] showed that the NiOOH/

FeOOH/CQD/BiVO$_4$ (NFCB) photoanode achieved a remarkable photo-current density of 5.99 mA cm^{-2} at 1.23 V vs. RHE under AM 1.5G in KH$_2$PO$_4$ aqueous solution without a hole scavenger (pH = 7) and a record high applied bias photon-to-current efficiency of 2.29% at 0.6 V vs. RHE for BiVO$_4$-based photoanodes. This novel NFCB photoanode could operate at the stable condition for 10 h with a Faraday efficiency of ~95%, demonstrating the great potential of using CQDs for solar water splitting.

6.3 COMPUTING

QDs provide a newer direction to quantum computers, which may be powerful supercomputers. These computers can operate and store infor-mation using quantum bits (qubits), which can simultaneously exist in two states; on and off. This particular phenomenon is quite important as it leads to highly improved information processing speeds and memory capacity as compared to computers. Quantum computing will become realistic based a day with a deep understanding of semiconductor physics at the nanometric scale in depth supported by the development of quantum information devices.

Kloeffel and Loss [18] have reviewed experimental and theoretical progress toward quantum computation with spins in QDs, with stress on QDs formed in GaAs heterostructures, on NW-based QDs, and self-assembled QDs. They reported a remarkable evolution in this field. One of the existing challenges for the realization of quantum computers, i.e., decoherence, will no longer remain a problem now. They have explained general concepts, relevant quantities, and basic requirements for spin-based quantum computing, along with opportunities and challenges of spin-orbit interaction and nuclear spins.

A new implementation of a universal set of one- and two-qubit gates has been proposed by Loss and Vincenzo [19] for quantum computation using the spin states of coupled single-electron quantum dots. It was proposed that desired operations are found to be affected by the gating of the tunneling barrier between neighboring dots. They computed several measures of the gate quality within derived spin master equation, where decoherence was incorporated, which was caused by a prototypical magnetic environment. Dot-array experiments were also proposed that would provide an initial demonstration of the desired non-equilibrium spin dynamics.

Imamoglu [20] discussed that QDs have some desired unique features, making them quite useful in quantum information processing. It has been widely accepted that spin of an electron forms an almost ideal elementary unit of quantum information, which is termed as a qubit. QDs can be used to localize and address single spins by both the methods; electrical or optical methods. A QDs embedded in a photonic nanostructure may also be considered as an ideal deterministic source of single photons. It has been shown theoretically that single photon pulses generated by a single QD, along with photodetectors and linear optical elements, may permit us for efficient quantum computation.

Quantum logic gates have been reported to be key elements in quantum computing. Wei and Deng [21] investigated the possibility of achieving a scalable and compact quantum computing, which is based on stationary electron-spin qubits. For this purpose, they made use of giant optical circular birefringence induced by quantum-dot spins in double-sided optical microcavities as a result of cavity quantum electrodynamics. They designed such compact quantum circuits for creating universal and deterministic quantum gates for electron-spin systems, which includes two-qubit Controlled-NOT (CNOT) gate and the three-qubit Toffoli gate. These are compact and economic, and did not require additional electron-spin qubits. These devices have good scalability and are quite attractive also as both of them are based on solid-state quantum systems, and the qubits are stationary.

Klein tunneling restricts electrostatic confinement of massless charge carriers. This problem in GR can be overcome by carving out nanostructures or applying electric displacement fields to open a band gap in bilayer GR. Presently, these approaches suffer from edge disorder or insufficiently controlled localization of electrons. An alternative strategy was designed by Freitag et al. [22] by combining a homogeneous magnetic field and electrostatic confinement in monolayer GR. They induced a confining potential in the Landau gaps of bulk GR without the requirement for physical edges. If the localized states are gated toward the Fermi energy, it will lead to regular charging sequences with more than 40 Coulomb peaks exhibiting typical addition energies of 7–20 meV. Orbital splittings (4–10 meV) and valley splitting (~3 meV) for the first orbital state can be deduced. This confinement approach seems to be suitable for creating QDs with well-defined wave function properties, which is far from the reach of traditional techniques.

Kamada and Gotoh [23] discussed some potential applications of elementary excitation in semiconductor quantum dot to quantum computation. They have proposed a scalable hardware and optical implementation of a logic gate, exploiting the discrete nature of electron-hole states as well as their well-concentrated oscillator strength for ultrafast gate operation. Here, a multiple-bit gate function is based on the nearest neighbor dipole-dipole coupling. Rabi population oscillation and long-lived coherence of an excitonic two-level system in an isolated $In_xGa_{1-x}As$ quantum dot guarantee quantum gate functions.

The control of solid-state qubits is needed to have a detailed understanding of the decoherence mechanisms. Although considerable progress has been made in uncovering the qubit dynamics in strong magnetic fields, decoherence at very low magnetic fields still remains difficult as the role of quadrupole coupling of nuclear spins is not properly understood. Phenomenological models of decoherence include two basic types of spin relaxation for spin qubits in semiconductor QDs. These are:

- Fast dephasing due to static but randomly distributed hyperfine fields (~2 ns); and
- Much slower process (>1 µs) of irreversible monotonic relaxation either due to nuclear spin co-flips or other complex many-body interaction effects.

Bechtold et al. [24] showed that this is the first simplification; as the spin qubit relaxation is determined by three rather than two distinct stages. The third additional stage corresponds to the effect of coherent precession processes that is there in the nuclear spin bath itself, leading to a relatively fast, but incomplete non-monotonic relaxation at intermediate timescales (~750 ns).

As silicon carbide is a promising candidate for single photon sources, quantum bits and nanoscale sensors based on individual color centers. Radulaski et al. [25] developed a scalable array of nanopillars, which incorporate single silicon vacancy centers in 4H-SiC. These vacancy centers are readily available for efficient interfacing with free-space objective and lensed-fibers. A substrate was irradiated with 2 MeV electron beams to create such vacancies. Then lithographic process forms 800 nm tall nanopillars with 400–1400 nm diameters. They obtained high collection efficiency of up to 22 k counts s^{-1} optical saturation rates from a single silicon vacancy center while single photon emission and the optically induced electron-spin polarization properties are preserved.

Lagoudakis et al. [26] demonstrated all-optical ultrafast complete, coherent control of a qubit formed by the single-spin/trion states of a charged site-controlled NW quantum dot. They showed that site-controlled QDs in NWs are quite promising hosts of charged exciton qubits. These qubits can be clearly manipulated in the same fashion as was in the case of randomly positioned quantum dot samples.

6.4 LIGHT-EMITTING DIODES (LEDS)

Quantum dot-based light-emitting diodes (QD-LEDs) are a form of light-emitting technology. These are considered the next generation of display technology after the organic light-emitting diodes (OLEDs) display. These are different from liquid crystal displays (LCDs), OLEDs, and plasma displays being an ideal blend of more brightness, flexibility efficiency with a long lifetime, as well as low-processing cost.

Solution-processed optoelectronic and electronic devices are interesting because of their potential for low-cost fabrication of large area devices, compatibility with lightweight, and flexible plastic substrates. Recently, the use of QDs in solution-processed LEDs as emitters have attracted attraction all over the globe.

Dai et al. [27] reported a solution-processed, multilayer QD-LED with good performance and reproducibility. This exhibited color-saturated deep-red emission, sub-bandgap turn-on at 1.7 volts, high external quantum efficiencies (20.5 %), low-efficiency roll-off (up to 15.1 %) of the external quantum efficiency at 100 mA cm^{-2}, and a long operational lifetime of more than 10^5 hours at 100 cd m.$^{-2}$ All these properties make this device the best-performing solution-processed red LED as-compared to vacuum-deposited organic LEDs. Such optoelectronic performance was achieved by inserting an insulating layer between the quantum dot layer and the oxide electron-transport layer, so as to optimize charge balance in the device and preserve the superior emissive properties of the QDs.

Bae et al. [28] demonstrated highly efficient green-LEDs based on CdSe@ZnS QDs with a chemical-composition gradient. They achieved excellent device performance by the moderate control of QD coverage in multilayered devices. The color-saturated green-light emission was mostly from the QD layers.

Zorn et al. [29] prepared QD/polymer hybrids by grafting a block copolymer containing thiol-anchoring moieties (poly(para-methyl triphenylamine-b-cysteamine acrylamide)) onto the surfaces of QDs through the ligand exchange procedure just to combine the optical properties of CdSe@ZnS QDs with the electrical properties of semiconducting polymers. As-prepared QD/polymer hybrids possess improved processability like enhanced solubility in different organic solvents and also the film formation properties with the improved colloidal stability, which was derived from the grafted polymer shells. They showed that LEDs based on QD/polymer hybrids, exhibited the improved device performance (almost 3-fold increase in the external quantum efficiency) as compared to devices prepared by pristine (unmodified) QDs.

Song et al. [30] reported novel quantum-dot light-emitting diodes based on all-inorganic perovskite $CsPbX_3$ (X = Cl, Br, I) NCs. The well-dispersed, single-crystal QDs exhibited high QYs, and tunable light emission wavelength. These novel perovskite QDs will open a new avenue toward designing various optoelectronic devices (displays, photodetectors, solar cells, lasers, etc.)

Highly bright and efficient inverted structure quantum dot based LEDs were reported by Kwak et al. [31] using solution-processed ZnO nanoparticles as the electron injection/transport layer. They also optimized energy levels with the organic hole transport layer. Highly bright red, green, and blue QLEDs were demonstrated showing maximum luminances up to 23 040, 218 800, and 2250 cd m^{-2}, and external quantum efficiencies of 7.3, 5.8, and 1.7%, respectively. It is also interesting to note that turn-on voltages as low as the band gap energy of each QD were shown along with long operational lifetime. It was attributed to the direct exciton recombination within QDs through the inverted device structure.

Ippen et al. [32] considered indium phosphide (InP) QDs as good alternatives to Cd-containing QDs for application in light-emitting devices. It was observed that multishell coating with ZnSe/ZnS improved the photoluminescence (PL) QY of InP QDs more strongly as compared to conventional ZnS shell coating. QY values (50–70%) along with peak widths (45–50 nm) can be routinely achieved, so as to make the optical performance of InP/ZnSe/ZnS QDs comparable to that of Cd-based QDs.

QDs-based backlight greatly enhances the color performance for LCDs. Luo et al. [33] reviewed QD backlight and discussed introduction and advantages of QD backlight over the white LED, like higher optical efficiency, much wider color gamut, enhanced ambient contrast ratio, and smaller color shift. They have also discussed some popular LC modes, including twisted nematic, fringing field switching (FFS) for touch panels, multi-domain vertical alignment (MVA) for TVs, and blue phase liquid crystal (BPLC) for next-generation displays. Out of these, QD-enhanced BPLC had the advantages of FFS and submillisecond response time.

6.5 PHOTODETECTORS

Efficient light detection is highly desirable in modern science and technology. Present day photodetectors are mainly photodiodes, which are based on crystalline inorganic semiconductors, such as silicon, III–V semiconductors, etc. Nowadays, photodetectors are made of solution-processed semiconductors including organic materials, metal-halide perovskites, QDs, etc. and these are also considered promising candidates for next-generation light sensing. They have certain benefits like ease of processing, tunable optoelectronic properties, facile integration with complementary metal–oxide–semiconductors, and compatibility with flexible substrates apart from good performance.

Photodetectors with response in the infrared region are more desirable, which have a wide range of applications ranging from cell phones, cameras, and home electronics to airplanes and satellites. Traditional fluorescent dyes lack long term stability and the ability to detect multiple signals simultaneously, but QDs overcome such obstacles, and therefore, they have a great potential for use as *in vivo* and *in vitro* fluorophores.

Li et al. [34] reported the fabrication of high-performance ultraviolet photodetectors, which are based on a heterojunction device structure having ZnO QDs decorated Zn_2SnO_4 NWs. They showed ultrahigh light-to-dark current ratio (up to 6.8×10^4), specific detectivity (up to 9.0×10^{17} Jones), photoconductive gain (up to 1.1×10^7), fast response, and excellent stability. A quantum dot decorated Zn_2SnO_4 NW had about 10 times higher photocurrent and responsivity as compared with a pristine Zn_2SnO_4 NW. Their high performance may be due to rational energy band engineering, which allows efficient separation of electron-hole pairs at the interfaces between ZnO QDs and Zn_2SnO_4 NW. Here, holes migrate to ZnO QDs increasing electron concentration and lifetime in the NW conduction channel; thus, leading to improved photoresponse. They fabricated flexible ultraviolet photodetectors and integrated into a 10×10 device array constituting a high-performance flexible ultraviolet image sensor.

Two-dimensional (2D) mono-elemental bismuth (Bi) crystal has recently attracted attention because of its interesting properties. Xing et al. [35] fabricated uniformly sized 2D Bi quantum dots (BiQDs) with an average diameter of 4.9 ± 1.0 nm and thickness (2.6 ± 0.7 nm) via a facile liquid-phase exfoliation (LPE) method. They evaluated photoresponse using photoelectrochemical measurements. As-fabricated BiQDs-based photodetector exhibited an appropriate capacity for self-driven broadband photoresponse, and high-performance photoresponse under low bias potentials ranging from UV to visible light and long-term stability of the on/off switching behavior.

Cu_2SnS_3 (CTS) QDs were synthesized by Dias et al. [36] via a solvo-thermal technique using poly(vinylpyrrolidone) (PVP) as a surfactant. Its electronic band gap was measured using cyclic voltammetry. The infrared photoresponse of the CTS QDs-based device was measured under infrared lamp, 1550 and 1064 nm lasers. The characteristics of the photodetector device were also determined that is, responsivity, external quantum efficiency, and specific detectivity.

CQDs have a potential in optoelectronics due to low-cost solution processing and high absorption efficiency. CdS CQD and Au NCs (NCs) were synthesized by Kan et al. [37] through a hot injection method. The absorbing layer was constructed by layer-by-layer spin-coating method treated with hexadecyltrimethylammonium bromide (CTAB) at room temperature. The response of the photodetector with and without Au NCs 1% by weight 1.27×10^{-4} and 4.96×10^{-5} A/W, respectively. The CdS CQDs photodetector (with Au NCs) had detectivity up to 1.42×10^9 Jones, which was almost two times than pure CdS CQDs photodetector. This was all because of the fact that Au NCs can effectively suppress the radiation recombination of the photogenerated carriers.

Chandan et al. [38] discussed the classic and current protocols used to synthesize QDs, along with adaptability of QD surfaces for versatile bioconjugation. An energy transfer mechanism is the main basis for the strong attraction of QDs to the biosensing community, and hence, efficient fluorescence resonance energy transfer, bioluminescence resonance energy transfer, and chemiluminescence resonance energy transfer are to be examined. They have also discussed recent advances in the detection of heavy metal ions, pathogens, and cancer biomarkers.

Zhang et al. [39] designed and fabricated PDs based on air-stable α-CsPbI$_3$ QDs and an up-conversion material (NaYF$_4$:Yb, Er QDs) using a facial low-temperature spin-coating method. Optical response of α-CsPbI$_3$ QDs is extended to the NIR region to allow broadband application from the UV to visible to NIR region (260 nm–1100 nm), on its surface-modification with NaYF$_4$:Yb, Er QDs. These PDs are capable of broad bandwidth photodetection from the deep UV to NIR region (260 nm–1100 nm) with good photoresponsivity (R, 1.5 A W^{-1}), high on/off ratio (up to 10^4) and very short rise/decay time (less than 5 ms). The performance of the device showed very little degradation with 60 days storage under ambient conditions.

Exploring the potentiality of enhancing the performance of avalanche photodiodes (APDs) using novel nanoscale structures is highly attractive for overcoming the bottleneck of avalanche probability. Ma et al. [40] reported large enhancement of electron-initiated photocurrent, may be due to impact ionization in InAs QDs within a GaAs APD structure. A five-layer stacked 2.25 MLs InAs QD/50 nm GaAs spacer multiplication structure integrated into a separated absorption, charge, and multiplication GaAs homo APD results in about six times higher multiplication factors, as compared to a reference device without QD over a temperature range of 77–300 K. Extremely low excess noise factor in close proximity to that of silicon was observed with an effective k$_{eff}$ factor less than 0.1.

A simple spin-coating process was adopted by Zheng et al. [41] to incorporate MAPbI$_3$ QDs (QDs) onto the surface of TiO$_2$ NTs. This heterostructure extends the response of TiO$_2$ NT from ultraviolet to visible range. An improvement of response in visible range by three orders of magnitude was observed as compared with pure TiO$_2$ NTs. It is also maintained in the UV range. These photodetectors showed a relatively fast and stable response in the 300–800 nm range. The photoelectric performance of the hybrid photodetector based on TiO$_2$ NTs is maintained even if it is exposed to moist air for 72 h or heated from room temperature to 100°C. Such a TiO$_2$ NTs/MAPbI$_3$ QDs heterostructure device possesses excellent flexibility and 85% transparency in the 400–800 nm range, as well as their photo detecting performance is also retained even after 200 cycles of repeated bending at 90°C. The proposed strategy combination of facile electrospinning and solution-processed QDs may open a new pathway for wide range response and flexible devices construction.

The fields of solution-processed photodetectors have been reviewed by Arquer et al. [42] from the angle of materials, device, the potential of the synergistic combination of materials and device engineering. The advances in metal-halide perovskite photodiodes and their recent applications in narrowband filterless photodetection have also been discussed.

Lead sulfide QDs are considered as promising materials for next-generation light, low-cost, and flexible photodetectors because of wide tunable band gaps, high absorption coefficients, easy solution processabilities, and high stabilities in the air. Normally, current single-layer PbS-QD photodetectors have disadvantages like large dark currents, low on-off ratios, and slow light responses, but if these are integrated with some metal nanoparticles, organics, and high-conducting GR/nanotube to form hybrid PbS-QD devices, then these are capable of enhancing photoresponsivity. Use of these approaches is associated with other problem like hampering the improvement of the overall device performance severely. Ren et al. [43] designed a bilayer QD-only device to overcome hurdles like current single-layer and hybrid PbS-QD photodetectors face, which may be integrated on some flexible polyimide substrate, so that it can outperform the conventional single-layer devices in response speed, detectivity, linear dynamic range, and signal-to-noise ratio, along with comparable responsivity.

Mihalache et al. [44] achieved significant improvement of the classical silicon nanowire (SiNW)-based photodetector via the realization of core-shell structures using newly designed GQDs-polyethyleneimines (GQDPEIs) via simple solution processing. Poly(ethyleneimine) (PEI)-assisted synthesis successfully tuned both the properties of GQDs; optical and electrical to fulfill the desired strong yellow PL emission along with a large band gap and the introduction of electronic states inside the band gap. This GQDPEI/SiNW photodetector exhibited a large photocurrent to the dark current ratio (Iph/Idark up to ~0.9×10^2 under 4 V bias), and the external quantum efficiency was also improved significantly. These GQDPEIs have the ability to control both; charge-carrier photogeneration and transport inside a heterojunction, which leads to various mechanisms simultaneously.

- Efficient suppression of the dark current governed by the type I alignment in energy levels,
- Charge photomultiplication determined by the presence of the PEI-induced electron trap levels, and
- Broadband ultraviolet-to-visible down conversion effects.

Tang et al. [45] prepared CsPbBr$_3$/reduced graphene oxide (RGO) nanocomposites via a facile hot-injection method. It was indicated that CsPbBr$_3$/

RGO nanocomposites displayed strong quenching phenomenon as compared to pure $CsPbBr_3$; maybe because of the fast separation and transfer of the photogenerated hole-electron pair. PL decay results of the $CsPbBr_3$/RGO nanocomposites (τ_{ave} = 4.52 ns) was relatively shorter almost one third than the corresponding $CsPbBr_3$ nanoparticles (τ_{ave} = 14.72 ns), which confirms the charge transfer between the $CsPbBr_3$ nanoparticles and GR.

In the last decade, a synergistic effect between the QDs and graphene in QDs/GR composite materials has drawn the attention of scientists all over the globe. The electron-hole pair's separation in QDs was enhanced due to the photogating effects in the composites, because of rapid transfer of electrons or holes to the surface of the GR sheets. It resulted in an excellent optical gain, so that these composites become promising materials for applications in optoelectronic and light harvesting devices. Tong et al. [46] reviewed a number of synthetic self-assembly methods like hydrothermal, solvothermal, atomic layer deposition, ion beam sputtering deposition, along with other methods like drop-casting, spin-coating, layer by layer, ultrasonication, polymethylmethacrylate (PMMA) aid transfer, and electrochemical methods. They have also discussed the applications of QDs/GR composites in photocatalysis, photodetectors, supercapacitors, lithium-ion batteries, solar cells, electrochemiluminescence (ECL) sensors, etc.

6.6   BIOLOGICAL APPLICATIONS

One of the most exciting frontline areas of nanobiotechnology is QDs. Luminescent semiconductor NCs or QDs possess desired optoelectronic properties so as to make them ideal for biological sensing applications. Synthesis of QDs significantly promoted fluorescent imaging for *in vitro* as well as *in vivo* applications. QDs have already established their role in the developing fields of nanomedicine and also in nanobiotechnology. If it is possible to prepare high-quality QDs from non-toxic materials, then these are quite useful in the biological field. It has been predicted that QDs may become dominant fluorescent reporters in biology and medicine in years to come. A new class of protean multifunctional nanoparticles may emerge from the combination of QDs with some targeting agents and photosensitizers.

Electrogenerated chemiluminescence (ECL) of QDs is very useful in the fabrication of biosensor, but high-toxicity of heavy metal ion QDs severely limits their further applications. Heavy metal detection technologies are very important in monitoring environment and human health protection; but most of the present technologies are quite a time to consume, expensive with sophisticated equipment, and require complicated sample pre-treatment limiting their use.

Some recent advances in Förster resonance energy transfer (FRET) sensing and bioimaging have been reported by Cheng et al. [47] using non-toxic silicon quantum dots (SiQDs). They prepared SiQDs-dye conjugates, which involves SiQDs as the donor and covalently attached organic dye as acceptors via self-assembled monolayer linkers. Enzymatic cleavage of the peptide will lead to changes in FRET response, which is then monitored using fluorescence lifetime imaging microscopy (FLIM-FRET). The combination of interfacial design and optical imaging may open up new avenues for bio-applications using non-toxic silicon quantum dots.

Susumu et al. [48] reported the synthesis of a series of $CdxZn_{1-x}Se/CdyZn_{1-y}S/ZnS$ and $ZnSe/CdyZn_{1-y}S/ZnS$ multishell alloyed luminescent semiconductor QDs, which showed fluorescence maxima in the range 410–530 nm covering the purple, blue, and green portions of the spectrum. They modified these to obtain water-soluble blue-emitting QDs and used for ratiometric pH sensing in aqueous buffers. These QDs were synthesized starting from ZnSe cores, and the peak positions of fluorescence were tuned by:

- Cation exchange with cadmium ions and/or
- Overcoating with $CdyZn_{1-y}S$ layers.

As-prepared QDs had high fluorescence QYs (about 30–55%), narrow fluorescence bands (FWHM~25–35 nm), and monodispersed semispherical shapes. They carried out successfully ligand exchange with hydrophilic compact ligands to prepare a series of water-soluble blue-emitting QDs, which retained intrinsic photophysical properties and showed excellent colloidal stability in aqueous buffers for more than a year. These blue-emitting QDs were conjugated with fluorescein isothiocyanate (FITC), a pH-sensitive dye, to construct a ratiometric pH sensing platform based on fluorescence resonance energy transfer. pH monitoring with the QD-FITC conjugates was successfully carried out in the pH range of 3.0–7.5. Assembly of such QD-FITC conjugates with membrane localization peptides permitted monitoring of the pH in extracellular environments.

A simple one-pot microwave-assisted green synthesis route has been reported by Kumawat et al. [49] for the fabrication of bright red-luminescent GQDs. Here, ethanolic extracts of *Mangifera indica* (mango) leaves were used and hence, these were called mGQDs. The size of mGQDs was in the range of 2–8 nm and they exhibited excitation-independent fluorescence emission in the NIR region in the range of 650–750 nm. A 100% cellular uptake and excellent biocompatibility on L929 cells were shown by these mGQDs even at high concentration (0.1 mg mL^{-1}) even after 24 h with post-treatment. Increased proliferation in L929 cells has been reported upon mGQDs treatment. These were also demonstrated as NIR-responsive fluorescent bioimaging probes, self-localizing themselves selectively in the cell cytoplasm. The temperature dependence of fluorescence intensity of these mGQDs proved them as a promising temperature sensing probe (at 10–80°C). It was found that the temperature signal remained stable even after a number of cycles of temperature switching between 30–80°C. A decrease in the fluorescence intensity of the mGQDs on increasing temperature makes it a suitable probe for temperature sensing.

Atchudana et al. [50] synthesized nitrogen-doped carbon dots (N-CDs) from *Chionanthus retusus (C. retusus)* fruit extract via a simple hydro-thermal-carbonization method. The ability of N-CDs to sense metal ions, and their biological activity (cell viability and bioimaging applications) were evaluated. The average size of the N-CDs was found to be approximately 5 ± 2 nm with an interlayer distance of 0.21 nm. The presence of phytoconstituent functionalities and the percentages of components in the N–CDs were also confirmed along with a nitrogen content of 5.3%. These demonstrated highly durable fluorescence properties and low cytotoxicity with a QY of about 9%. N–CDs were also used as probes for the detection of metal ions. A high sensitivity and selectivity of N–CDs was exhibited towards Fe^{3+} with a linear relationship between 0 and 2 μM and a detection limit of 70 μM. The synthesized N–CDs can have promising biomedical applications like bioimaging due to high fluorescence, excellent water solubility, good cell permeability, and negligible cytotoxicity. The potential of N–CDs as biological probes was determined using fungal (*Candida albicans* and *Cryptococcus neoformans*) strains via fluorescent microscopy. It was found that N–CDs were suitable candidates for differential staining applications in yeast cells as they have good cell permeability and localization with negligible cytotoxicity. They opined that N–CDs may have utility as probes for the detection of cellular pools of metal ions (Fe^{3+}) as well as early detection of opportunistic yeast infections in biological samples.

Dong et al. [51] reported ECL behavior of eco-friendly silicon quantum dots (SiQDs) in neutral aqueous condition. Stable and intense cathodic ECL emission was observed in the phosphate buffer solution using potassium persulphate as coreactant. A new ECL biosensor (ECL resonance energy transfer; ECL-RET system) was prepared by using SiQDs ECL as energy donor and gold nanoparticles (AuNPs) as an energy acceptor. AuNPs was connected at the terminal of hairpin DNA to form a signal probe. When this probe was modified on SiQDs, ECL-RET system was there due to the short distance between AuNPs and SiQDs, with the result that there is an apparent decrease of the ECL signal. The target DNA can open loop of hairpin DNA, and move the AuNPs away from the electrode surface. As a consequence, the ECL-RET process was hampered, and the ECL emission resumed. It was observed that increased ECL signals varied linearly with the target DNA concentrations in the range of 0.1 fM to 1 pM with a detection limit of 0.016 fM (3σ). This proposed ECL sensor was highly sensitive for the detection of target DNA.

A strategy was developed by Lee et al. [52] for synthesizing immediately activable, water-soluble, and compact (~10–12 nm hydrodynamic diameter) QDs, which have a small number of stable and controllable conjugation handles for long distance delivery and subsequent biomolecule conjugation. The sample resulted in a population consisting of low valency QDs on covalent conjugation with engineered monovalent streptavidin. They also synthesized a quantum dot with a small number of biotin molecules, which can self-assemble with engineered divalent streptavidin via high-affinity biotin-streptavidin interaction. These low-valency QDs (compact, stable, and highly specific against biotinylated proteins of interest) were considered ideal for labeling and tracking single molecules on the cell surface with a high spatiotemporal resolution for different biological systems and applications.

Velusamy et al. [53] synthesized fluorescent nitrogen-doped carbon dots (N-CDs) via a simple hydrothermal route. They used *Hylocereus undatus* (*H. undatus*) extract and aqueous ammonia as carbon and nitrogen source, respectively. As-synthesized N-CDs emitted strong blue fluorescence at 400 nm under the excitation of 320 nm. They also observed excitation-dependent emission properties from the fluorescence of synthesized N-CDs. It was revealed that synthesized N-CDs were spherical in shape with an average diameter of 2.5 nm. The graphitic nature of synthesized N-CDs was confirmed by XRD pattern. The doping of nitrogen was ascertained by EDS and FT-IR studies. The cytotoxicity and biocompatibility of N-CDs were evaluated through MTT assay on L–929 (Lymphoblastoid–929) and MCF–7 (Michigan Cancer Foundation–7) cells. It was found that the fluorescent N-CDs show less cytotoxicity and good biocompatibility on both the cells; L–929 and MCF–7. These N-CDs show excellent catalytic activity towards the reduction of methylene blue by sodium borohydride.

Amjadi et al. [54] presented a very simple and rapid method for the synthesis of CQD/silver (CQD/Ag) nanocomposites at room temperature. They prepared glucose-derived CQDs via a facile microwave-assisted method and used as both; reducing and stabilizing agents for synthesis of CQD/Ag nanocomposites. It was observed that a unique interaction between as-prepared nanocomposites and antithyroid drug methimazole occurred in the solution, resulting in a sharp color change from yellow to red. The intensity of surface plasmon resonance (SPR) peak of CQD/Ag nanocomposites at 400 nm decreases and a new peak appeared at a longer wavelength. It provides a basis for developing a new colorimetric detection method for methimazole. The calibration curve for this drug was found to be linear in the concentration range of 2.0–40 μg L^{-1} with a detection limit of 1.0 μg L.$^{-1}$ The method was successfully applied for the determination of methimazole in urine samples.

A combination of advantages of fluorescence and colorimetry has been designed by Liu et al. [55] to provide a bidimensional optical sensing platform for arginine (Arg) detection. This system was established by monitoring the influence of Arg on the growth of gold nanoparticles/carbon QDs (Au/CQDs) composite. CQDs were synthesized by using ethylene glycol as the reducing as well as a stabilizing agent. Arg is an amino acid containing guanidine group with the highest isoelectric point (pI) value at 10.76, and it would carry positive charges at pH 7.4 with positively charged guanidine group. Arg can attract AuCl$_4^-$ and CQDs through electrostatic interaction, which adversely affected the growth of Au/CQDs composite. As a result, the color of the system did not change, and the fluorescence quenching of CQDs was prevented in the presence of Arg. A low detection limit for Arg (37 nM) and a detection limit of 450 nM was obtained by fluorescence spectroscopy. This dual-signal sensor also revealed excellent selectivity toward Arg over other amino acids. Arg can be detected in urine samples with satisfactory results by this sensor.

Niu et al. [56] reported the development of a "turn-on" fluorescence sensor for Pb^{2+} detection, which is based on GQDs and gold nanoparticles. They achieved an extremely broad detection range of Pb^{2+} from 50 nM to 4 μM, with a detection limit of 16.7 nM. This system is highly sensitive and selective for determination of lead ions. This strategy can be considered for detection of other heavy metal, antigen, or DNAs by slight modification of sensing molecules, and also for fast examination in chemical and biological applications.

Water-soluble $AgInS_2$ quantum dots (AIS QDs) were synthesized in an electric pressure cooker in the presence of polyethyleneimine by Wang et al. [57]. As - prepared QDs exhibited yellow emission with a PL QY up to 32%. They also showed excellent water/buffer stability. The highly luminescent AIS QDs were used for the detection of hydrogen peroxide/glucose and cell imaging. The amino-functionalized AIS QDs show high sensitivity and specificity for glucose and hydrogen peroxide with detection limits of 0.90 and 0.42 µM, respectively. A linear correlation was observed between PL intensity and concentration of H_2O_2 in the ranges of 0.5–10 µM and 10–300 µM, while these linear ranges were 1–10 µM and 10–1000 µM for detection of glucose. These AIS QDs had negligible cytotoxicity on HeLa cells. The luminescence of AIS QDs gives the function of optical imaging.

6.7 PHOTOCATALYSTS

A highly efficient and eco-friendly oxidation process is always required for purification of air or water. Coupling of a semiconducting photocatalyst with GQDs has been found effective in enhancing the photocatalytic as well as photoelectrical conversion performances of the composites, but the preparation of such composites is quite a time consuming and involves difficult post-treatment steps.

Different sized GQDs have been synthesized by Chinnusamy et al. [58] via an inexpensive wet chemical method. They used bird charcoal as a precursor. As-prepared GQDs were found to have luminescence and visible light absorption. Then these GQDs were further coupled with titanium dioxide to form TiO_2–GQDs nanocomposites. Such GQD nanostructures exhibited band gap tunability and had the potential to increase the photoabsorption in TiO_2. The hybrid combination of the nanomaterial decreases the recombination of charge carriers, but increases charge carrier mobility; thus, improved the overall photoconversion efficiency. These composites showed higher photocatalytic activity and rate constants value as compared to pure TiO_2.

Copper (Cu)-doped tin oxide QDs were prepared by Babu et al. [59] using energy-efficient solution combustion synthesis. As - synthesized nanoparticles were duly characterized. The X-ray diffraction (XRD) pattern showed that the QDs have tetragonal rutile-like structure. Modi-fication in the growth inhibition of the SnO_2 lattice parameters with Cu doping was confirmed by HR-TEM. The optical band gap energy of SnO_2 was found to decrease on increasing the Cu concentration. X-ray photo-electron spectroscopy (XPS) revealed the presence of copper and tin as Cu^{2+} and Sn^{4+}, respectively. The photocatalytic degradation of methyl orange dye was observed under visible light irradiation using the QDs as catalysts. The QDs with 0.03 mol % Cu doping showed the highest photocatalytic performance, where 99% dye was degraded under visible light within 180 min.

Wang et al. [60] synthesized a novel full spectrum light-driven CQDs/Bi_2WO_6 (CBW) hybrid materials via a facile hydrothermal method. It was indicated that CBW heterojunctions were an assembly of CQDs on m-BWO and these had high separation efficiency of photo-generated carriers and full light spectrum absorption. The photocatalytic mechanism of CBW hybrid materials was revealed, which suggests that their excellent photocatalytic activity was due to upconverted PL and electron reservoir properties of CQDs. Density functional theory (DFT) calculations indi-cated that complementary conduction and valence band-edge hybridiza-tion between CQDs and m-BWO may increase separation efficiency of electron-hole pairs of CBW hybrid materials. O_2^-, ·OH, and h^+ were considered the main active species during the photocatalytic process as evident from ESR measurement and quenching experiments.

Photocatalytic reduction of Cr(VI) has attracted enormous research interest because it provides a clean and low-cost way for the removal of toxic Cr(VI). Despite tremendous efforts, the present day great chal-lenge in material science is to develop highly efficient photocatalysts for Cr(VI) reduction at low cost. A new composite material consisting of SnS_2 quantum dots (QDs) grown on RGO has been reported by Yuan et al. [61] These act as highly-efficient photocatalysts for photocatalytic reduction of Cr(VI) under visible-light irradiation. They prepared these nanostructured materials by a simple and cost-effective one-step hydrothermal process using stannic chloride pentahydrate, L-cysteine, and GR oxide were used as precursors. The photocatalytic performance of SnS_2 QDs/RGO composite photocatalysts in the reduction of Cr(VI) has been investi-gated. The results showed that the photocatalysts exhibited significantly enhanced photocatalytic activities for Cr(VI) reduction. The 1.5% SnS_2 QDs/RGO photocatalyst achieved the highest photoreduction efficiency of about 95.3% under visible-light irradiation. This enhanced photocatalytic activity of SnS_2 QDs/RGO photocatalysts was mainly attributed to excellent electron transportation ability of GR that impedes the recombination of electron-hole pairs, and the effective charge transfer from SnS_2 to GR was confirmed by the significant reduction of photoluminescence intensity in SnS_2 QDs/RGO photocatalysts.

GQDs/Mn-N-TiO$_2$/g-C$_3$N$_4$ (GQDs/TCN) composite photocatalysts have been synthesized by Nie et al. [62] GQDs/TCN catalysts have been successfully used for the photodegradation of organic pollutants like p-nitrophenol (4-NP), diethyl phthalate (CIP) and ciprofloxacin (DEP) coupled with simultaneous photocatalytic production of hydrogen. It was observed that 5%GQDs/TCN–0.4 sample showed the best photocatalytic hydrogen production and organic pollutant degradation rate under simulated solar irradiation. The photocatalytic hydrogen production rates in the solution of all the three pollutants (4-NP, CIP, and DEP) were larger than pure water system. However, H_2 evolution rate in 4-NP solution was found to be smaller than the solutions of CIP and DEP, which indicates that the photodegradation rate of 4-NP is larger than that of CIP and DEP. It was opined that photogenerated electrons were used in the photodegradation process of 4-NP but not in that of CIP and DEP.

It is known that CQDs synthesized by hydrothermal method possess upconversion properties that could transfer low energy photons to high energy photons, which gives an enhanced visible light response. CQDs modified TiO_2 photocatalysts were successfully prepared by Ke et al. [63] via a facile sol-gel method. They investigated the photophysical and surficial properties of the as-prepared composite photocatalyst. The photocatalytic performance was tested by degradation of methylene blue under visible light irradiation, and its efficiency was as high as 90% within 120 min, which is 3.6 times higher than pure TiO_2.

Liu et al. [64] fabricated hydroxyl-GQDs modified mesoporous graphitic carbon nitride (mpg-C$_3$N$_4$) composites through electrostatic interactions. It was reported that 0.5 wt% GQDs/mpg-C$_3$N$_4$ composites exhibited higher photocatalytic activity than pure mpg-C$_3$N$_4$. Rhodamine B and tetracycline hydrochloride were selected as modal pollutants. It was indicated that uniform dispersion of GQDs on the surface of mpg-C$_3$N$_4$ and an intimate contact between the two materials was responsible for the enhanced activity. More O_2^- species and a small fraction of holes (h$^+$) were generated for photocatalytic degradation in the presence of GQDs/mpg-C$_3$N$_4$ composites.

Sulfur-doped graphene oxide quantum dots (S-GOQDs) were synthesized by Gliniak et al. [65]. They used it for efficient photocatalytic generation of hydrogen. As-synthesized S-GOQDs exhibited three absorption bands at 333, 395, and 524 nm, C=S and C−S stretching vibration signals at 1075 and 690 cm^{-1}, and two excitation-wavelength-independent emission signals with maxima at 451 and 520 nm, respectively, which confirmed that doping of S atom into the GOQDs was successfully completed. It was also suggested that the S-GOQDs has conduction band and valence band levels suitable for water splitting. An initial rate of 18,166 μmol h^{-1} g^{-1} in pure water and 30,519 μmol h^{-1} g^{-1} in 80% aqueous ethanol solution were obtained in the presence of direct sunlight. Thus, metal-free, and inexpensive S-GOQDs have a great potential in the development of sustainable and environmental friendly photocatalysts for water splitting.

Cu NPs were modified by Zhang et al. [66] with CQDs for an improvement of their photocatalytic ability. Cu/CQDs composites were prepared by facile *in situ* photoreductions. Higher hydrogen generation rate was achieved with it than pure Cu NPs. It was observed that the highest H$_2$ evolution rate was 64 mmol g^{-1} h^{-1} using a sample containing 15.6 wt% of CQDs. CQDs were found to act as electron reservoir to trap electrons generated from Cu NPs and as a result, hinder the recombination of an electron-hole pair. Broad spectrum photocatalytic activities of these samples were achieved under monochromatic light irradiation due to the SPR effect of Cu nanoparticles.

A facile one-step hydrothermal route has been reported by Min et al. [67] for the preparation of GQDs coupled TiO$_2$ (TiO$_2$/GQDs) photocatalysts. Here, 1,3,6-trinitropyrene (TNP) was used as a sole precursor of GQDs. During the process, TNP molecules undergo an intramolecular fusion to form GQDs, which was simultaneously decorated on the surface of TiO$_2$ nanoparticles. This effective coupling of GQDs on TiO$_2$ can extend the light absorption to visible region, and the charge separation efficiency of TiO$_2$/GQDs composites was enhanced. Thus, GQDs act as a photosensitizer and an excellent electron acceptor. These advances make the TiO$_2$/GQDs photocatalyst highly active towards the hydrogen generation, resulting in seven and three-time rate and photocurrent response at optimal GQDs content than TiO$_2$ alone, respectively.

Huang et al. [68] reported a novel CQDs /ZnFe$_2$O$_4$ composite photocatalyst via a facile hydrothermal process. The CQDs/ZnFe$_2$O$_4$ (15 vol

%) composite had stronger transient photocurrent response, almost eight times higher than that of $ZnFe_2O_4$ alone, indicating the better transfer of photogenerated electrons and separation efficiency of photogenerated electron-hole pairs. $CQDs/ZnFe_2O_4$ composite displayed enhanced photocatalytic activities on gaseous NOx removal and high selectivity for nitrate formation under visible light ($\lambda > 420$ nm) irradiation as compared to pristine $ZnFe_2O_4$ nanoparticles. It was indicated that the reactive species contributing to NO elimination were O_2^- and $\cdot OH$ radicals. The CQDs are known to act as an electron reservoir and transporter as well as a powerful energy-transfer component. It was reported that this composite has good biocompatibility and low cytotoxicity.

Que et al. [69] reported a facile fabrication of $Bi_2O_3/CQDs$ photocatalysts. Initially, irregular β-Bi_2O_3 nanosheets were obtained by a thermal treatment of $Bi_2O_2CO_3$ precursor at 350°C for 3 h in air. Different contents of CQDs were then incorporated onto as-obtained β-Bi_2O_3 nanosheets in ethanol solution with vigorous stirring. It was observed that visible light absorption of β-Bi_2O_3 photocatalyst was greatly enhanced with the incorporation of CQDs. As-prepared heterostructure photocatalysts exhibited an enhanced photocatalytic activity for the photodegradation of rhodamine B under visible light exposure.

Use of QDs in different fields makes then quite a potential candidate for water splitting, photocatalytic degradation, bioimaging, photodetectors, LEDs, etc. and time is not far off, when these QDs will find a prominent position in the majority of these fields.

KEYWORDS

- laser
- LEDs
- photocatalyst
- photodetectors
- photovoltaics
- quantum computers
- quantum dots (QDs)
- solar cells

REFERENCES

1. Brus, L., (1984). Electron-electron and electron-hole interactions in small semiconductor crystallites: The size dependence of the lowest excited electronic state. *J. Chem. Phys., 80,* 4403–4409.

2. Brus, L., (1986). Electronic wave functions in semiconductor clusters: Experiment and theory. *J. Phys. Chem., 90*(12), 2555–2560.

3. Ekimov, A. I., & Onushchenko, A. A., (1981). Quantum size effect in three-dimensional microscopic semiconductor crystals. *JETP Lett., 34*(6), 345–349.

4. Yang, Z., Fan, J. Z., Proppe, A. H., Arquer, F. P. G., Rossouw, D., Voznyy, O., et al., (2017). Mixed-quantum-dot solar cells. *Nature Commun., 8,* doi: 10.1038/s41467–017–01362–1.

5. Vyskočil, J., Hospodková, A., Petříček, O., Pangrác, J., Zíková, M., Oswald, J., et al., (2017). GaAsSb/InAs/ (In)GaAs type II quantum dots for solar cell applications. *J. Crystal Growth, 464,* 64–68.

6. Wang, Y., Su, W., Zang, S., Li, M., Zhang, X., & Liu, Y., (2017). Bending-durable colloidal quantum dot solar cell using a ZnO nanowire array as a three-dimensional electron transport layer. *Appl. Phys. Lett., 110,* doi.org/10.1063/1.4980136.

7. Zhang, X., Santra, P. K., Tian, L., Johansson, M. B., Rensmo, H., & Johansson, E. M. J., (2017). Highly efficient flexible quantum dot solar cells with improved electron extraction using MgZnO nanocrystals. *ACS Nano., 11*(8), 8478–8487.

8. Tang, Q., Zhu, W., He, B., & Yang, P., (2017). Rapid conversion from carbohydrates to large-scale carbon quantum dots for all-weather solar cells. *ACS Nano., 11*(2), 1540–1547.

9. Li, H., Shi, W., Huang, W., Yao, E. P., Han, J., Chen, Z., et al., (2017). Carbon quantum dots/TiOx electron transport layer boosts the efficiency of planar heterojunction perovskite solar cells to 19%. *Nano Lett., 17*(4), 2328–2335.

10. Zhou, D., Liu, D., Pan, G., Chen, X., Li, D., Xu, W., et al., (2017). Cerium and ytterbium Co-doped halide perovskite quantum dots: A novel and efficient down converter for improving the performance of silicon solar cells. *Adv Mater., 29*(42), doi:10.1002/adma.201704149.

11. Kisslinger, R., Hua, W., & Shankar, K., (2017). Bulk heterojunction solar cells based on blends of conjugated polymers with II-VI and IV-VI inorganic semiconductor quantum dots. *Polymers, 9*(2), doi: 10.3390/polym9020035.

12. Meinardi, F., Ehrenberg, S., Dhamo, L., Carulli, F., Mauri, M., Bruni, F., et al., (2017). Highly efficient luminescent solar concentrators based on earth-abundant indirect-bandgap silicon quantum dots. *Nature Photonics, 11,* 177–185.

13. Muthalif, M. P. A., Lee, Y. S., Sunesh, C. D., Kim, H. J., & Choe, Y., (2017). Enhanced photovoltaic performance of quantum dot-sensitized solar cells with a progressive reduction of recombination using Cu-doped CdS quantum dots. *Appl. Surface Sci., 396,* 582–589.

14. Teymourinia, H., Salavati-Niasari, M., Amiri, O., & Farangi, M., (2018). Facile synthesis of graphene quantum dots from corn powder and their application as down-conversion effect in quantum dot-dye-sensitized solar cell. *J. Mol. Liq., 251,* 267–272.

15. Wang, R., Wu, X., Xu, K., Zhou, W., Shang, Y., Tang, H., et al., (2018). Highly efficient inverted structural quantum dot solar cells. *Adv Mater., 30*(7), doi.org/10.1002/adma.201704882.

16. Veerathangam, K., Pandian, M. S., & Ramasamy, P., (2017). Photovoltaic performance of Ag-doped CdS quantum dots for solar cell application. *Mater. Res. Bull., 94*, 371–377.

17. Ye, K. H., Wang, Z., Gu, J., Xiao, S., Yuan, Y., Zhu, Y., et al., (2017). Carbon quantum dots as a visible light sensitizer to significantly increase the solar water splitting performance of bismuth vanadate photoanodes. *Energy Environ. Sci., 10*, 772–779.

18. Kloeffel, C., & Loss, D., (2013). Prospects for spin-based quantum computing in quantum dots. *Annu. Rev. Condens. Matter. Phys., 4*, 51–81.

19. Loss, D., & Vincenzo, D. P. D., (1998). Quantum computation with quantum dots. *Phys. Rev. A., 57*(1), doi.org/10.1103/PhysRevA.57.120.

20. Imamoglu, A., (2003). Are quantum dots useful for quantum computation? *Phys. E: Low-dimens. Syst. Nanostruct., 16*(1), 47–50.

21. Wei, H. R., & Deng, F. G., (2014). Scalable quantum computing based on stationary spin qubits in coupled quantum dots inside double-sided optical microcavities. *Sci. Rep., 4*, doi: 10.1038/srep07551.

22. Freitag, N. M., Chizhova, L. A., Nemes-Incze, P., Woods, C. R., Gorbachev, R. V., Cao, Y., et al., (2016). Electrostatically confined monolayer graphene quantum dots with orbital and valley splitting. *Nano Lett., 16*(9), 5798–5805.

23. Kamada, H., & Gotoh, H., (2004). Quantum computation with quantum dot excitons. *Semicond Sci. Technol., 19*, S392–S396.

24. Bechtold, A., Rauch, D., Li, F., Simmet, T., Ardelt, P. L., Regler, A., et al., (2015). Three-stage decoherence dynamics of an electron spin qubit in an optically active quantum dots. *Nature Phys., 11*, 1005–1008.

25. Radulaski, M., Widmann, M., Niethammer, M., Zhang, J. L., Lee, S. Y., Rendler, T., et al., (2017). Scalable quantum photonics with single color centers in silicon carbide. *Nano Lett., 17*(3), 1782–1786.

26. Lagoudakis, K. G., McMahon, P. L., Dory, C., Fischer, K. A., Müller, K., Borish, V., et al., (2016). Ultrafast coherent manipulation of trions in site-controlled nanowire quantum dots. *Optic., 3*(12), 1430–1435.

27. Dai, X., Zhang, Z., Jin, Y., Niu, Y., Cao, H., Liang, X., et al., (2014). Solution-processed, high-performance light-emitting diodes based on quantum dots. *Nature, 515*, 96–99.

28. Bae, W. K., Kwak, J., Park, J. W., Char, K., Lee, C., & Lee, S., (2009). Highly efficient green-light-emitting diodes based on CdSe@ZnS quantum dots with a chemical-composition gradient. *Adv. Mater., 21*, 1690–1694.

29. Zorn, M., Bae, W. K., Kwak, J., Lee, H., Lee, C., Zentel, R., et al., (2009). Quantum dot–block copolymer hybrids with improved properties and their application to quantum dot light-emitting devices. *ACS Nano., 3*(5), 1063–1068.

30. Song, J., Li, J., Li, X., Xu, L., Dong, Y., & Zeng, H., (2015). Quantum dot light-emitting diodes based on inorganic perovskite cesium lead halides ($CsPbX_3$). *Adv. Mater., 27*, 7162–7167.

31. Kwak, J., Bae, W. K., Lee, D., Park, I., Lim, J., Park, M., et al., (2012). Bright and efficient full-color colloidal quantum dot light-emitting diodes using an inverted device structure. *Nano Lett., 12*(5), 2362–2366.

32. Ippen, C., Greco, T., & Wedel, A., (2012). InP/ZnSe/ZnS: A novel multishell system for InP quantum dots for improved luminescence efficiency and its application in a light-emitting device. *J. Inform Display, 13*(2), doi.org/10.1080/15980316.2012.68 3537.

33. Luo, Z., Xu, D., & Wu, S. T., (2014). Emerging quantum-dots-enhanced LCDs. *J. Display Technol., 10*(7), 526–539.

34. Li, L., Gu, L., Lou, Z., Fan, Z., & Shen, G., (2017). ZnO quantum dot decorated Zn_2SnO_4 nanowire heterojunction photodetectors with drastic performance enhancement and flexible ultraviolet image sensors. *ACS Nano., 11*(4), 4067–4076.

35. Xing, C., Huang, W., Xie, Z., Zhao, J., Ma, D., Fan, T., et al., (2018). Ultrasmall bismuth quantum dots: Facile liquid-phase exfoliation, characterization, and application in high-performance UV–Vis photodetector. *ACS Photonics., 5*(2), 621–629.

36. Dias, S., Kumawat, K., Biswas, S., & Krupanidhi, S. B., (2017). Solvothermal synthesis of Cu_2SnS_3 quantum dots and their application in near-infrared photodetectors. *Inorg. Chem., 56*(4), 2198–2203.

37. Kan, H., Liu, S., Xie, B., Zhang, B., & Jiang, S., (2017). The effect of Au nanocrystals applied in CdS colloidal quantum dots ultraviolet photodetectors. *J. Mater. Sci. Mater. Electron., 28*(13), 9782–9787.

38. Chandan, H. R., Schiffman, J. D., & Balakrishna, R. G., (2018). Quantum dots as fluorescent probes: Synthesis, surface chemistry, energy transfer mechanisms, and applications. *Sens. Actuators B: Chem., 258*, 1191–1214.

39. Zhang, X., Wang, Q., Jin, Z., Zhang, J., & Liu, S., (2017). Stable ultra-fast broad-bandwidth photodetectors based on α-CsPbI$_3$ perovskite and NaYF$_4$:Yb, Er quantum dots. *Nanoscale, 9*, 6278–6285.

40. Ma, Y. J., Zhang, Y. G., Gu, Y., Chen, X. Y., Wang, P., Juang, B. C., et al., (2017). Enhanced carrier multiplication in InAs quantum dots for bulk avalanche photodetector applications. *Adv. Optic. Mater., 5*(9), doi: org/10.1002/adom.201601023.

41. Zheng, Z., Zhuge, F. W., Wang, Y. G., Zhang, J. B., Gan, L., Zhou, X., et al., (2017). Decorating perovskite quantum dots in TiO$_2$ nanotubes array for broadband response photodetector. *Adv. Funct. Mater., 27*, doi: org/10.1002/adfm.201703115.

42. Arquer, P. F. G. D., Armin, A., Meredith, P., & Sargent, E. H., (2017). Solution-processed semiconductors for next-generation photodetectors. *Nature Rev. Mater., 2*, doi: /10.1038/natrevmats.2016.100.

43. Ren, Z., Sun, J., Li, H., Mao, P., Wei, Y., Zhong, X., et al., (2017). Bilayer PbS quantum dots for high-performance photodetectors. *Adv. Mater., 29*, doi: org/10.1002/adma.201702055.

44. Mihalache, I., Radoi, A., Pascu, R., Romanitan, C., Vasile, E., & Kusko, M., (2017). Engineering graphene quantum dots for enhanced ultraviolet and visible light p-Si nanowire-based photodetector. *ACS Appl. Mater. Interfaces, 9*(34), 29234–29247.

45. Tang, X., Zu, Z., Zang, Z., Hu, Z., Hu, W., Yao, Z., et al., (2017). CsPbBr$_3$/Reduced graphene oxide nanocomposites and their enhanced photoelectric detection application. *Sens. Actuators B: Chem., 245*, 435–440.

46. Tong, L., Qiu, F., Zeng, T., Long, J., Yang, J., Wang, R., et al., (2017). Recent progress in the preparation and application of quantum dots/graphene composite materialism. *RSC Adv., 7*, 47999–48018.

47. Cheng, X., Benjamin, F. P. M., Robinson, A. B., Longatte, G., O'Mara, P. B., Tan, V. T. G., et al., (2017). Colloidal silicon quantum dots: From preparation to the modification of self-assembled monolayers for bioimaging and sensing applications. *Proc. Colloidal Nanoparticles for Biomedical Applications*, 10078, doi: 10.1117/12.2249592.

48. Susumu, K., Field, L. D., Oh, E., Hunt, M., Delehanty, J. B., Palomo, V., et al., (2017). Purple-, blue-, and green-emitting multishell alloyed quantum dots: Synthesis, characterization, and application for ratiometric extracellular pH sensing. *Chem. Mater., 29*(17), 7330–7344.

49. Kumawat, M. K., Thakur, M., Gurung, R. B., & Srivastava, R., (2017). Graphene quantum dots from *Mangifera indica*: Application in near-infrared bioimaging and intracellular nanothermometry. *ACS Sustain Chem. Eng., 5*(2), 1382–1391.

50. Atchudana, R., Nesakumar, T., Edison, J. I., Chakradhar, D., Perumal, S., Shim, J. J., et al., (2017). Facile green synthesis of nitrogen-doped carbon dots using *Chionanthus retusus* fruit extract and investigation of their suitability for metal ion sensing and biological applications. *Sens. Actuators B: Chem., 246*, 497–509.

51. Dong, Y. P., Wang, J., Peng, Y., & Zhu, J. J., (2017). Electrogenerated chemilumines-cence of Si quantum dots in neutral aqueous solution and its biosensing application. *Biosens. Bioelectron., 89*, 1053–1058.

52. Lee, J., Feng, X., Ou, C., Bawendi, M. G., & Huang, J., (2018). Stable, small, specific, low-valency quantum dots for single-molecule imaging. *Nanoscale, 10*, 4406–4414.

53. Velusamy, A., Thomas, N. J., Immanuel, E., Lee, Y. R., & Mathur, G. S., (2017). Biological and catalytic applications of green synthesized fluorescent N-doped carbon dots using *Hylocereus undatus*. *J. Photochem. Photobiol. B.: Biol., 168*, 142–148.

54. Amjadi, M., Hallaj, T., Asadollahi, H., Song, Z., Frutos, M. D., & Hildebrandt, N., (2017). Facile synthesis of carbon quantum dot/silver nanocomposite and its application for colorimetric detection of methimazole. *Sens. Actuators B: Chem., 244*, 425–432.

55. Liu, T., Li, N., Dong, J. X., Zhang, Y., Fan, Y. Z., Lin, S. M., et al., (2017). A colorimetric and fluorometric dual-signal sensor for arginine detection by inhibiting the growth of gold nanoparticles/carbon quantum dots composite. *Biosens. Bioelectron., 87*, 772–778.

56. Niu, X., Zhong, Y., Chen, R., Wang, F., Liu, Y., & Luo, D., (2018). A "turn-on" fluores-cence sensor for Pb^{2+} detection based on graphene quantum dots and gold nanoparticles. *Sens. Actuators B: Chem., 255*, 1577–1581.

57. Wang, L., Kang, X., & Pan, D., (2017). Gram-scale synthesis of hydrophilic PEI-coated $AgInS_2$ quantum dots and its application in hydrogen peroxide/glucose detection and cell imaging. *Inorg. Chem., 56*(11), 6122–6130.

58. Chinnusamy, S., Kaur, R., Bokare, A., & Erogbogbo, F., (2018). Incorporation of graphene quantum dots to enhance photocatalytic properties of anatase TiO_2. *MRS Commun., 8*(1), 137–144.

59. Babu, B., Kadam, A. N., Ravikumar, R. V., & Byon, C., (2017). Enhanced visible light photocatalytic activity of Cu-doped SnO_2 quantum dots by solution combustion synthesis. *J. Alloys Compd., 703*, 330–336.

60. Wang, J., Tang, L., Zeng, G., Deng, Y., Dong, H., Liu, Y., et al., (2018). 0D/2D interface engineering of carbon quantum dots modified Bi_2WO_6 ultrathin nanosheets with enhanced photoactivity for full spectrum light utilization and mechanism insight. *Appl. Catal. B: Environ., 222*, 115–123.

61. Yuan, Y. J., Chen, D. Q., Shi, X. F., Tu, J. R., Hu, B., Yang, L. X., et al., (2017). Facile fabrication of "green" SnS_2 quantum dots/reduced graphene oxide composites with enhanced photocatalytic performance. *Chem. Eng., J., 313*, 1438–1446.

62. Nie, Y. C., Yu, F., Wang, L. C., Xing, Q. J., Liu, X., Pei, Y., et al., (2018). Photocatalytic degradation of organic pollutants coupled with simultaneous photocatalytic H_2 evolution over graphene quantum dots/Mn-N-TiO_2/g-C_3N_4 composite catalysts: Performance and mechanism. *Appl. Catal. B: Environ., 227*, 312–321.

63. Ke, J., Li, X., Zhao, Q., Liu, B., Liu, S., & Wang, S., (2017). Upconversion carbon quantum dots as visible light responsive component for efficiency enhancement of photocatalytic performance. *J. Colloid Interface Sci., 496*, 425–433.

64. Liu, J., Xu, H., Xu, Y., Song, Y., Lian, J., Zhao, Y., et al., (2017). Graphene quantum dots modified mesoporous graphite carbon nitride with significant enhancement of photocatalytic activity. *Appl. Catal. B: Environ., 207*, 429–437.

65. Gliniak, J., Lin, J. H., Chen, Y. T., Li, C. R., Jokar, E., Chang, C. H., et al., (2017). Sulfur-doped graphene oxide quantum dots as photocatalysts for hydrogen generation in the aqueous phase. *Chem. Sus. Chem., 10*(16), 3260–3267.

66. Zhang, P., Song, T., Wang, T., & Zeng, H., (2017). In-situ synthesis of Cu nanoparticles hybridized with carbon quantum dots as a broad spectrum photocatalyst for improvement of photocatalytic H_2 evolution. *Appl. Catal. B: Environ., 206*, 328–335.

67. Min, S., Hou, J., Lei, Y., Ma, X., & Lu, G., (2017). Facile one-step hydrothermal synthesis toward strongly coupled TiO_2/grapheme quantum dots photocatalysts for efficient hydrogen evolution. *Appl. Surface Sci., 396*, 1375–1382.

68. Huang, Y., Liang, Y., Rao, Y., Zhu, D., Cao, J. J., Shen, Z., et al., (2017). Environment-friendly carbon quantum dots/$ZnFe_2O_4$ photocatalysts: Characterization, biocompatibility, and mechanisms for NO removal. *Environ. Sci. Technol., 51*(5), 2924–2933.

69. Que, Q., Xing, Y., He, Z., Yang, Y., Yin, X., & Que, W., (2017). Bi_2O_3/Carbon quantum dots heterostructured photocatalysts with enhanced photocatalytic activity. *Mater. Lett., 209*, 220–223.

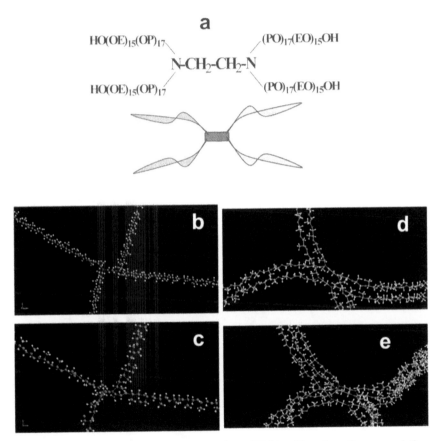

FIGURE 10.2 (a) Structural formula of star shaped T904 and its schematic representation. H (white), C (grey), O (red), and N (blue). (b) A graphical model of the structure and (c) its relaxed structure. (d) A graphical model of an aggregate of three molecules and (e) their relaxed structure.

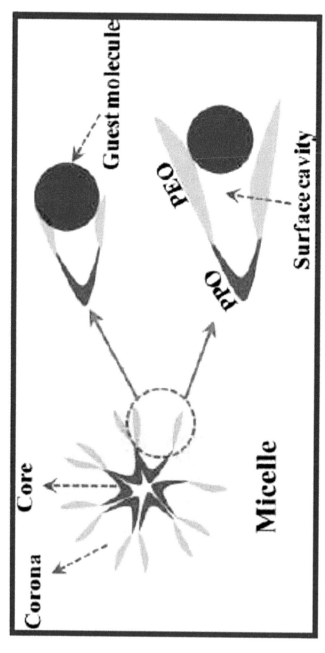

a Red dotted circle shows a possible surface cavity whose size is related to the number of PEO and PPO units. A larger cavity can easily accommodate a guest molecule in comparison to a smaller cavity.

FIGURE 10.3   A TBP Micelle with the Core Occupied by PPO Units and the Corona Constituted by PEO Unitsa

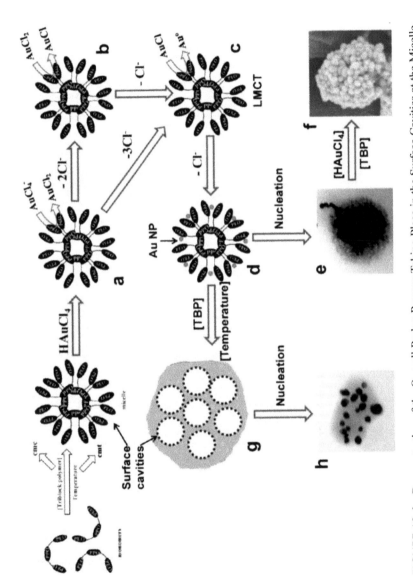

FIGURE 10.4 Demonstration of the Overall Redox Process Taking Place in the Surface Cavities at the Micelle-Solution Interface of TBP Micelles.

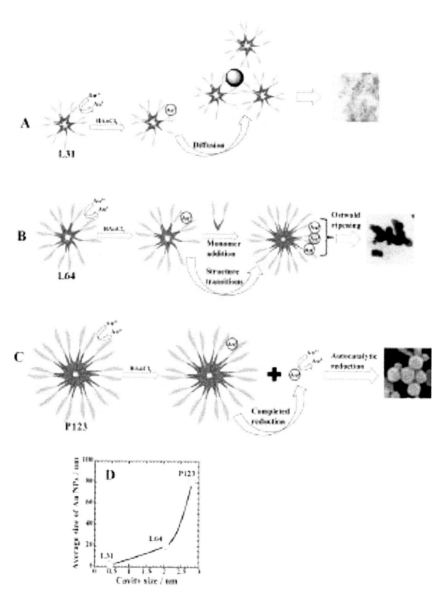

FIGURE 10.5   Schematic Representation of the Proposed Mechanism for the Synthesis of Au NPs by Using Micelles of L31, L64, and P123, Respectively (see details in the text). (D) Plot of the Average Size of NPs Estimated from TEM Images versus Cavity Size for L31, L64, and P123.

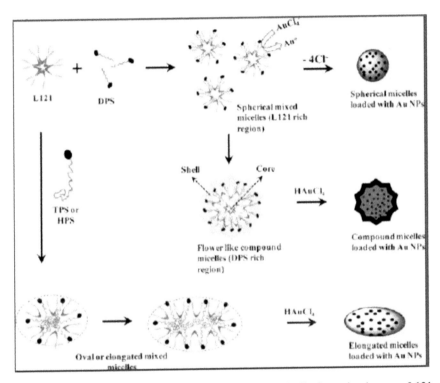

FIGURE 10.6 Schematic representation of the mixed micelle formation between L121 and zwitterionic surfactants and their subsequent use as micelle templates for the self-assembled Au NPs. Top reaction shows the formation of spherical L121 + DPS mixed micelles in the L121-rich region of the mixture by incorporating the hydrocarbon chains of DPS molecules in the L121 micelles. In the DPS-rich region, predominantly hydrophobic L121 is solubilized by the hydrocarbon tails of DPS in the form of a typical compound micelle where DPS molecules occupy the shell while L121 resides in the core, resulting in the formation of flower-like morphologies with Au NPs mainly accommodated in the core due to the presence of L121 surface cavities. Lower reaction shows that using TPS or HPS instead of DPS induces their longer hydrocarbon tails in the L121 micelles, thus causing structure transitions with the formation of oval or elongated morphologies that are visualized by the self-assembled Au NPs.

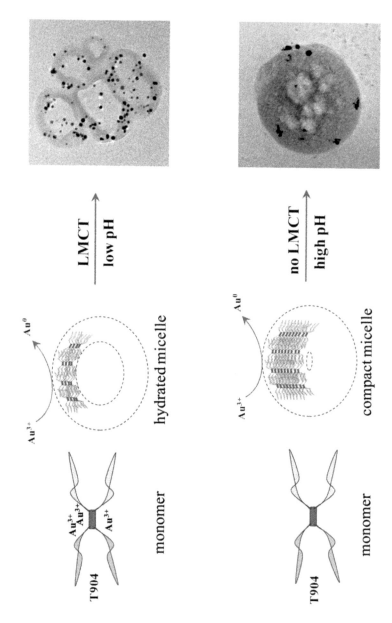

FIGURE 10.15 Schematic representation of a T904 monomer and its vesicles at low and high pH. Hydrated vesicle with core–shell type morphology is formed at low pH while compact compound micelle with no clear core–shell regions is produced at high pH.

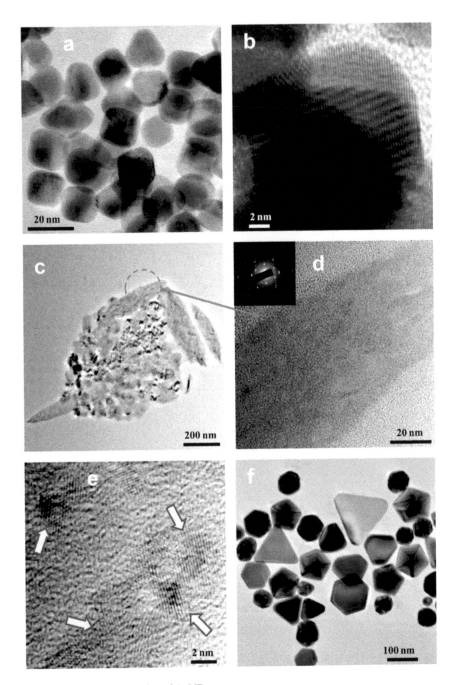

FIGURE 10.16   Geometries of AuNPs.

CHAPTER 7

POLYMER NANOCOMPOSITES, A SMART MATERIAL: SYNTHESIS, PREPARATION, AND PROPERTIES

SONIA KHANNA

Department of Chemistry, School of Basic Sciences and Research, Sharda University, Greater Noida, India, E-mail: sonia.khanna@sharda.ac.in

ABSTRACT

Nanocomposites are the materials that are made up of two materials of which one is in the form of sheets, fibers or particles, and other is matrix phase. A polymer composite is a combination of polymer (matrix) and filler (reinforcement). The additives can be one-dimensional (nanotubes and fibers), two-dimensional (layered materials like clay) or three-dimensional (spherical particles). Polymeric nanocomposites are attracting researchers in light of its extraordinary properties gaining considerable attention both in academia and in industries, due to their outstanding mechanical properties such as lightweight, high elastic stiffness, high durability, corrosion resistance, and strength. These materials have a big potential for applications in the automotive, aerospace industry as well as in the biotechnological field. Various methods of synthesis and fabrication of polynanocomposites are discussed in this chapter

7.1 INTRODUCTION

Nanotechnology is the latest area of research encompassing a wide range of technologies, processing techniques, and measurements dealing with nanostructures with components with the size between 1 to 100-nanometer dimensions [1]. These particles and materials are referred to as nanoparticles or nanomaterials and exhibit unusual and exotic properties that are not present in traditional bulk materials. It entails the ability to build molecular systems with atom-by-atom precision, yielding a variety of nanomaterials. Nanomaterials possess a large surface area, a high aspect ratio, and a high surface-to-mass ratio and significantly influence the physical, chemical, biological, mechanical, and electrical properties. Nanotechnology is a wide interdisciplinary field of research, innovation, development, and industrial activity that has been growing in a speedy way worldwide for the last few years [2, 3]. Nanotechnology has benefitted any aspects of life by its applications in the energy storage and production, information technology, medical purposes, manufacturing, food, and water purification, instrumentation, biomedical, and environmental uses [4].

Composites consist of reinforcing phase in the form of sheets, fibers or particles, and other a tough or ductile matrix phase. Generally, the reinforcing phase have low densities and provides the strength of the resulting composites if combined properly with the matrix material in correct proportion. These materials have unprecedented flexibility and hardness. The reinforcing material can be made up of particles (e.g., minerals), sheets (e.g., exfoliated clay stacks) or fibers (e.g., carbon nanotubes or electrospun fibers) [5]. The nanocomposites are characterized by solid structures dispersed at the nanoscale in a matrix, for example, inorganic nanoparticles in an organic matrix. For example, adding carbon nanotubes improves the electrical and thermal conductivity. The behaviors of nanocomposites are dependent on their sizes. A reduction in size results in dramatical emergence of new properties such as electrical conductivity, high strength, and insulating behavior not observed in its macrostate. For example, nanocomposites with size less than 5nm show catalytic activity and zinc oxide and TiO_2 colloids become invisible when below 15nm [6]. Nanocomposites display exceptionally high surface to volume ratio of the reinforcing phase and/or its exceptionally high aspect ratio. The percentage by weight (called mass fraction) of the nanoparticulates introduced can remain very low (on the order of 0.5% to 5%) due to the low filler percolation threshold, especially for the most commonly used non-spherical, high aspect ratio fillers (e.g., nanometer-thin platelets, such as clays, or nanometer-diameter cylinders, such as carbon nanotubes). The orientation and arrangement of asymmetric nanoparticles, thermal property mismatch at the interface, interface density per unit volume of the nanocomposite, and polydispersity of nanoparticles significantly affect the effective thermal conductivity of nanocomposites [7].

The properties of polymer like thermoset cure, polymer chain mobility, and polymer chain conformation, degree of polymer chain ordering or crystallinity has shown to affect the nature of nanocomposites [8]. The large amount of reinforcement surface area has an observable effect on the macroscale properties of the composite. Other kinds of nanoparticles may result in enhanced optical properties, dielectric properties, heat resistance or mechanical properties such as stiffness, strength, and resistance to wear and damage and are used accordingly.

7.2   TYPES OF NANOCOMPOSITES

There are three types of nanocomposites

1. ceramic nanocomposites;
2. metal matrix nanocomposites; and
3. polymer nanocomposites.

The properties of nanocomposites rely on a range of variables like degree of dispersion, size, shape, and orientation of the reinforcing phase in matrix and the interactions between the matrix and the second phase.

i) **Ceramic nanocomposites:** These nanocomposites consist of single ceramic layers (1nm thick) homogeneously dispersed in a continuous matrix. The host ceramic layer tends to orient themselves parallel to each other due to dipole-dipole interaction. Ceramic matrix nanocomposites having at least one phase in nano-dimension have outstanding electrical and mechanical properties. Generally, the common methods used in micro composites fabrications are conventional powder method, polymer precursor route, spray pyrolysis, and chemical methods. Some common ceramic matrix nanocomposites include Al_2O_3/SiO_2, $SiO_2/Ni.$, $Al_2O_3/TiO/SiC$ [9]. The pioneering studies have been reported on ceramic nanocomposites on Al_2O_3/SiC system by Niihara et al. [10, 11]. Addition of small amount of SiC particles has shown noticeable strengthening of the Al_2O_3 matrix. Likewise, the incorporation of high strength nanofibers into ceramic matrices has allowed the preparation of advanced nanocomposites with high toughness and superior failure characteristics compared to the sudden failures of ceramic materials.

ii) **Metal-matrix nanocomposites:** Metal-matrix nanocomposites consist of a ductile metal or alloy matrix in which some nanosized reinforcement material is implanted. These materials combine metal and ceramic features, i.e., ductility, and toughness with high strength and modulus. Metallic materials such as Al [12–14], Mg [15, 16], Ti [17, 18], Cu [19, 20] and their alloys have been investigated for their industrial applications. Thus, metal matrix nanocomposites are suitable for production of materials with high strength in shear/compression processes and high service temperature capabilities. They show an extraordinary potential for application in many areas, such as aerospace and automotive industries and development of structural materials. Both metal and ceramic nanocomposites have potential applications, but the low wettability of molten metal does not allow the synthesis by conventional casting methods.

Nanocomposites based on a metal oxide matrix with metal oxide nanoparticles, nanowires as reinforcing phase, makes them useful as a sensor by virtue of sensing properties of metal oxide. These metal oxides have also been used for the fabrication of chemical sensors and biosensors. They are simple to be prepared and have good stability and high sensitivity.

The advantages of these nanocomposites are their stability in air, relative inexpensiveness, and easy preparation in the ultra-dispersed state.

iii) **Polymer nanocomposites:** Usually, the synthesis of polymer nano-composites applies bottom-up or top-down methodologies. Polymer nanocomposites have been investigated with various additives. Nanocomposites reinforced with carbon nanotubes have also been extensively reviewed [22, 23].

This chapter concentrates on polymer nanocomposites and highlights their synthesis, properties, and applications.

7.3 POLYMER NANOCOMPOSITES

Polymer composite is a combination of polymer (matrix) and a filler (reinforcement). The additives can be one-dimensional (nanotubes and fibers), two-dimensional (layered materials like clay) or three-dimensional (spherical particles). The choice of reinforcement material depends on the applications. The polymer matrix and fillers are generally bonded by weak intermolecular forces; however, in some cases, chemical bonding is employed, as well. The dispersion of the filler material on an atomic or molecular level (nanometer level) and chemical bonding with the matrix material results in remarkable improvements in the mechanical properties of the resulting composite materials. Clay minerals (montmorillonite, saponite, hectorite, etc.) are used as filler materials for achieving high strength.

Polymer nanocomposites have gained popularity due to their outstanding mechanical properties such as lightweight, high elastic stiffness, high durability, corrosion resistance, and strength. They also show properties of barrier resistance, flame retardancy, wear resistance, magnetic, electrical, and optical properties. Nanoparticle addition in polynanocomposites bring about remarkable changes in the material properties like high surface area, high surface energy. An anisotropic geometry in the polymer matrix is generated, decreasing the interparticle distance and increasing polymer matrix interaction strength. Various filler materials such as exfoliated clay and carbon nanotubes, exfoliated graphite and nanocrystalline metals as reinforcing materials have also been reported [21]. Such modified polymer nanocomposites with completely new set of properties have emerged as suitable for new applications extending from conventional composite materials [22, 24]. Vollenberg and Heikens reported nanocomposites samples by thoroughly mixing filler particles-polystyrene (PS), styrene–acrylonitrile copolymer (SAN), polycarbonate (PC) and polypropylene (PP) with alumina beads having 35nm and 400nm dimension as filler materials with polymer matrix. [25]. In this study, polymers were dissolved in a polar solvent, mixed in the beads for several hours and then poured over a large surface container to allow the solvent to evaporate. It was subsequently dried under a vacuum at 100°C. Pure polymer was then added to samples to achieve the desired particle volume fractions.

7.3.1 PROCESSING OF POLYMER NANOCOMPOSITES

The processing of nanocomposites depends on various factors during its preparation [26, 27]. These factors are:

1. types and orientations of filler materials.
2. degree of mixing of two phases.
3. type of adhesion at the matrix interface.
4. volume fraction of nanoparticles.
5. nanoparticle characteristics.
6. nature of the interphase developed at the matrix interface.
7. size and shape of nanofiller materials.
8. morphology of the system.

In order to achieve enhanced properties of nanocomposites, appropriate reinforcing materials and matrix should be selected. The nanosized particles should be dispersed and distributed in the matrix material properly to prevent agglomeration of particles else the properties of nanocomposites will deteriorate. These aggregates can act as defects and limits the property enhancement of nanocomposite. The nanoparticles should be homogeneously dispersed in the matrix to achieve maximum property enhancement. [28]. The nature of interface between matrix and filler material is another important factor affecting the property of nanocomposite. If the interface region has good, the overall properties of the nanocomposite will be excellent. Most of the interphase properties depend on the bound surface and can be modified by optimizing the interfacial bond between the nanofiller and polymer matrix. The interaction between the interconnecting phases depends on the ratio of surface energy of filler and matrix. The total surface area of a nanoparticle determines the extent of interface phenomenon contributing to the properties of nanocomposites [29, 30].

7.3.2   PREPARATION OF POLYMER NANOCOMPOSITES

There are reports of various methods for the preparation of polymer nanocomposites. These are: (1) in situ polymerization; (2) intercalation of polymer from solution; (3) direct mixing of polymer and fillers; (4) melt intercalation; (5) template synthesis; and (6) sol-gel process. The melt blending is a relatively new process of nanocomposite fabrications. This process involves the melting of polymer to form a viscous liquid. The nanoparticles are dispersed into polymer matrix by means of high shear rate along with diffusion at high temperature. The nanocomposites are then fabricated by either compression molding or injection molding.

7.3.3   FABRICATION OF POLYMER MATRIX NANOCOMPOSITE

Nanoreinforcements need to be surface modified before incorporating them into polymer matrices. Polymer matrix nanocomposites can be fabricated either by chemical or mechanical process. Uniform and homogeneous dispersion of nanoparticles in the polymer matrix is one of the major problems encountered in polymer nanocomposite fabrication. The nanofillers have a tendency to aggregate and form micron size filler cluster, which limit the dispersion of nanoparticles in the polymer matrix thereby deteriorating the properties of nanocomposites. There are various ways of dispersing nanofillers uniformly and homogeneously in the polymer matrix by complicated polymerization reactions or surface modification of filler materials [31]. Mostly polymer nanocomposites are fabricated by the following five methods.

1. intercalation method;
2. chemical/mechanical method;
3. in situ polymerization;
4. sol-gel method; and
5. direct mixing of polymer and nanofillers.

1.   Intercalation Method

Intercalation method generally involves the dispersion of nanoplatelets of nanomaterials into the polymer matrix. Polymer chains are inserted into layered structures such as clays, irrespective of the ratio of polymer to layered structure. It is followed by flocculation of intercalated and stacked layers and separation of the individual layers in the polymer matrix. The material prepared has better high storage modulus, increased tensile and flexural properties, heat distortion temperature and reduced gas permeability as compared to matrix material or conventional micro and macro-composite materials. This technique improves the bulk properties such as stiffness, shrinkage, and flammability. Intercalation is a top-down approach and requires surface modification of nanoplatelets for homogeneous dispersion of plate-like nanofillers in the polymer matrix. Intercalated morphology occurs when polymer chains diffuse into the spacing of layered structure [32, 33].

2. Chemical/Mechanical Techniques

In chemical technique, nanoparticles are dispersed into polymer followed by additional polymerization process. The nanoparticles are swollen in monomer solution, and the polymer formation occurs between the intercalated sheets by polymerization method. In the mechanical technique, the polymer is dissolved in a co-solvent, and nanoplatelets sheets are swollen in the solvent, and these two solutions are mixed together, the polymer chains in the solution intercalate into the nanoplatelets layers and displace the solvent.

3. Melt Intercalation Method

Melt intercalation involves mixing the nanofillers (clays) into the polymer matrix at its molten temperature. In this method, mixture of polymer and nanofibers are annealed either statically or under shear above softening point of polymer and it allows the use of polymers, which are not suitable for in situ polymerization or solution intercalation. The final shape of components can be fabricated by compression molding, injection molding or fiber production technique [34–36].

4. In Situ Polymerization Method

In this method, polymer is generated in situ between the layers, by swelling the layer hosts within the liquid monomer or monomer solution. The nanofillers are swollen within the liquid monomer solution (initiation of polymerization by heating or radiation or by diffusion). The low molecular weight monomer solution can easily seeps in between layers causing swelling. The resulting mixture is polymerized either using radiation, heat, initiator diffusion or by organic initiator [37]. Another technique is in situ template synthesis. In this method, the clay layers are synthesized in the presence of polymer chains. Both polymer matrix and clay layers are dissolved in an aqueous solution and gel is generally refluxed at high temperature. The polymer serves as a nucleating agent and promotes the growth of the inorganic filler crystals. As those crystals grow, the polymer is trapped within the layers and thus forms the nanocomposite. The growth of clay layers take place on the polymer chains at high temperature. The only drawback of this process is that high-temperature synthesis causes decomposition of polymer.

5. Sol-Gel Method

Sol-gel method is a bottom-up approach, and it is based on an opposite principle than all the previous methods [38]. In this method, solid nanoparticles are dispersed in the monomer solution, forming a colloidal suspension of solid nanoparticles (sol), forming interconnecting network between phases (gel) by polymerization reactions. The polymer nanoparticle 3D network extends throughout the liquid [40]. The polymer serves as a nucleating agent and promotes the growth of layered crystals. As the crystals grow, the polymer is seeped between layers, and thus nanocomposite is formed.

6. Direct Mixing of Polymer and Nanofillers

Direct mixing of a polymer matrix and nanofillers is a top-down approach of nanocomposite fabrication, and it is based on the breakdown the aggregated nanofillers during mixing process [31, 39]. This method is suitable for fabricating polymer matrix nanocomposites, and it involves two general ways of mixing the polymer and nanofillers. One way is mixing a polymer, in the absence of any solvents, with nanofillers above the glass transition temperature of the polymer, generally called melt-compounding method. The other way involves mixing of polymer and nanofillers in solution employing solvents, generally called solvent method/solution mixing [39].

7. Melt Compounding

In this method, nanofibers are added to the polymer above the glass transition temperature with vigorous mixing. These nanofiller are broken and aggregated, promoting homogeneous and uniform nanofiller dispersion in the polymer matrix.

8. Solvent Method

In this method, nanoparticles, dispersed in solvent and polymer dissolved in a co-solvent are mixed together forming nanocomposites. In this method, the shear stresses induced in the polymer matrix are lowered compared to that in melt compounding. The nanofillers are pre-dispersed in the solvent by sonication in order to breakdown the nanofiller aggregates [31]. The polymer nanocomposites fabricated by one of the above methods are finally processed by conventional manufacturing methods like injection molding, calendaring, casting, compression molding, blow molding, rotational molding, extrusion molding, thermoforming, etc. [31].

9. Applications of Polymer Nanocomposites

The remarkable properties of polynanocomposites like high mechanical strength, thermal stability and distortion resistance, chemical resistance, electrical conductivity and many more, has made polymer nanocomposites an ideal candidate for various applications in numerous automotive and electronic applications. The utility of polymer-based nanocomposites in these areas is quite diverse involving many potential applications and have been proposed for their use in various applications. Some of the applications are given below:

i) A number of polymer nanocomposites derived from polymers such as styrene, butadiene rubbers have been used as covers against various gases such as CO_2, N_2, O_2 and chemicals like HNO_3, H_2SO_4, etc.

ii) They have been utilized in making door handles, engine covers, fuel cells, solar cell and covers for different electronic devices.

iii) Nanotechnology is deeply embedded in the design of advanced devices for electronic and optoelectronic applications.

iv) Another potential application includes photovoltaic (PV) cells and photodiodes, supercapacitors, printable conductors, light emitting diodes (LEDs) and field effect transistors [40].

v) The electrical conductivity of carbon nanotubes in insulating polymers has been applied in electromagnetic interference shielding, transparent conductive coatings, electrostatic dissipation, supercapacitors, electromechanical actuators and various electrode applications [41].

vi) Polymer-based solar cells have been used to make cheap large flexible panels for optical displays, catalysis, PV, gas sensors, electrical devices, mechanics, photoconductors, and superconductor devices [42].

7.4   CONCLUSION

Polymer nanocomposites have been researched in recent years in light of its remarkable improvement in properties when compared to conventional microcomposites. In this chapter, the classification of nanocomposites based on matrix materials has been discussed along with their processing techniques. The methods used for fabricating the nanocomposites before applied are also discussed along with problems related to the dispersion of nanophase in polymer matrix. This can be researched to devise ways overcome this problem being encountered with nanosized materials. Nevertheless, there is hope that polymer/carbon nanotube nanocomposites might emerge has an amazing nanomaterial which helps in miniaturization of electronic devices and other applications.

KEYWORDS

- **applications**
- **fabrication**
- **polymer nanocomposites**
- **preparation**

REFERENCES

1. Roy, R., Roy, R. A., & Roy, D. M., (1986). "Alternative perspectives on 'quasi crystallinity': non-uniformity and nanocomposites." *Materials Letters, 4*(8 & 9), 323–328.
2. Stander, L., & Theodore, L., (2011). "Environmental implications of nanotechnology: An update." *Int. J. Environ. Res. Public Health, 8*, 470–479.
3. Waseem, S., Khan, C. M., & Asmatulu, R., (2012). "Effects of nanotechnology on global warming." *ASEE Midwest Section Conference, Rollo, MO, 19–21*, p. 13.
4. Ratner, M., & Ratner, D., (2002). *"Nanotechnology: A Gentle Introduction to the Next Big Idea."* Prentice Hall. ISBN-13: 978-0131014008.
5. Kamigaito, O., (1991). "What can be improved by nanometer composites?" *J. Jpn. Soc. Powder Metall., 38*(3), 315–321. doi: 10.2497/jjspm.38.315.in.
6. Mulvaney, P., (2001). *MRS Bulletin.*
7. Zhiting, T., & Han, H. Y., (2013). "A molecular dynamics study of effective thermal conductivity in nanocomposites." *Int. J. Heat Mass Transfer, 61*, 577–582. doi: 10.1016/j.ijheatmasstransfer.2013.02.023.
8. Ajayan, P. M., Schadler, L. S., & Braun, P. V., (2003). *Nanocomposite Science and Technology.* Wiley. ISBN 3–527–30359–6.

9. Parameswaranpillai, J., Kuria, H. N., & Yingfeng, T. Y., (2016). *Nanocomposite Materials*. Taylor and Francis, USA. ISBN-13: 978-0131014008.

10. Niihara, K., (1991). "New design concept of structural ceramics-ceramic nano-composite." *Journal of the Ceramic Society of Japan (Nippon Seramikkusu Kyokai Gakujutsu Ronbunshi), 99*(6), 974–982.

11. Nakahira, A., & Niihara, K., (1992). "Structural ceramics-ceramic nanocomposites by sintering method: Roles of nanosize particles." *Journal of the Ceramic Society of Japan, 100*(4), 448–453.

12. Deng, C. F., Wang, D. Z., Zhang, X. X., & Ma, Y. X., (2007). "Damping characteristics of carbon nanotube reinforced aluminum composite." *Mater. Lett., 61,* 3229–3231.

13. Shehata, F., Fathy, A., Abdelhameed, M., & Mustafa, S. F., (2009). "Preparation and properties of Al_2O_3 nanoparticle reinforced copper matrix composites by in situ processing." *Mater. Design, 30,* 2756–2762.

14. Goussous, S., Xu, W., & Xia, K., (2010). "Developing aluminum nanocomposites via severe plastic deformation." *J. Phys. Conf. Ser., 24,* 012106.

15. Wang, Z., Wang, X., Zhao, Y., & Du, W., (2010). "SiC nanoparticles reinforced magnesium matrix composites fabricated by ultrasonic method." *Trans. Nonferrous Met., 20,* s1029–s1032.

16. Nie, K. B., Wang, X. J., Xu, L., Wu, K., Hu, X. S., & Zheng, M. Y., (2012). "Influence of extrusion temperature and process parameter on microstructures and tensile properties of a particulate reinforced magnesium matrix nanocomposites." *Mater. Design, 36,* 199–205.

17. Luo, P., McDonald, D. T., Zhu, S. M., Palanisamy, S., Dargusch, M. S., & Xia, K., (2012). "Analysis of microstructure and strengthening in pure titanium recycled from machining chips by equal-channel angular pressing using electron backscatter diffraction." *Mater. Sci. Eng. A., 538,* 252–258.

18. Stolyarov, V. V., Zhu, Y. T., Alexandrov, I. V., Lowe, T. C., & Valiev, R. Z., (2001). "Influence of ECAP routes on the microstructure and properties of pure Ti." *Mater. Sci. Eng. A., 299,* 59–67.

19. Bozic, D., Stasic, J., Dimcic, B., Vilotijevic, M., & Rajkovic, V., (2001). "Multiple strengthening mechanisms in nanoparticle-reinforced copper matrix composites." *J. Mater. Sci., 34,* 217–226.

20. Quang, P., Jeong, Y. G., Yoon, S. C., Hong, S. H., & Kim, H. S., (2007). "Consolidation of 1 vol.% carbon nanotube reinforced metal matrix nanocomposites via equal channel angular pressing." *J. Mater Proc. Techn., 187–188,* 318–320.

21. Paul, D. R., & Robeson, L. M., (2008). "Polymer nanotechnology: Nanocomposites," *Polymer, 49,* 3187–3204.

22. Karen, I. W., & Richard, A., (2007). Vaia, "polymer nanocomposites." *MRS Bulletin, 32,* 314–322.

23. Schmidt, D., Shah, D., & Giannelis, E. P., (2002). "New advances in polymer/layered silicate nanocomposites." *Current Opinion in Solid State & Materials Science, 6*(3), 205–212.

24. Giannelis, E. P., (1996). "Polymer-layered silicate nanocomposites." *Advanced Materials, 8*(1), 29–35.

25. Jordon, J., Jacob, K. I., Tannenbaum, R., Sharaf, M. A., & Jasiuk, I., (2005). "Experimental trends in polymer nanocomposites- a review." *Material Science and Engineering A., 393*, 1–11.

26. Kumar, S. K., & Krishnamoorti, R., (2010). "Nanocomposites: Structure, phase behavior, and properties." *Annu. Rev. Chem. Biomol. Eng., 1*, 37–58.

27. Jeon, I. Y., & Baek, J. B., (2010). "Nanocomposites derived from polymers and inorganic nanoparticles."*Materials, 3*, 3654–3674.

28. Ajayan, P. M., Schadler, L. S., & Braun, P. V., (2003). *"Nanocomposite Science and Technology" WILEY-VCH.* ISBN 3–527–30359–6.

29. Ciprari, D., Jacob, K., & Tannenbaum, R., (2006). "Characterization of polymer nanocomposite interphase and its impact on mechanical properties." *Macromolecules, 39*, 6565–6573.

30. Miller, S. G., (2008). "Effects of nanoparticle and matrix interface on nanocomposite properties." *PhD dissertation* (pp. 21, 22). University of Akron.

31. Tanahashi, M., (2010). "Development of fabrication methods of filler/polymer nanocomposites: With focus on simple melt-compounding-based approach without surface modification of nanofillers." *Materials, 3*(3), 1593–1619.

32. Ogata, N., Jimenez, G., Kawai, H., Ogihara, T., & Ogata, N., (1997). "Structure and thermal/mechanical properties of poly (l-lactide)–clay blend." *Journal of Polymer Science Part B: Polymer Physics, 35*, 389–396.

33. Zhao, X., Urano, K., & Ogasawara, S., (1989). "Adsorption of polyethylene glycol from aqueous solutions on montmorillonite clays." *Colloid and Polymer Science, 267*, 899–906.

34. Vaia, R. A., & Giannelis, E. P., (1997). "Lattice of polymer melt intercalation in organically modified layered silicates." *Macromolecules, 30*, 7990–7999.

35. Gilmann, J. W., (1999). "Flammability and thermal stability studies of polymer-layered –silicate (clay) nanocomposites." *Applied Clay Science, 15*(1 & 2), 31–49.

36. Aymonier, C., Bortzmeyer, D., Thomann, R., & Lhaupt, R. M., (2003). "Poly (methyl methacrylate)/palladium nanocomposites: Synthesis and characterization of the morphological, thermomechanical, and thermal properties." *Chemistry of Materials, 15*(25), 4874–4878.

37. (a) Mittal, V., (2009). "Polymer-layered silicate nanocomposites: A review." *Materials, 2*, 992–1057. (b) Ray, S. S., & Okamoto, M., (2003). "Polymer/layered silicate nanocomposites: A review from preparation to processing." *Prog. Polym. Sci., 28*, 1539–1641.

38. Vorotilov, K. A., Yanovskaya, M. I., Turevskaya, E. P., & Sigov, A. S., (1999). "Sol-gel derived ferroelectric thin films: Avenues for control of microstructural and electric properties. *Journal of Sol-Gel Science and Technology, 16*(2), 109–118.

39. Reddy, R. J., (2010). "Preparation, characterization, and properties of injection molded graphene nanocomposites." *Master's Thesis, Mechanical Engineering.* Wichita State University, Wichita, Kansas, USA. Page no 37 of thesis.

40. Baibarac, M., & Gomez-Romero, P., (2006). "Nanocomposites based on conducting polymers and carbon nanotubes from fancy materials to functional applications." *J. Nanosci. Nanotechnol., 6*, 1–14.

41. (a) Baughman, R. H., Zakhidov, A. A., & De Heer, W. A., (2002). "Carbon nanotubes: the routes towards applications." *Science, 297*, 787–792. (b) Moniruzzaman, M.,

& Winey, K. I., (2006). "Polymer nanocomposites containing carbon nanotubes." *Macromolecules, 39,* 5194–5205.

42. Kothurkar, N. K., (2004). "*Solid State, Transparent, Cadmium Sulfide-Polymer Nanocomposite.*" University of Florida.

CHAPTER 8

THEORETICAL BASES OF SIMULATION OF FORMATION PROCESSES OF NANOSTRUCTURES IN A GAS MEDIUM AND RESULTS OF NUMERICAL STUDIES OF NANOSYSTEMS

A. YU. FEDOTOV

Udmurt Federal Research Center of the Ural Branch of the Russian Academy of Sciences, Institute of Mechanics, Izhevsk, Russia, E-mail: alezfed@gmail.com

ABSTRACT

This chapter considers the methodology and the universal multi-level mathematical model with the possibility of applying to a wide range of problems and technologies for a multidimensional study of the processes of condensation, motion, formation, and interaction of nanostructures. On the basis of a multilevel mathematical model and adapted numerical schemes, a problem-adapted software package with a block-modular structure and with a graphical interactive user interface was created. Numerical simulation of the mechanisms of formation and growth of nanostructures used to feed plants from the gaseous medium has been carried out. The intensity of condensation, the composition and structure of the nanoparticles formed are established. The main properties of nanoelements are calculated. Comprehensive studies of the formation of metal nanoparticles for the technology of thermal evaporation and subsequent condensation under various regimes, compositions, and ratios of parent metals have been performed. The internal structure of nanoobjects is revealed depending on the types of raw materials. With the help of statistical analysis, the laws of distribution of the basic properties of nanoclusters are established. The processes of formation of special-purpose aerosol nanostructures used in fire-extinguishing systems have been studied. On the basis of the features of the structure and composition of nanoelements, the possibility of efficient operation of an aerosol gas generator for fire extinguishing in rooms with computer equipment and electrical equipment was tested. The technological process of molecular beam epitaxy on solid substrates has been numerically studied in order to identify various variants of nanostructure formation. A multistage numerical modeling of the formation of nanofilms on porous aluminum oxide matrices by the method of discrete thermal evaporation of powder under high vacuum conditions is carried out. The main dimensional and structural properties of the nanoclusters formed are calculated; the degree of pore filling in the substrates is determined.

8.1   INTRODUCTION

Nanostructure and nanoscale elements, structures, object have taken a firm place among the new promising materials used in various areas of human activity. These materials attract attention due to their features and properties at the nanoscale, thanks to which their performance characteristics and functional behavior in a favorable way distinguish this class of substances from ordinary and familiar macro objects.

The interest of world industry and science in nanomaterials is confirmed by the growing number of publications, the volume of investments and the number of innovative projects in this field. In 2010, in the field of physics, the Nobel Prize was awarded for the discovery of such nanomaterial as graphene and experiments on it [1]. Nanotechnologies are widely used in machine building [2] and metalworking [3, 4], in the space industry and aircraft building [5–7], in medicine and biotechnology [8–12], in the food and textile industries [13, 14], in the automotive industry [15–17], agriculture [18–20], ecology [21], solar energy and energy saving [22–25], and, of course, electronics [26–28].

Expansion of the scope of application of products to new areas and the creation of high technology products in the nanoindustry is possible only in conjunction with the development of fundamental science in this industry. Theoretical studies due to modeling allow to reduce the amount of expensive and laborious experiments conducted, to determine and formulate the main regularities in the behavior of nanomaterials. A separate perspective direction of mathematical modeling is the discovery of new properties of nanomaterials, the prediction of their parameters and behavior in the process of operational impact.

The formation of a scientific and technical reserve and the accumulation of potential in the field of theoretical research for nanotechnological productions is an integral part of the creation of the final commercial product. Accelerated development of nanotechnology is possible only if there is a strong knowledge base of modeling methods and a foundation from the mathematical apparatus in this field. Based on the foregoing, it can be concluded that the improvement of the theory of mathematical modeling in the tasks of nanotechnology and the nanoindustry occupies a key position in the development of science and industry.

Directly or indirectly, the processes of manufacturing and functioning of nanostructured materials must be considered together with gaseous media. Often in the vacuum or gaseous media, their formation takes place, the main structural features and operational properties are laid. A generalization of the problems arising in the field of nanotechnology and related to air media is demonstrated in Figure 8.1. The source of atoms and molecules of the gaseous phase in such nanosystems is a gas generator, which can be a separate or combined high-temperature furnace, evaporator, and particle atomizer, laser radiation, ionized plasma or other energy-intensive devices. The basic mechanisms for the formation of nanostructures can be realized directly in gaseous media, on surfaces, for example, in the case of deposition of nanofilms, and in bulk media inside nanocomposites.

The variety of research tasks for nanostructured objects associated with gas media requires the development of a knowledge base and the accumulation of information in this field. A special place in the study of the properties of nanostructures is occupied by mathematical models and algorithms, since they do not require significant costs for fabrication of samples and expensive equipment; they allow preliminary unsuccessful natural experiments and predict potentially claimed properties of new materials. In addition, mathematical modeling allows us to generalize the results of research and formulate fundamental laws and mechanisms for the formation, interaction, and functioning of nanoobjects. The construction of theoretically grounded multiscale models with the ability to adjust to specific technological processes and detail the results at each level is necessary in many areas of nanotechnology and nanoindustry, including for gas-containing nanosystems.

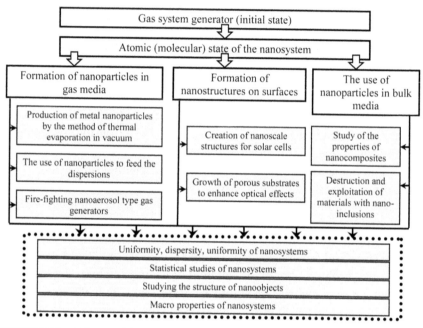

FIGURE 8.1   Scheme of the generalization of simulation tasks for nanosystems formed or functioning in a gaseous medium.

The aim of the work is to develop a methodology and a universal multi-level mathematical model with the ability to apply to a wide range of tasks and technologies for multidimensional research of the processes of condensation, motion, formation, and interaction of nanostructures.

8.2   MATHEMATICAL MODELS AND THEORETICAL BASES

A multilevel mathematical model of the problem of the formation of nano-structures in a gaseous medium, combines several approaches at once: quantum mechanics, molecular dynamics, mesodynamics of particles [29, 30]. The general structural diagram of the model is shown in Figure 8.2. With the help of the apparatus of quantum mechanics, certain energy characteristics and charges of atoms in molecules were calculated. If the nanosystem under consideration was molecular, equilibrium configurations of molecules were also determined: bond lengths, angles, etc.

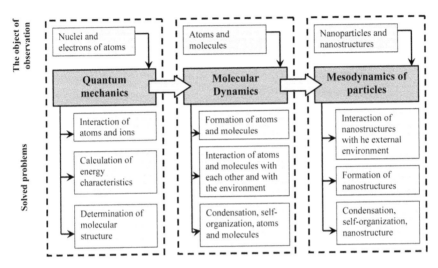

FIGURE 8.2 Structure of a mathematical model for studying the formation of nanoobjects in a gaseous medium.

To calculate the equilibrium configurations, we solved the stationary Schrödinger equation

$$\hat{H}\,\Psi\,(\mathbf{R},\,\mathbf{r}) = E\,\Psi\,(\mathbf{R},\,\mathbf{r}), \tag{1}$$

where E – the total energy of the system under consideration; $\hat{H}$ – Hamiltonian consisting of the sum of operators of potential and kinetic energies; $\Psi\,(\mathbf{R},\mathbf{r})$ – the full-electron wave function of the system, which depends on the coordinates of all the nuclei $\mathbf{R}$ and the positions of the electrons $\mathbf{r}$.

In this chapter, the Born-Oppenheimer approximation was used, which makes it possible to separate the variables and to represent the total wave function in terms of the product of nuclear and electronic functions.

$$\Psi\,(\mathbf{R},\,\mathbf{r}) = \Psi_{nuc}\,(\mathbf{R})\,\Psi_{el}(\mathbf{R},\,\mathbf{r}), \tag{2}$$

where $\Psi_{nuc}(\mathbf{R})$ – multinuclear wave function;
$\Psi_{el}(\mathbf{R},\mathbf{r})$ – many-electron wave function.

Taking into account the separation of energies and variables into nuclei and electrons, the Schrödinger equation for the particle system (1) is given

$$by \left[-\frac{\hbar^2}{2} \sum_{k=1}^{N_{nuc}} \frac{1}{M_k} \nabla_k^2 - \frac{\hbar^2}{2m_e} \sum_{i=1}^{N_{el}} \nabla_i^2 + U_{nuc,nuc}(\mathbf{R}) + U_{el,nuc}(\mathbf{R},\mathbf{r}) + U_{el,el}(\mathbf{r}) \right] \times$$

$$\times \Psi_{nuc}(\mathbf{R}) \Psi_{el}(\mathbf{R},\mathbf{r}) = E \Psi_{nuc}(\mathbf{R}) \Psi_{el}(\mathbf{R},\mathbf{r}), \qquad (3)$$

where $U_{nuc,nuc}(\mathbf{R})$ – nuclear repulsion energy;

$U_{el,el}(\mathbf{r})$ – electron repulsion energy;

$U_{el,nuc}(\mathbf{R},\mathbf{r})$ – energy of attraction of nuclei and electrons;

M_k – mass of the k-th nucleus;

m_e – electron mass; and

operators ∇_k^2 and ∇_i^2 act on the desired wave functions of the nuclei and electrons. The first two terms in equation (3) are operators of the kinetic energy of electrons and nuclei.

Molecular dynamics was based on the results obtained using the equations of quantum mechanics and detailed the further evolution of the nanosystem. With the help of molecular dynamics studies, the processes of condensation of atoms and molecules into nanostructures, their self-assembly, and self-organization, the formation and growth of nanofilms on substrates of various types, the interaction of nanoelements were considered. A wide range of interaction potentials, both pairwise and multiparticle, was used. The exchange of energy with the external environment occurred through the application of algorithms of thermostats and barostats.

The method of molecular dynamics is realized through the solution of the differential equation of motion for a set of atoms in conjunction with the initial conditions.

$$m_i \frac{d^2 \mathbf{r}_i(t)}{dt^2} = \mathbf{F}_i(t,\mathbf{r}) \ , t_0 = 0, \ \mathbf{r}_i(t_0) = \mathbf{r}_{i0}, \ \frac{d\mathbf{r}_i(t_0)}{dt} = \mathbf{V}_i(t_0) = \mathbf{V}_{i0}, \ i = 1,2,..,N_a, \quad (4)$$

where N_a – number of atoms forming the nanosystem;

t_0 – initial time;

m_i – atomic mass at number i;

$\mathbf{r}_{i0}, \mathbf{r}_i(t)$,– the radius vectors of the i-th atom at the initial and current time, respectively, the particle coordinates are determined within the computational domain;

$\mathbf{r} = \{\mathbf{r}_1(t), \mathbf{r}_2(t), K, \mathbf{r}_{Na}(t)\}$ – the set of coordinates of all particles;

$\mathbf{F}_i(t,\mathbf{r})$ – the resulting force acting from the whole set of particles on the i-th atom; and

$\mathbf{V}_{i0}, \mathbf{V}_i(t)$ – speed of the i-th atom, at the initial and current time.

The magnitude of the forces $\mathbf{F}_i\,(t,r)$ in Eq. (4) is determined by the potential of the nanosystem $U(r)$. For uninsulated nanosystems, there is an exchange of energy with the external environment by means of an additional external force, $\mathbf{F}_i^{\,ex}\,(t,r)$.

$$\mathbf{F}_i\left(t,\mathbf{r}\right)=-\nabla U\left(\mathbf{r}\right)+\mathbf{F}_i^{ex}\left(t,\mathbf{r}\right)=-\frac{\partial U\left(\mathbf{r}\right)}{\partial \mathbf{r}_i\left(t\right)}+\mathbf{F}_i^{ex}\left(t,\mathbf{r}\right),\quad i=1,2,..,N_a \qquad (5)$$

Molecular dynamics was used in studying the processes of formation and interaction of nanostructures at the initial stage of condensation. The further process of aggregation of nanoobjects was considered with the aid of the apparatus of mesodynamics of particles. For a certain period of time, the simulation was carried out jointly by both molecular dynamics and the mesodynamics of the particles. In this interval, a comparison was made of the results obtained with the aid of various models and the adjustment of the parameters of the nanosystem.

Elements of the model in mesodynamics of particles are the formed nanostructures shown in Figure 8.3. In this method, their movement, enlargement, and interaction are considered. The positions and masses of nanoclusters are calculated by formulas.

$$m_j=\sum_{i=1}^{N_j} m_i,\quad \mathbf{r}_j=\frac{\displaystyle\sum_{i=1}^{N_j} m_i\mathbf{r}_i}{m_j},\quad i=1,2,..,N_a,\ j=1,2,..,N_p,\ \sum_{j=1}^{N_p} N_j=N_a, \qquad (6)$$

where m_i and m_j, $\mathbf{r}_i$ and $\mathbf{r}_j$–masses and radius vectors for the i-th atom and the j-th nanoparticle, respectively;

N_p–total number of nanoparticles in the system;

N_j–the number of atoms entering the j-th nanoparticle.

The velocities of the nanoparticles are calculated from the law of conservation of momentum. The resultant force acting on each nanostructure

$$\mathbf{V}_j=\frac{\displaystyle\sum_{i=1}^{N_j} m_i\mathbf{V}_i}{m_j},\quad \mathbf{F}_j=\sum_{i=1}^{N_j}\mathbf{F}_i,\ j=1,2,..,N_p, \qquad (7)$$

where $\mathbf{V}_i$ and $\mathbf{V}_j$–atomic velocities and nanoparticles, respectively;

$\mathbf{F}_i$ and $\mathbf{F}_j$–the forces acting on the atom and on the nanostructure; the index j indicates the belonging to the number of the nanoparticle, the index i to the number of the atom.

Due to the uneven application of forces to the atoms of nanoparticles and the heterogeneity of the potential field in which they are located, nanostructures appear rotational motion. Rotational motion is described by an additional vector equation of the system. The resulting system of equations for the motion of nanoparticles will have the form.

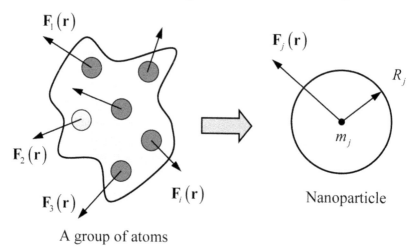

A group of atoms

FIGURE 8.3   Combining atoms in nanostructures.

$$\begin{cases} m_j \dfrac{d^2 \mathbf{r}_j(t)}{dt^2} = \mathbf{F}_j(t,\mathbf{r}) \\[2mm] I_j \dfrac{d^2 \varphi_j(t)}{dt^2} = \mathbf{M}_j(t,\varphi) \end{cases}, \quad j = 1,2,\ldots,N_p, \tag{8}$$

where $\mathbf{F}_j(t,\mathbf{r})$–force acting on the j-th nanoparticle;
I_j–moment of inertia nanoparticle;
$\varphi_j(t)$–vector of rotation angles, for the three-dimensional case $\varphi = (\alpha, \beta, \gamma)$, contains a set of three angles: precession, nutation, and the angle of proper rotation;
$\mathbf{M}_j(t,\varphi)$–moment of strength of j-th nanostructure;
$\varphi = \{\varphi_1(t), \varphi_2(t),\ldots \varphi_{N_p}(t)\}$.

The moment of inertia and the rotational moment of the nanostructure in the system (8) are calculated using the following expressions:

$$I_j = \sum_{i=1}^{N_j} m_i r_i^2, \quad \mathbf{M}_j(t,\varphi) = \mathbf{r}_j(t) \times \mathbf{F}_j(t,\mathbf{r}), \quad j = 1,2,..,N_p \tag{9}$$

In comparison with the apparatus of classical molecular dynamics, in the method of mesodynamics of particles, a certain modification was obtained from Eq. (8)

$$F_j(t,\mathbf{r}) = -\frac{\partial U(\mathbf{r})}{\partial \mathbf{r}_j} - m_j \mathbf{g} + \mathbf{\eta}_j(t) - m_j \zeta \frac{d\mathbf{r}_j(t)}{dt}, \quad j = 1, 2, \ldots, N_p, \qquad (10)$$

where g–acceleration of gravity;

ς–coefficient of friction;

$\eta_j(t)$–random force applied to the j-th nanostructure; and

$\mathbf{r} = \left\{ \mathbf{r}_1(t), \mathbf{r}_2(t), \ldots, \mathbf{r}_{N_p}(t) \right\}$ – indicates the dependence of the values on the set of radius vectors of all nanoparticles.

The first term in (10) is the potential gradient. The nature of the potential forces acting in the nanosystems remains the same as in the MD method. For a nanoparticle, the potential gradient is formed by averaging over all atoms entering the nanostructure. Gravity forces are introduced in (10) in connection with the growth of the sizes and mass of nanoclusters over time. The $\eta_j(t)$ is stochastic in nature and reflects the interaction of nanostructures with the free atoms of the nanosystem. Since the nanoparticle simultaneously interacts with a large number of atoms, it is possible to consider with high accuracy the force $\eta_j(t)$ distributed according to the normal law with the following properties. The average value of the random force should be chosen equal to zero in order to exclude the directed movement of nanostructures under the influence of a random factor. The force $\eta_j(t)$ links the mobility of the nanoparticle with the coefficient of friction and temperature, so it is theoretically justified to select the standard deviation by the quantity.

$$\sigma = \sqrt{2k_B T_t \zeta m_j}, \quad j = 1, 2, \ldots, N_p, \qquad (11)$$

where k_B–Boltzmann's constant; and

T_t–target temperature of the nanosystem. The relationship between the root-mean-square deviation and the coefficient ς is due to the fact that the random force should partially compensate for the friction losses

The last element of the sum in (10) is the frictional force, which is proportional to the mass and velocity of each nanoparticle. The coefficient ς is related to the instantaneous and target temperatures of the nanosystem by the following equation

$$\zeta = \vartheta\left(\frac{T}{T_t} - 1\right), \tag{12}$$

where ϑ–empirical parameter.

The agglomeration of nanoelements in the method of mesodynamics of particles is determined by the acting forces of interaction and energetically favorable rearrangements. As shown by the studies carried out, in condensation problems the fundamental role is played by the processes of combining nanostructures (Figure 8.4), the separation occurs in rare cases and can be taken into account by the parameters of the model. Therefore, with a sufficient degree of accuracy, the mechanisms of the decay of nanoparticles in the method of mesodynamics of particles can be neglected. The combination of nanostructures is influenced by the following factors: the size and mass of the nanoparticles; the forces applied to them; distance between interacting objects; module and direction of nanoelement velocities.

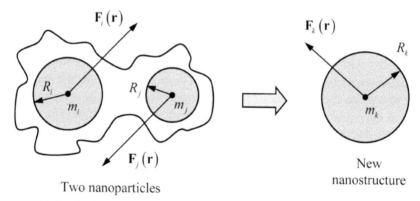

Two nanoparticles　　　　　New nanostructure

FIGURE 8.4　Combining several nanostructures.

The condition for combining two nanoparticles into one is the achievement of a minimum value of the potential.

$$U_{ij}(r) = U_{min}, \quad i, j = 1, 2, \ldots, N_p, \tag{13}$$

where r–distance between nanostructures.

The distance of the action of the potential is calculated taking into account the size of the nano-objects. In the case of approaching the nanoelements at a distance closer than the r_{min}, they combine. The amount of nanoparticles is reduced by one unit. The distance r_{min} corresponds to the minimum value of the potential energy U_{min}. The assumption of the model is the spherical shape of the nanoparticles.

The coordinates and velocities of the new nanostructures when combined are determined through the equations of the center of mass. Since in the model of mesodynamics nanoparticles are assumed to be incompressible, the size of the new nanoobjects is calculated in accordance with the principle of conservation of volume.

$$m_k = m_i + m_j, \quad \mathbf{r}_k = \frac{m_i \mathbf{r}_i + m_j \mathbf{r}_j}{m_k}, \quad \varphi_k = \frac{m_i \varphi_i + m_j \varphi_j}{m_k}, \quad \mathbf{V}_k = \frac{m_i \mathbf{V}_i + m_j \mathbf{V}_j}{m_k},$$

$$\omega_k = \frac{m_i \omega_i + m_j \omega_j}{m_k}, \quad R_k = \sqrt[3]{R_i^3 + R_j^3}, \quad \mathbf{F}_k = \mathbf{F}_i + \mathbf{F}_j, \quad i, j = 1, 2, .., N_p, \quad (14)$$

where the indices i and j belong to the colliding nanostructures, the index k indicates a new nanoparticle, N_p after the transformations it decreased by one.

In the model of mesodynamics, spatial, and temporal scales increased with allowance for energy control and parameters of the nanosystem. The increase in the spatial dimension was realized through the union of symmetrically mapped images of the computational domain of modeling. The space was rescaled many times.

The integration step was defined as part of the period of oscillations of the fastest nanoobjects.

$$\Delta t = n_v \cdot \min_j \left(\frac{1}{v_j} \right) = n_v \cdot \min_j \left(\frac{h}{E_j} \right) = n_v \cdot \min_j \left(\frac{2h}{m_j V_j^2 + I_j \omega_j^2} \right),$$

$$\lambda_j = \frac{h}{m_j V_j}, \quad v_j = \frac{E_j}{h}, \quad 0 < n_v \leq 1, \, j = 1, 2, .., N_p, \quad (15)$$

where h–Planck's constant;

E_j–energy of the j-th nanoparticle;

λ_j and v_j–length and frequency of the de Broglie wave of the nanostructure;

n_v–parameter, indicates which fraction of the inverse frequency of the nanoparticles is the time integration step.

The choice of Δt is made primarily with allowance for the motion of the fastest nanostructures. To optimize and reduce additional computations, the integration step was recalculated in time through a number of iterations.

The work was adapted numerical schemes of different accuracy order to describe the motion of nanostructures in a gaseous medium, including the equations of rotational motion. Numerical methods of higher accuracy order are used in the theory of mesodynamics of particles, where the increased value of the integration step requires the use of stable computational schemes. In contrast to the previously known condensation models, the mesodinamic equations of particles make it possible to trace the detailed behavior of each nanostructure in the system throughout the entire condensation process, to investigate the composition and mechanisms of growth of the generated nanoobjects, and not only to observe the change in the overall characteristics of the material.

On the basis of a multilevel mathematical model, a program-adapted complex was developed for theoretical studies and modeling of the fundamental processes of formation, motion, and interaction of nanostructures in a gaseous medium. The general software package is shown in Figure 8.5. The software package includes a source data preparation unit, a computational module, a visualization and analysis module, and an adapted inter-program coordination unit. The solution of the problems under consideration is carried out sequentially, the direction of the transitions between the blocks and modules is indicated by arrows. For the developed problem-oriented complex, the author's certificate of registration of the software resource was obtained. The proposed organization of the complex covers various programming languages and has a graphical interactive user interface.

The main algorithms for visualization and analysis of results implemented in the software complex include: the identification of atoms grouped in nanostructures; determination of the uniformity of nanofilms and nanocomposites; establishment of the chemical composition and proportions of the initial elements of nanostructures; determination of the fraction of condensed atoms and molecules; search for structural and dimensional properties of nanoobjects; clarification of the internal structure of nanoparticles, nanofilms, substrates, and analysis of a typical nanoparticle for a given material. Conducted studies on convergence, stability, analysis of the reference structure, demonstration of the work

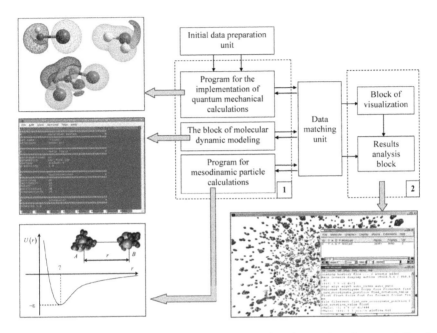

FIGURE 8.5 General structure of the software and tool complex: 1–computational module, 2–analysis and visualization module.

of thermostats and barostats and comparison of the results of solving test problems with previously known theoretical and experimental data at different levels confirm the adequacy and correctness of the mathematical model and indicate the reliability of the obtained modeling results.

The value of the created software product lies in the wide potential of studying the properties and parameters of new homogeneous and nano-composite materials, as well as the possibility of predicting their properties and recreating a particular technological process. The complex is adapted for nanosystems containing gas phase media and supports various existing data storage formats for similar tasks.

8.3 THE RESULTS OF A NUMERICAL STUDY OF THE PROCESSES OF FORMATION OF NANOSTRUCTURES USED TO FEED PLANTS FROM THE GASEOUS MEDIUM

The first task was to study the processes of nanostructures used to feed plants from the gaseous medium. The introduction of fertilizers into the soil is not always effective because of the insufficient penetration of minerals into the stems and leaves of plants. Therefore, the question arises of finding new approaches for feeding plants and increasing their yields. The process of the introduction of mineral salts in the form of nanoparticles through the pores of plants is realized as follows: initially, the fertilizers together with the combustible material are compressed into a tablet. The tablet is ignited, which leads to the transfer of mineral substances into the gaseous state. In the gas phase, nanoparticles are formed, which subsequently penetrate and feed the plants.

According to the results of the research, the composition of the gas, after burning the mineral tablet, includes more than 20 types of molecules. The main elements of the gas mixture and the number of molecules are shown in Figure 8.6. For each of the six types of molecules forming the gas phase and shown in Figure 8.6, equilibrium configurations were calculated using the equations of quantum mechanics.

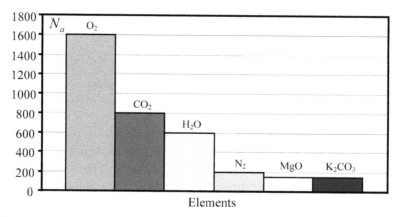

FIGURE 8.6 Numerical content of molecules in the nanosystem used to feed plants from the gaseous medium.

The study showed that the formed nanoclusters include molecules of water, magnesium oxide, and potassium carbonate. Molecules of nitrogen, oxygen, and carbon dioxide remain at all times in the gaseous state and are present in the nanosystem as components of the air mixture, as well as combustion products.

The processes of agglomeration and growth of nanoformations for different initial components proceeded with different intensities. At the cooling stage, K_2CO_3 molecules are more active in the condensation, the molecules of magnesium oxide and water are agglomerated with less intensity. The fluctuation character of the change in mass fraction is traced, which is associated with insignificant forces of attraction between certain atoms and molecules. Because of small forces, unstable nanoobjects can lose molecules and decay into smaller nanoclusters.

The inhomogeneities and the disturbances in the homogeneity of the composition in the nanosystem that were present in the mass distribution at the initial moments are leveled off over time. In the future, the unevenness of the composition decreases and does not exceed 5% of the total mass of the composite gas-composition mixture.

The method of mesodynamics of particles made it possible to analyze the processes of condensation and growth of nanostructures at spatial and temporal scales close to real ones, and also to identify the type and nature of the change in the average size and number of nanoobjects per unit volume. The graphs of the average diameter of nanoclusters and the number of nanoelements per unit volume are presented jointly in Figure 8.7. When considering smaller time scales, almost linear behavior of these parameters was observed with a deviation in a small range of values. The extended time scale established a decrease in the intensity of the change in functions and the approximation of their form to the asymptotic form. Due to the growth of the mass of nanoobjects, their mobility and degree of interaction decreased.

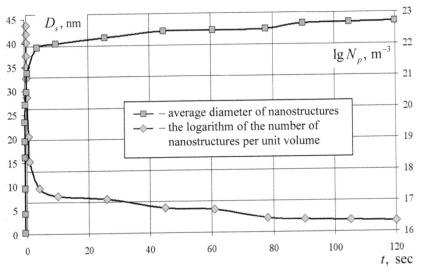

FIGURE 8.7 Change in the average size and number of nanostructures per unit volume, for fertilizing plants from the gaseous medium.

The calculated final values of the average size of nanoparticles and their number per unit volume indicate the concentration of nanoelements in the air medium formed after burning a tablet with mineral fertilizers equal to 1.6–2.2 mg/m³, which is close to the maximum allowable concentration of dust particles.

Experimental studies have shown the presence of a two-modal particle size distribution with peak frequency values in the micro- and nanoscale region. The micron distribution of the particle sizes predominantly corresponds to the combustion products formed after burning the mineral solid fuel pellet. The distribution in the nanometer interval, shown in Figure 8.8, determines the spread of the nanostructure diameter of the target fertilizing of plants.

Comparison of experimental data and modeling results in the form of dimensional distributions of nanoparticles indicate a satisfactory agreement of the values obtained. One-type character of the behavior of the constructed dependences with a pronounced peak and attenuation is observed when the distribution in the form of the average diameter of the nanoelements is removed from the value of the mathematical expectation. The average value of the size of nanoobjects, obtained by modeling, corresponds to a value of 44.1 nm. A similar experimental parameter is equal to 37.15 nm.

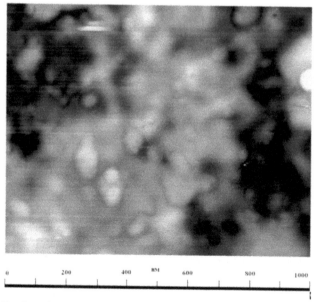

FIGURE 8.8 Snapshot of the atomic force microscope of nanostructures used to feed plants from the gaseous medium.

The difference between the obtained characteristics is about 15%. The introduction of silver atoms as a coagulating agent in a nanoscale is an effective way to control the mechanisms of agglomeration and growth of nanostructures used to feed plants from the gaseous medium. The addition of silver atoms leads to stimulation of the formation of new condensation centers, an increase in the number of nanoelements, and contributes to the coarsening of already existing particles. The intensity of condensation of the molecules of potassium carbonate, water, and magnesium oxide after the introduction of silver atoms into the nanosystem increased two times or more.

8.4 RESULTS OF RESEARCH OF RECEPTION PROCESSES OF THE METAL NANOSTRUCTURES RECEIVED BY A METHOD OF EVAPORATION AND CONDENSATION

The next problem was built on the modeling of the synthesis of metal nanostructures by the method of evaporation, thermal saturation, and subsequent condensation. The scheme of the technology is illustrated in Figure 8.9. This technology is easy to operate, cost-effective, and is used in the production of metal nanoparticles on an industrial scale. Various combinations of proportions and initial compositions were considered, but the main attention was paid to nanostructures Ag-Cu, Ag-Zn, Ag-Au-Zn. The process of formation and growth of metallic nanostructures was divided into two main stages. In the first stage, condensation centers are formed in the form of nanoelements formed by free metal atoms. For the second stage, the unification of already formed nanoobjects is more characteristic. The duration and intensity of the stages is determined by the type and properties of the parent metals. The number of nanoclusters grows significantly at the first stage and gradually decreases to the second, as a consequence of the enlargement of nanoformations.

With equal conditions for cooling the nanosystem to normal temperature, the speed and intensity of agglomeration of silver-copper nanoobjects is approximately ten times higher than that of silver-zinc nanoclusters formed similarly. An indicator of growth is the dynamics of the average size of nanostructures. In the first case, the average diameter increased from 0.4 nm to 0.66 nm in 10 nsec, in the second—from 0.4 nm to 0.86 nm in 170 nsec. The behavior of the total volume and surface area of the nanoelements for these compositions was the same: the initial rapid increase and the subsequent slight change.

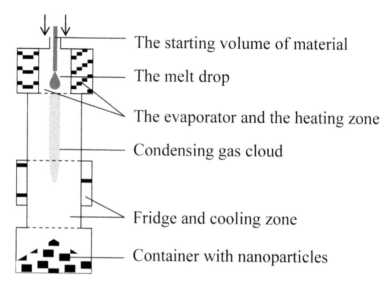

The starting volume of material

The melt drop

The evaporator and the heating zone

Condensing gas cloud

Fridge and cooling zone

Container with nanoparticles

FIGURE 8.9 The scheme of the plant for producing nanoparticles, implementing the principle of thermal saturation and condensation.

Studies of internal organization and component content of nanoelements have shown that the composition of the formed nanoparticles depends on the physicochemical properties of metals and their quantitative share in the nanosystem. For silver-copper nanostructures, the overwhelming majority had a composition corresponding to the proportion of the initial materials in the initial state. The silver-zinc nanoclusters were formed on a different principle, and the variations in their composition did not have an explicit dependence on the initial proportion of the metals. The cooling rate of the nanosystems does not significantly affect the total content of the two-component silver-copper nanostructures.

In the computational experiments, no homogeneity and uniformity of the composition of metallic nanosystems were observed. The heterogeneities of the total nanocomposite composition were insignificant and did not exceed 10% of the mass in total. The resulting mass fluctuations of the medium appear in the final stages of condensation and are caused by the movement of particularly large nanoformations.

On the structure, internal homogeneity and homogeneity of the formed metallic nanoobjects is significantly influenced by the type of initial materials. Nanocluster silver-copper has a mixed structure, silver-zinc nanoelements have a shell structure, three-component silver-gold-zinc nanoparticles are formed multilayered, which confirms the distribution of their relative density, shown in Figure 8.10. In all cases, cavities within the nanoformation are not observed.

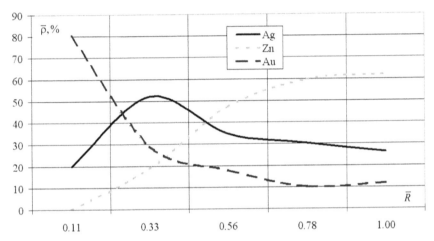

FIGURE 8.10 Dynamics of the relative density of silver-gold-zinc nanostructures as a function of the dimensionless radius.

For the purpose of statistical research, the problem of forming silver-copper nanostructures was solved many times with different distributions of initial coordinates and velocities. For each computational experiment, the properties of the nanoparticles were determined, and a variational series was constructed.

Statistical analysis of the formation and growth of silver-copper nano-clusters has shown that the main structural and quantitative parameters of nanoobjects, demonstrated in Figure 8.11, such as the average density, the average weight fraction of silver, the average size and the number of nanoelements in the calculation area according to the Pearson criterion at the significance level of 0.8 are distributed according to the normal law. The mathematical expectation and standard deviation for these properties of nanoparticles, respectively, were: $\mu = 9.06$ and $\sigma = 0.30$; $\mu = 49.96$ and $\sigma = 0.88$; $\mu = 6.14$ and $\sigma = 0.12$; $\mu = 182.14$ and $\sigma = 6.84$.

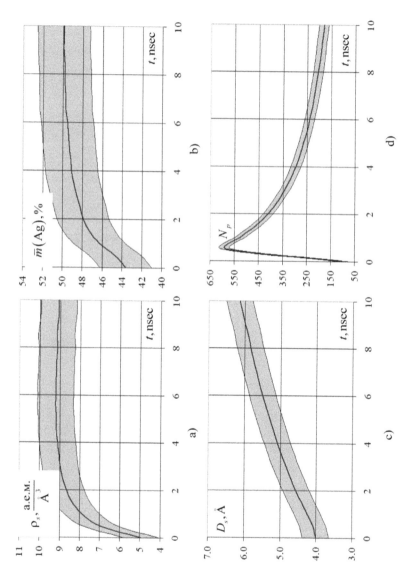

FIGURE 8.11 Change in the parameters of Ag-Cu nanostructures with the error limit: (a) the average density, (b) the average mass fraction of silver, (c) the average diameter, (d) the number of nanoobjects in the calculated cell.

Verification of the hypothesis using the Pearson criterion requires a considerable sample size and depends nonlinearly on it. If data volumes are insufficient, statistics can increase, in the case of a representative sample size, Pearson's statistics decrease. The statistical functions of the variational series of the parameters under consideration had a linear form, were well approximated by a straight line and, therefore, confirmed Pierson's hypothesis of the normal distribution law. Comparison of the probability functions of the normal law and selective empirical functions on the distribution histogram shows a good coincidence of the sample and theoretical data.

8.5 RESULTS OF MODELING THE GROWTH OF NANOSTRUCTURES USED IN FIRE EXTINGUISHING AEROSOL SYSTEMS

As a next series of computational experiments, the task of forming nanostructures, which are used in nanoaerosol fire extinguishing systems, was solved. The appearance of the fire-extinguishing optically transparent aerosol generator is shown in Figure 8.12. The gas generator is triggered by the response of a special sensor, and the aerosol-gas fire extinguishing agent enters the air. Such fire extinguishers are designed to contain and extinguish fires of a wide range of materials, universal, as a rule, automated, and widely used due to economic efficiency. For greater efficiency, fire extinguishing gas generators can be used together with a filter, as shown in Figure 8.12.

Modeling of multicomponent heterophase processes arising during the operation of nanoaerosol extinguishers was divided into several stages. At the initial stage, experimental methods established the composition and components of the aerosol-air medium formed after the fire-extinguishing gas generator was activated. For the molecular elements obtained, equilibrium optimal geometric configurations were determined, and data were prepared for further simulation of the operation of the gas generator. The immediate operating time of the aerosol gas generator corresponded to an elevated temperature of the nanosystem (600 K), at which the gas-air mixture components were uniformly distributed within the design area. The boundary conditions along the directions of all the coordinate axes were periodic. As further studies have shown, the ongoing processes of condensation of nanostructures are fairly homogeneous, and the choice of periodic boundary conditions in order to reduce computational costs is justified. After triggering the gas generator, the atoms, molecules, and nanoaerosol particles appeared in the gas-air medium under conditions close to normal (310 K), as a result of which the molecules and atoms were condensed into nanoobjects.

FIGURE 8.12   Appearance of the nanoaerosol extinguishing generator.

The stage of condensation of nanostructures was the longest and was carried out using apparatuses of molecular dynamics and mesodynamics of particles. At the stage of molecular-dynamic modeling, the temperature was maintained using thermostatic algorithms. The mesodynamic mathematical model includes temperature correction mechanisms and does not require additional computational processing. The results of the simulation were subsequently compared with the experimental data obtained by means of optical-electronic measurements.

Conducting theoretical studies was based on a certain form of gas generator nanoaerosol with a previously approved composition. As the most optimal solid-fuel composition as a result of literature and patent searches, thermodynamic calculations, and a large number of tests, a 3-component formulation was considered, consisting of fuel (carbon black), oxidizer (potassium nitrate) and inorganic binder. The ratio of the fuel-oxidizer components is close to the stoichiometric, which allows the final combustion products to be produced during the operation of the gas generator: CO_2, N_2, K_2CO_3. Toxic components in the form of CO, CH, and any other under-oxidized compounds are absent. In addition, the stoichiometric ratio makes the resulting composition flameless, which prevents the generator of the fire-extinguishing nanoaerosol itself from becoming a secondary source of ignition.

Based on the results of experimental studies, the composition of the gas-air mixture formed after the activation of the nanoaerosol gas generator was established. Solid-phase elements are molecules of magnesium oxide and calcium, potassium carbonate. The constituent components of the gaseous medium include the molecules of oxygen, nitrogen, and carbon dioxide present in the composition also as a combustion product. In addition, there are water molecules in the gas-air mixture.

Studies of the processes of formation of special-purpose aerosol nanostructures, shown in Figure 8.13, showed that the molecules of potassium carbonate, water, magnesium, and calcium oxides actively condense into agglomerates. The greatest intensity of association in nanoclusters is potassium carbonate. The molecules of oxygen, carbon dioxide and nitrogen in the condensation processes were not significantly involved. These elements remained predominantly in the gaseous medium throughout the cooling stage and were present in the calculated volume as components of the air-to-air mixture. The increased content of carbon dioxide is explained by the ongoing combustion processes when the gas generator is activated.

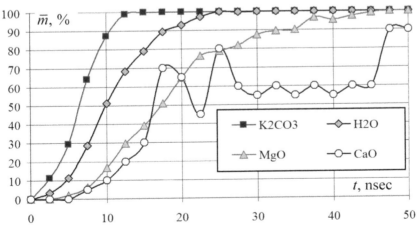

FIGURE 8.13 Dynamics of the mass fraction of solid- and liquid-phase components of a fire extinguishing gas generator for grouped molecules in nanostructures.

The active condensation of water molecules in a nanosystem allows one to talk about the possibility of effective operation of an aerosol gas generator for fire extinguishing in rooms with computer equipment and electrical equipment, where the ingress of moisture into electronic devices is undesirable. Also in the process of functioning, the gas generator additionally binds about 5% of the molecules of free oxygen, adding to the working environment an excess amount of carbon dioxide, which is one of the proven and reliable approaches to suppress the combustion and extinguishing processes.

Using the methods of multilevel mathematical modeling, the main characteristics of special purpose aerosol systems at a time of 120 sec have been obtained, such as the average size of nanostructures –4.33 μm and the particle concentration –$10^{15,27}$ m^{-3}. These parameters are shown in Figure 8.14. A series of computational experiments demonstrated that with the passage of time the intensity of the change in size and quantity functions for nanoclusters decreases, the dependencies reach a regime close to stationary and vary insignificantly.

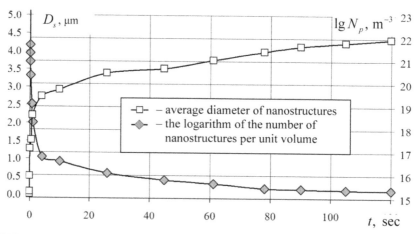

FIGURE 8.14   Change in the average size and number of nanostructures per unit volume, for gas extinguishing aerosol generator.

Compared with the modeling of nanostructures used to feed crops from the gaseous medium, the agglomeration intensity in aerosol systems is an order of magnitude higher. The final concentration of nanoclusters, in turn, has lower values than in plants for plant nutrition.

Nanoelements for a fire-extinguishing gas generator are formed larger, but in a smaller amount.

The experimental procedure involved the steps of the gas generator's operation in a special exhaust chamber, the deposition of nanoparticles, microparticles, and various agglomerates onto a slide and subsequent complex analysis of the characteristics of the structures obtained. As a hardware base, a complex system for measuring the properties of Nanotest 600 materials and a non-contact optical profilometer NewView 6300 was used. The results of experimental images of deposited nanostructures on pre-met glass are shown in Figure 8.15 and 8.16. Carrying out experimental studies (Figures 8.15 and 8.16) with the help of a hardware-software system made it possible to establish that the final diameter of nanoformations varied in the range 0.3–12.1 μm and corresponded to an average value of 3.87 μm.

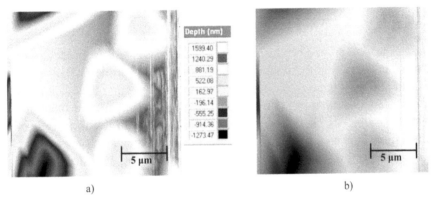

FIGURE 8.15 Scanning pattern (a) and image of the sample site (b) special-purpose nanostructures obtained with the help of Nanotest 600.

The discrepancy with the analogous parameter obtained with the help of mathematical modeling is 10.4%. Comparison of experimental and theoretical data in the form of dimensional distributions of nanoparticles indicate a satisfactory agreement of the values obtained. Constructed histograms have the same type of character with closely spaced values of probabilistic frequencies. In both cases, the number of clusters increased significantly, and their number decreased significantly.

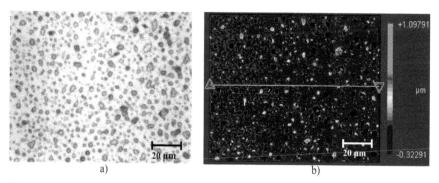

FIGURE 8.16   Image (a) and the color diagram of heights (b) of the sample obtained for the composition of the fire extinguishing gas generator.

8.6   RESULTS OF MODELING THE GROWTH PROCESSES OF NANOFILMS FOR THE TECHNOLOGY OF MOLECULAR BEAM EPITAXY

Another important and promising task is the modeling of the processes of molecular beam epitaxy of nanofilms on continuous templates. This principle of applying nanolayer coatings consists in depositing the evaporated elements in a vacuum or atmosphere of inert gases on a heated substrate. This technology is one of the most frequently used in the formation of films, since it makes it possible to observe the mechanisms of sputtering directly in the growth chamber and allows developing nanostructures of the required thickness and with given relief and surface properties. The technology is applied in various fields, including for the creation of photovoltaic films and solar cells.

The general scheme of molecular-beam epitaxy, used to solve the problem under consideration, is shown in Figure 8.17. The source of the deposited particles is the evaporator, which is located at the top of the simulated system. The substrate is located in the lower region.

The conducted series of computational experiments made it possible to establish various variants for the formation of nanostructures and nanofilms on silicon templates. Gold was collected in spherical and arbitrary nanoclusters directly on the surface of the substrate, followed by the filling of free regions with silicon in the next stage of epitaxy (Figure 8.18). Gallium, in combination with gold, formed nanofilms by regions and regions. Indium was characterized by agglomeration into nanoparticles above the template with further precipitation on it (Figure 8.19). In this case, gallium formed a relatively even surface layer.

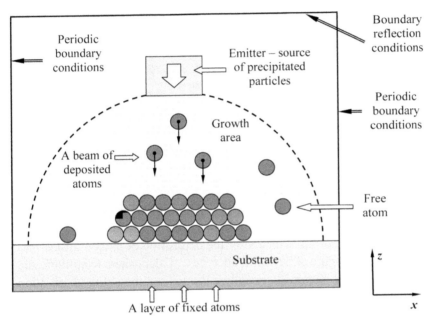

FIGURE 8.17   Modeling scheme for the problem of molecular-beam epitaxy of nanofilms on solid substrates.

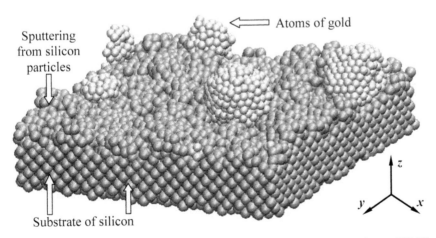

FIGURE 8.18   Results of two-layer deposition of gold atoms and silicon on substrate Si(100).

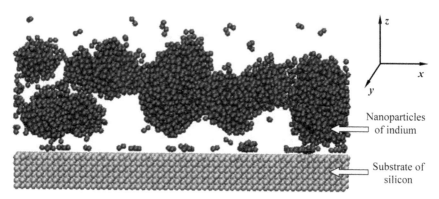

FIGURE 8.19   Formation of indium nanoparticles above the surface the Si(100).

Two-component antimony-silicon nanocomposites had a layered structure with a small mixing of atoms in the intermediate regions. With the deposition of gallium and antimony, the formation of high-grade nanostructures occurred, which after the deposition of silicon turned into quantum dots (Figure 8.20). For all types of deposited nanofilms, the substrate structure remained crystalline throughout the entire observation process. Diffusion of sedimented particles inside the template did not occur. The internal structure of the deposited nanofilms and nanostructures was predominantly amorphous. Exceptions are nanoclusters of gold and indium, and their structure was close to crystalline.

Modes of molecular-beam epitaxy, including the substrate temperature, have a significant effect on the mechanisms of nanostructure formation. For gold nanoparticles, a decrease in temperature from 800 K to 300 K led to a restructuring of the structure from amorphous to crystalline. A rise in temperature from 300 K to 800 K in the formation of antimony-gallium-silicon nanocomposite contributed to a decrease in the number of nanoparticles and an increase in their sizes. The shape of nanoclusters has also changed to streamlined and spherical, and the structure has become more dense.

The technology of producing photovoltaic nanofilms and nanostructures of antimony, gallium, and silicon on silicide substrates by molecular-beam epitaxy was experimentally tested, which led to the confirmation of the island structure of nanoobjects, which was previously obtained by mathematical modeling.

Using mathematical modeling methods, the type, properties, and structure of a new promising nanomaterial of special purpose with the introduction of quantum dots based on gallium and antimony in the region of silicon nanofilm were predicted. Integration of quantum dots leads to an increase in the sensitivity of the photodiode to 4–5%, which is due to a significant increase in the photoresponse in the wavelength range 1200–1600 nm.

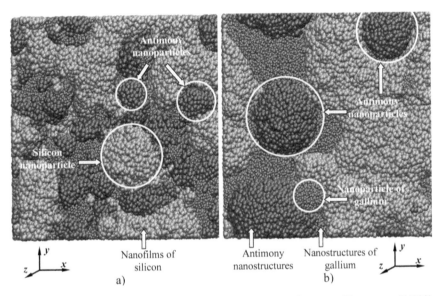

FIGURE 8.20 Formation of nanostructures of silicon-antimony-gallium on a Si(100) substrate heated to (a) 300 K and (b) 800 K.

The formation of silicon nanocomposites with embedded nanostructures of gallium-antimony with high density makes it possible to create new materials with controlled optical, thermoelectric, and photoluminescent properties. Such systems are used in microelectronics as active elements of light-emitting diodes, photodetectors, and thermoelectric converters.

8.7 RESULTS OF MODELING THE FORMATION OF NANOFILMS ON POROUS ALUMINUM OXIDE MATRICES BY THE METHOD OF DISCRETE THERMAL EVAPORATION

In the next problem, step-by-step modeling of the formation of nanofilms on porous aluminum oxide matrices by the method of discrete thermal evaporation of a powder under high vacuum conditions was considered. Materials based on porous alumina have become widespread as templates and templates for the synthesis of nanoelements of various shapes and forms, including nanotubes, nanocells, nanowires, nanotubes, and many other structures. With the help of porous substrates, ordered arrays of nanostructured objects filled with, for example, a phosphor or semiconductor elements of the same size and shape, can be obtained. Each such nanoelement is a separate light emitter, and their combined integration leads to a coherent addition of radiation from all sources, which in turn causes a significant increase in the intensity of light.

Modeling of the formation of nanofilms and nanostructured coatings on porous alumina substrates was carried out in several stages. The sequence of stages is shown schematically in Figure 8.21. The first step in solving the problem is to form a substrate of amorphous alumina. For this purpose, oxygen, and aluminum atoms in a 3: 2 ratio are placed in the calculated region with periodic boundary conditions in the horizontal directions (Figure 8.21a). Under normal thermodynamic conditions and the absence of external influences on the nanosystem, the solid substrate passes into an amorphous energetically more favorable state and stabilizes (Figure 8.21b). At the second stage, an opening of the required diameter and depth-pore was cut in the substrate. This hole is shown in Figure 8.21 in. In the long run, it was covered by atoms and molecules, as shown in Figure 8.21d. The last stage is the longest and is the most interesting, since during it the formation of nanostructures and nanofilms occurs.

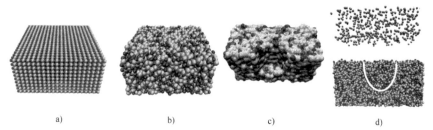

a)   b)   c)   d)

FIGURE 8.21   Sequence of stages in modeling the formation of nanofilms on porous substrates.

As a result of a series of computational experiments, various variants of the formation of nanostructures and nanolayer coatings were identified.

Iron deposition demonstrated that the agglomeration of atoms begins in a vacuum, and then the grouped structures are assembled on the surface of the substrate. The relief of a nanofilm is formed not even, with differences of several nm. When the atoms of silver and gold are epitaxed, a nanofilament is formed uniform with a small subsidence in the pore region, large structures in a vacuum are not formed (Figure 8.22). The deposition of germanium and gallium atoms led to the island formation of nanofilms, areas. Separated on the substrate were grouped gallium nanoparticles. During the deposition of palladium and platinum, the hole did not overgrow during the entire deposition time, and the nanofilm was formed uniformly. For all types of atoms, single particles reached the bottom of the cavity.

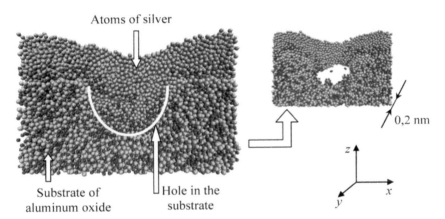

FIGURE 8.22 Results of the formation of silver nanofilms on a porous substrate, deposition time 0.2 ns.

The most complete and stable pore contraction, recorded when gallium atoms were deposited, showed that the active growth of the number of particles in the pore occurs in the initial period of time and is accompanied by an active rearrangement of the atomic structure. For different pore sizes, the temperature is characterized by the variable position of the center of mass of the gallium nanostructure formed inside it (Figure 8.23). For pores of radius 2 and 3 nm, the center of mass is formed above the middle of the depth ($H = 4$ nm). An increase in the radius of the hole to 5 nm or more led to a shift of the center of mass to the middle of the depth of the pore, which indicates a fairly dense filling of the deposited atoms.

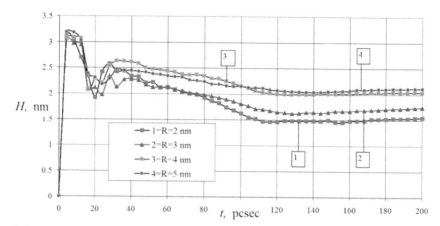

FIGURE 8.23   Change in the position of the mass center of nanostructures inside pores of various sizes when gallium deposition on alumina templates.

As shown in Figure 8.24, the growth processes of nanofilms and nano-clusters of zinc sulfide indicate incomplete overgrowth of cavities within the substrate at a level of 3.5% of the total number of molecules deposited. Partial pore filling is in good agreement with qualitative and quantitative experimental results and requires the search for control capabilities for the formation of ordered single-type nanostructured matrices based on optical elements and semiconductors.

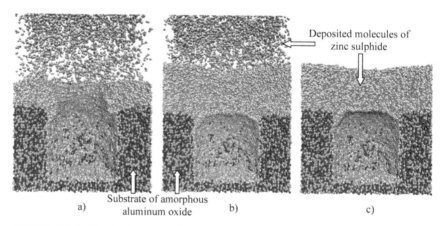

FIGURE 8.24   Result of overgrowing a porous alumina substrate with zinc sulfide for the deposition time a) 0.2 ns, b) 0.4 ns, and c) 0.6 ns.

The introduction of alloying additives in the form of copper and manganese sulfide with a mass fraction of 1–10% on the mechanisms of formation of nanofilms from zinc sulfide and their quantitative assessments did not significantly influence. These additives participate in epitaxy processes on a par with the main type of deposited atoms and are used in technological processes to impart yellow and green glow to nanoparticles and layers.

Analysis of the structure of materials based on zinc sulfide indicates a predominantly amorphous structure of templates and formed nanofilms with small areas of crystallization having different spatial directions. The end of the precipitation is the key to changing the structure, after which a redistribution of atoms and molecules to an energetically more favorable state was observed.

By the example of the deposition of silver atoms, it is shown that the phase transition from gas to liquid and solid state can be identified by a change in energy and temperature and occurs mixed between the melting points and the boiling point of silver at a temperature $T = 1520$ K.

In studying possibilities of process control nanoplenki growth and formation of similar arrays of ordered nanostructures and found that the deposition angles 0–60 degrees and sintering nanosystems to 293–593 K temperature do not lead to a significant change of properties and size of nanoclusters. It occurs minor rebuilding nanoelements structure due to temperature fluctuations, but more dense filling of the pores and change nanofilms properties not observed. This result may be associated with increased melting point of oxide elements in nanoscale and high potential barriers occurring near the openings on the substrate.

8.8   MAIN RESULTS AND CONCLUSIONS

A multilevel mathematical model based on the methods of quantum mechanics, molecular dynamics and mesodynamics of particles is developed and allows one to observe in the dynamics the processes of formation, interaction and condensation of nanostructures in the gas medium throughout the entire simulation cycle, from the atomic-electronic state to the macrolevel of the system. The constructed model of mesodynamics of particles describes the processes of condensation and agglomeration of nanostructures with allowance for their rotational motion and includes the effect of potential fields, gravity forces, the external medium in the form of a stochastic balanced component and frictional force for adjusting thermodynamic parameters.

A problem-oriented software package with a graphical interactive user interface consisting of a computational module, a visualization and data analysis module, an initial information preparation unit, and a program for coordinating control variables was created. The main algorithms for visualization and analysis of results include: the identification of atoms grouped in nanostructures; determination of the uniformity of nanofilms and nanocomposites; establishment of the chemical composition and proportions of the initial elements of nanoclusters; determination of the fraction of condensed atoms; search for structural and dimensional properties of nanoobjects; clarification of the internal structure of nanoparticles, nanofilms, substrates and analysis of a typical nanoparticle for a given material.

Investigation of the mechanisms of formation and growth of nanostructures used to feed plants from the gaseous medium showed that the formed nanoclusters include molecules of water, magnesium oxide and potassium carbonate. Molecules of nitrogen, oxygen and carbon dioxide remain at all times in the gaseous state and are present in the nanosystem as components of the air mixture, as well as combustion products. The processes of agglomeration and growth of nanoforms for different initial components proceeded with different intensities, the most active in the condensation involved the molecules K_2CO_3.

Complex studies of the method of thermal evaporation and subsequent condensation of metallic nanostructures showed that the process of their formation and growth is divided into two main stages. In the first stage, condensation centers are formed in the form of nanoelements formed by free metal atoms. For the second stage, the unification of already formed nanoobjects is more characteristic. On the structure, internal homogeneity and homogeneity of nanoobjects is significantly influenced by the type of raw materials. Nanoclusters of silver-copper have a mixed structure, silver-zinc nanoelements possess a shell structure, three-component silver-gold-zinc nanoparticles are formed multilayered. In all cases, cavities within the nano-formation are not observed.

Observation of the processes of formation of special-purpose aerosol nanostructures revealed that molecules of potassium carbonate, water, magnesium and calcium oxides are actively condensed into agglomerates.

The molecules of oxygen, carbon dioxide and nitrogen in the condensation processes were not significantly involved. The active condensation of water molecules allows us to talk about the possibility of efficient operation of an aerosol gas generator for fire extinguishing in rooms with computer equipment and electrical equipment.

The results of a step-by-step simulation of the formation of nanofilms on porous aluminum oxide matrices by the method of discrete thermal evaporation of the powder under high vacuum conditions made it possible to identify various variants for the formation of nanostructures and nanolayer coatings. For all types of atoms, single particles reached the bottom of the cavity in the template. The growth processes of nanofilms and nanoclusters of zinc sulphide indicate incomplete overgrowing of the cavities within the substrate at a level of 3.5 % of the total number of precipitated molecules.

When studying the possibility of controlling the growth processes of nanofilms and forming arrays of the same type and ordered nanostructures, it was obtained that the deposition angles of 0-60 degrees and the sintering of the nanosystem to temperatures of 293-593 K do not lead to a significant change in the properties and sizes of nanoclusters. It occurs minor rebuilding nanoelements structure due to temperature fluctuations, but more dense filling of the pores and change nanofilms properties not observed.

KEYWORDS

- mathematical modeling
- quantum mechanics
- molecular dynamics
- mesodynamics of particles
- molecular-beam epitaxy
- thermal evaporation of metals
- discrete evaporation of pock in high vacuum
- nutrition of plants from the gaseous medium
- photovoltaic films

- anodic porous alumina
- nanofilms
- nanoparticles
- nanostructures
- luminescent properties of zinc sulphide
- gas fire-extinguishing nanoaerosol generator

REFERENCES

1. Novoselov, K.S.; Geim, A.K.; Morozov, S.V.; Jiang, D.; Zhang, Y.; Dubonos, S.V.; Grigorieva, I.V.; Firsov, A.A. Electric Field Effect in Atomically Thin Carbon Films. *Science*. **2004**, 306(5696), 666-669.
2. Regan, B.C.; Aloni, S.; Jensen, K.; Ritchie, R.O.; Zettl, A. Nanocrystal-Powered Nanomotor. *Nano Letters*. **2005**, 5(9), 1730-1733.
3. Dutta, A.K.; Narasaiah, N.; Chattopadhyay, A.B.; Ray, K.K. Influence of Microstructure on Wear Resistance Parameter of Ceramic Cutting Tools. *Mater. and Manuf. Proc.* **2002**, 17(5), 651-670.
4. Gaitonde, V.N.; Karnik, S.R.; Figueira, L.; Paulo Davim, J. Analysis of Machinability During Hard Turning of Cold Work Tool Steel. *Mater. and Manuf. Proc.* **2009**, 23(4), 1373-1382.
5. Gkikas, G.; Sioulas, D.; Lekatou, A.; Barkoula, N.M.; Paipetis, A.S. Enhanced Bonded Aircraft Repair Using Nano-Modified Adhesives. *Materials and Design*. **2012**, 41, 394-402.
6. David, M.; Rawal, S.P.; Rummel, K. Multifunctional Structures for Advanced Spacecraft. *Journal of Spacecraft and Rockets*. **2001**, 38(2), 226-230.
7. Okpala, C.C. The Benefits and Applications of Nanocomposites. *International Journal of Advanced Engineering Technology*. **2014**, 5(4), 12-18.
8. Salata, O.V. Applications of Nanoparticles in Biology and Medicine. *J. Nanobiotechnology*. **2004**, 2(1), 15-20.
9. Fakruddin, M.; Hossain, Z.; Afroz, H. Prospects and Applications of Nanobiotechnology: a Medical Perspective. *J. Nanobiotechnology*. **2012**, 10(31), 2-8.
10. Emerich, D.F.; Thanos, C.G. Nanotechnology and Medicine. *Expert Opinion on Biological Therapy*. **2003**, 3(4), 655-663.
11. Parak, W.J.; Gerion, D.; Pellegrino, T.; Zanchet, D.; Micheel, C.; Williams, C.S.; Boudreau, R.; Le Gros, M.A.; Larabell, C.A.; Alivisatos, A.P. Biological Applications of Colloidal Nanocrystals. *Nanotechnology*. **2003**, 14, R15-R27.
12. Darroudi, M.; Ahmad, M.B.; Abdullah, A.H.; Ibrahim, N.A.; Shameli, K. Effect of Accelerator in Green Synthesis of Silver Nanoparticles. *Int. J. Mol. Sci.* **2010**, 11(10), 3898-3905.

13. Sawhney, A.P.S.; Condon, B.; Singh, K.V.; Pang, S.S.; Li, G.; Hui, D. Modern Applications of Nanotechnology in Textiles. *Textile Research Journal.* **2008**, 78(8), 731-739.

14. Wong, Y.W.H.; Yuen, C.W.M.; Leung, M.Y.S.; Ku, S.K.A.; Lam, H.L.I. Selected Applications of Nanotechnology in Textiles. *Autex Research Journal.* **2006**, 6(1), 1-10.

15. Presting, H.; Konig, U. Future Nanotechnology Developments for Automotive Applications. *Materials Science and Engineering C.* **2003**, 23, 737-741.

16. Louda, P. Applications of Thin Coatings in Automotive Industry. *Journal of Achievements in Materials and Manufacturing Engineering.* **2007**, 24(1), 51-56.

17. Wong, K.V.; Paddon, P.A. Nanotechnology Impact on the Automotive Industry. *Recent Patents on Nanotechnology.* **2014**, 8(3), 181-99.

18. Tze, W.T.Y. Nanoindentation of Wood Cell Walls: Continuous Stiffness and Hardness Measurements / W.T.Y. Tze, S. Wang, T.G. Rials, G.M. Pharr and S.S. Kelley // Compos. Part A: Appl. Sci. Manuf. – 2007. – Vol. 38. – Pp. 945-953.

19. Gindl, W.; Schöberl, T.; Jeronimidis, G. The Interphase in Phenol-Formaldehyde and Polymeric Methylene Diphenyl-di-Isocyanate Glue Lines in Wood. *Int. J. Adhes. Adhes.* **2004**, 24(4), 279-286.

20. Bhatnagar, A.; Sain, M. Processing of Cellulose Nanofiber-Reinforced Composites. *Journal of Reinforced Plastics and Composites.* **2005**, 24(12), 1259-1268.

21. Zaharov, R.S.; Glotov, O.G. Characteristics of Combustion of Pyrotechnic Compositions with Powdered Titanium. *Bulletin of NSU. Series: Physics.* **2007**, 2(3), 32-40.

22. Spinelli, P.; Ferry, V.E.; van de Groep, J.; van Lare, M.; Verschuuren, M.A.; Schropp, R.E.I.; Atwater, H.A.; Polman, A. Plasmonic Light Trapping in Thin-Film Si Solar Cells. *Journal of Optics.* **2012**, 4(2), 024002.1-11.

23. Ma, H.Y.; Bendix, P.M.; Oddershede, L.B. Large-Scale Orientation Dependent Heating from a Single Irradiated Gold Nanorod. *Nano Lett.* **2012,** 12(8) 3954-3960.

24. Sanchot, A.; Baffou, G.; Marty, R.; Arbouet, A.; Quidant, R.; Girard, C.; Dujardin, E. Plasmonic Nanoparticle Networks for Light and Heat Concentration. *ACS Nano.* **2012**, 6(4), 3434-3440.

25. Stupca, M.; Alsalhi, M.; Saud, T.A.; Almuhanna, A.; Nayfeh, M.H. Enhancement of Polycrystalline Silicon Solar Cells Using Ultrathin Films of Silicon Nanoparticle. *Appl. Phys. Lett.* **2007**, 91(6), 063107.1-3.

26. Ju, S.; Facchetti, A.; Xuan, Y.; Liu, J.; Ishikawa, F.; Ye, P.; Zhou, C.; Marks, T.; Janes D. Fabrication of Fully Transparent Nanowire Transistors for Transparent and Flexible Electronics. *Nat. Nanotechnol.* **2007**, 2, 378-384.

27. Boyer, D.; Tamarat, P.; Maali, A.; Lounis, B.; Orrit M. Photo-thermal Imaging of Nanometer-Sized Metal Particles among Scatterers. *Science.* **2002**, 297(5584), 1160-1163.

28. Shipway, A.N.; Katz, E.; Willner, I. Nanoparticle Arrays on Surfaces for Electronic, Optical, and Sensor Applications. *Chemphyschem.* **2000**, 1(1), 18-52.

CHAPTER 9

CARBON NANOTUBES AS CHEMICAL SENSORS AND BIOSENSORS: A REVIEW

RAKSHIT AMETA[1,2], KANCHAN KUMARI JAT[3], JAYESH BHATT[1], AVINASH RAI[1], TARACHAND NARGAWE[1], DIPTI SONI[1], and SURESH C. AMETA[1]

[1]Department of Chemistry, PAHER University, Udaipur–313003 (Raj), India

[2]Department of Chemistry, Faculty of Science, J. R. N. Rajasthan Vidyapeeth (Deemed-to-be University), Udaipur–313002 (Raj), India

[3]Department of Chemistry, Mohan Lal Sukhadia University, Udaipur–313002 (Raj), India

ABSTRACT

Carbon nanotubes (CNTs) are a promising material for quite sensitive chemical sensors as well as biosensors, because they have excellent material properties with the high surface-to-volume ratio. These can efficiently detect very low concentration of alcohol, ammonia, aldehyde, moisture, etc. CNTs are also utilized in detecting the presence of various pathogenic bacteria. Not only this, these are very good biomarkers for early detection of different types of cancer like prostate cancer, colon cancer, breast cancer, etc. Uses of CNTs, chemical sensors and biosensors have been reviewed.

9.1 INTRODUCTION

Smart sensors are needed for health, safety, environmental, and security purposes like environmental gas monitoring, detection of hazardous gases, control of air quality and chemical processes. Gas sensors are commonly utilized in various industries, and they are very important for environmental monitoring. Sensors are also important for the detection of hazardous gas leaks. Sensors based on flexible substrates are being increasingly developed, mainly because of the increasing proliferation of wearable, handheld, portable consumer electronics [1, 2]. Such a flexible technology will provide a promising alternative to expensive silicon technology. Most of research scientists focus their interest on improving sensitivity, selectivity, stability, and response-recovery times of sensing materials. Chemical sensors that are based on a change in resistance upon binding with analyses are the preferred candidates for use in a new generation of low-power, low cost, and high-precision portable sensing devices with interesting electrical & physicochemical properties and a high capacity for gas adsorption.

Materials with high specific surface areas have been found to exhibit even better sensitivities and response times. Numerous materials, including TiO_2 [3], ZnO [4] and WO_3 [5] are currently being studied for their possible use as the active components in sensing layers, but these materials are typically most effective at high temperatures. Carbon nanotubes (CNTs) are now commonly used in detection because of number of sites are available for adsorption due to hollow geometry, one-dimensional nanoscale morphology, and high specific surface area that promotes physical adsorption or chemical reaction with desired molecules for signal transformation with high efficiency and speed [6–8]. The sensing mechanism is based on the transfer of electrons to and from adsorbate molecule. It may still operate even in the presence of a surface-adsorbed species, with which electrons are transferred upon interaction with a different analyte.

A good sensor must be:

- sensitive to the material to be detected/measured;
- insensitive to other materials, which are likely to be encountered during application; and
- does not influence the measured material.

Sensors could be divided into:

- gas-phase sensor;
- liquid-phase sensor;
- solid-phase sensor;

- physical;
- chemical; and
- biosensors.

9.1.1 ADSORPTION MECHANISM

The sensor-related applications of CNTs to detect variant kinds of gases have never failed to be highlighted in recent years. CNTs are supposed to be a new type of adsorbent which hold a very prominent position in carbon-based sensor materials due to many reasons. The major ones are [9]:

- they possess chemically inert surfaces and high specific surface area for physical adsorption, directly; and
- diversity of well-defined adsorption sites are available for adsorbed molecules.

Apart from this, different charge distribution resulting from the charge transfer and different adsorption energy attributed to gas morphology coexisted in the adsorption process, which is responsible for the increasing or decreasing conductivity in gas adsorption on CNT sensors [10]. Therefore, the specific gases are differentiated from each other.

9.1.2 ADSORPTION SITES

There are four potential adsorption sites for the adsorption of diverse gases in the CNTs [11, 12]. These are:

- Internal sites in the hollow interior of every tube;
- Interstitial channels in the hollow channels between individual tubes;
- Grooves in the exterior surface of the tubes, where two adjacent parallel tubes meet; and
- Outside surface, the curved surface of tubes on the outside of the nanotube bundles.

The gas molecules are able to interact with CNTs through the outer surface of bundles, and the interstitial channels between the tubes and the inside of CNTs [13]. A number of studies [14–18] have been carried out just confirm the place, where gas molecules are adsorbed on CNTs (Figure 9.1).

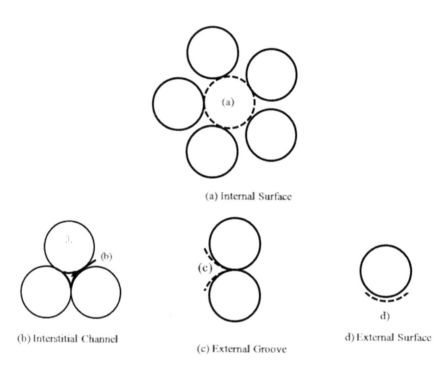

(a) Internal, (b) Interstitial channel, (c) External groove, and (d) External surface.

FIGURE 9.1　Adsorption sites of CNTs.

9.2　CHEMICAL SENSORS

A chemical sensor is an analytical device, which can provide information about the chemical composition of its environment. This information may be provided in the form of physical signal, which can be measured. It is correlated with the concentration of a particular chemical species (called analyte). There are two main steps in the functioning of a chemical sensor and these are:

i)　recognition; and
ii)　transduction.

In the first step, analyte molecules interact selectively with receptor molecules or certain sites included in the structure of the recognition element of the sensor. As a result, a characteristic physical parameter is varied, and this variation can be measured by means of an integrated transducer generating the output signal.

CNTs have been regarded as a promising material for highly sensitive gas sensors as they have excellent material properties combined with their one-dimensional structural advantages, i.e., a high surface-to-volume ratio. The capacitance of single-walled carbon nanotubes (SWCNTs) is highly sensitive to a broad class of chemical vapors. Because of transduction mechanism, they can form the basis for a fast, low-power sorption-based chemical sensor.

Molecular adsorbates are polarized by the fringing electric fields radiating from the surface of a SWCNT electrode in the presence of a dilute chemical vapor, which causes an increase in its capacitance. This effect was used by Snow et al. [19] to construct a high-performance chemical sensor. It is done by thinly coating the SWCNTs with chemoselective materials that provide a large, class-specific gain to the capacitance response. These SWCNT chemicapacitors are fast, highly sensitive, and completely reversible.

A high-sensitive humidity sensor based on using graphene oxide (GO) film with evenly dispersed MWCNTs on Au interdigitated electrodes (IDEs) has been reported by Li et al. [20]. GO was used for multiwalled carbon nanotubes (MWCNTs) as dispersant. The sensitivity of the sensor with GO/MWCNT film was found to be thirteen times more than with sensor with GO film only. SEM confirmed the presence of GO/MWCNT with the less stacking layers as compared to GO. This may be due to the addition of MWCNTs, which played a supporting role in the GO/MWCNT composite. This composite structure would lead to more accessible surface area and thus accelerate the diffusion of the water molecules. It was concluded that GO/MWCNT composite could be used for high-sensitive humidity detection.

Sheng et al. [21] reported a simple, low cost and effective route to fabricate CNT-based chemical sensors at ambient temperature. Silk fibroin was incorporated in vertically aligned CNT arrays (CNTA), which were obtained through a thermal chemical vapor deposition (CVD) method making feasible the direct removal of CNTA from substrates without any rigorous acid or sonication treatment. The functionalization of CNTA with polyaniline (PANI) significantly improved the sensing performance of CNT-based chemical sensors in detecting ammonia and hydrogen chloride vapors. These chemically modified CNTA also show responses to organic vapors such as menthol, ethyl acetate, and acetone. Although the detection limits of this chemical sensor is of the same orders of magnitudes but they show some advantages like simplicity, low cost, and energy efficiency in preparation and fabrication of devices. It was reported that a linear relationship between the relative sensitivity and concentration of analyte permits accurate determination of trace concentrations of trace chemical vapors.

Lupan et al. [22] investigated the influence of CNT hybridization on ultraviolet (UV) and gas sensing properties of individual and networked ZnO nanowires (NWs). Concentration of CNT was varied so as to achieve optimal conditions for the hybrid with improved sensing properties. In case of CNT decorated ZnO nanonetworks, the effect of relative humidity (RH) and applied bias voltage on the UV sensing properties was observed. It was observed that on increasing the CNT content to about 2.0 wt% (with respect to the entire ZnO network), the UV sensing response was also increased from 150 to 7300. The ZnO-CNT networks demonstrated an excellent selectivity and also a high gas response to NH_3 vapor. They reported a response of 430 to 50 ppm at room temperature with an estimated detection limit of about 0.4 ppm. The highest sensing performance for NH_3 vapor was obtained for the finest NW with 100 nm diameter and response of about 4 to 10 ppm at room temperature.

Liu et al. [23] made chemiresistive detectors for amine vapors from SWCNTs by noncovalent modification with cobalt meso-aryl porphyrin complexes. They showed that it is possible to improve the magnitude of the chemiresistive response to ammonia with the changes in the oxidation state of the metal, the electron-withdrawing character of the porphyrinato ligand, and the counter anion. Such devices exhibited sub-ppm sensitivity and high selectivity toward amines and also good stability to air, moisture, and time. The application of these chemiresistors was demonstrated in the detection of various biogenic amines (i.e., putrescine, cadaverine) and monitoring of spoilage in raw meat and fish samples (chicken, pork, salmon, cod) over several days.

A gas sensor, fabricated by the simple casting of SWCNTs on an IDE has been reported by Li et al. [24] for gas and organic vapor detection at room temperature. It was observed that sensor responses are linear for concentrations of sub-ppm to hundreds of ppm with detection limits of 44 ppb for NO_2 and 262 ppb for nitrotoluene. The time for the detection response was of the order of seconds and minutes for the recovery. It was found that variation of the sensitivity is less than 6% for all the tested devices, as with commercial metal oxide or polymer microfilm sensors. The room-temperature high sensitivity of the SWCNT transistor sensors and manufacturability of the commercial sensors was also retained. This detection capability from gas to organic vapors was attributed to direct charge transfer on individual semiconducting SWCNT conductivity supported by additional electron hopping effects on intertube conductivity through physically adsorbed molecules between SWCNTs.

A CNT-based gas sensor has been reported by Jeon et al. [25] They assembled highly purified, solution-processed 99.9% semiconducting CNT networks bridged by palladium source/drain electrodes in a field-effect transistor (FET) configuration with a local back-gate electrode. The gas responses of the CNT-FETs were investigated under different controlled operating regimes for the enhanced detection of H_2 and NO_2 gases.

Kaniyoor et al. [26] fabricated hydrogen gas sensors, which was based on noble nanometal decorated one-dimensional MWCNTs and two-dimensional graphene by a simple drop casting technique. Pt-decorated functionalized graphene sheets (Pt/f-G), and Pt-decorated functionalized multi-walled carbon nanotubes (Pt/f-MWCNT) were synthesized by them and employed as hydrogen sensors. Although (Pt/f-G) had a response time comparable to that of (Pt/f-MWNT) for hydrogen sensing, even at a low detection level of 4 vol% hydrogen in air, but there was a two-fold increase in the sensitivity at room temperature. It was interesting to note that these sensors were stable over repeated cycles of hydrogenation and dehydrogenation.

Jung et al. [27] investigated the effects of Co doping on the NO gas sensing characteristics of ZnO-carbon nanotube (ZnO-CNT) layered composites. These composites were fabricated by coaxial coating of single-walled CNTs with ZnO using pulsed laser deposition. It was confirmed that a CNTs coated with ZnO nanoparticles (NPs) had an average diameter as small as 10 nm and it showed little influence of doping 1 at.% Co into ZnO on the morphology of the ZnO-CNT composites. Co-doping into ZnO (with 1 at.%) gave rise to a significant improvement in the response of the ZnO-CNT composite sensor to NO gas exposure. This Co-doped ZnO-CNT composite sensor had high sensitivity and fast response to NO gas even at low temperatures and NO concentrations. An improvement of the NO gas sensing properties was attributed to an increase in the specific surface area and the role of the doped Co elements as a catalyst. It was suggested that Co-doped ZnO-CNT composites are useful as high-performance NO gas sensors.

Hashishin and Tamaki [28]. This was based on the concept that defects and carboxyl groups may contribute in increasing number of adsorption sites of gas molecules. TEM revealed that oxidized MWCNTs had a rough sidewall surface. The sensor response to dilute NO_2 (Ra/Rg) of an oxidized MWCNT-based sensor was about 50% higher than that of as-grown MWCNT-based sensor. This suggests that there was an increase in the number of NO_2 adsorption sites with an increase in the density of defects on the surface of MWCNTs. The p-n junction of a semiconductor generally contributes in increasing electric resistance with the formation of a space charge layer, which could also enhance NO_2 gas adsorption. Therefore, MWCNTs as p-type semiconductors were modified with an n-type semiconductor such as WO_3 and SnO_2 NPs, which naturally exhibited an excellent sensor response to dilute NO_2. MWCNT-WO_3 and MWCNT-SnO_2 composite sensors showed a fairly good sensor response (Ra/Rg) to dilute NO_2, compared to MWCNT-based sensor. Adding n-type oxide to MWCNTs results in formation of a large depletion layer within MWCNTs by the p-n junction on contact points between MWCNTs and oxides; thus, an increase in resistance in air and the enhancement of NO_2 adsorption was observed.

A reversible chemoresistive sensor has been developed by Esser et al. [29], which is able to detect even sub-ppm concentrations of ethylene. This method has high selectivity towards ethylene and is simply prepared in few steps from readily available materials. The sensing mechanism depends on the high sensitivity in resistance of SWCNTs to changes in their electronic surroundings.

Ishihara et al. [30] reported an amperometric detection of formaldehyde using hydroxylamine hydrochloride and SWCNTs. Hydroxylamine hydrochloride reacts with HCHO where HCl vapor evaluate, which injects a hole carrier into semiconducting SWCNTs. The increase of conductivity in SWCNTs can be monitored using an ohmmeter. It was observed that de-bundling of SWCNTs with a Metallo-supramolecular polymer (MSP) increases the active surface area in the SWCNTs network, leading to excellent sensitivity to HCHO with a limit of detection (LOD) as 0.016

ppm. The response of sensor is reversible, and it is reusable also. The selectivity to HCHO is more than hundred times higher than interferences with other volatile compounds like water, methanol, and toluene. False-positive responses caused by a significant change of humidity and/ or temperature were also successfully discriminated from true-positive responses by using two sensors: (i) with hydroxylamine hydrochloride, and (ii) without hydroxylamine hydrochloride.

Neumann et al. [31] compared chemical sensing properties of single wire and mat form sensor structures fabricated from the same CNTs materials. Sensing properties of these sensors were evaluated upon electrical response in the presence of five compounds like acetone, acetic acid, ethanol, toluene, and water. It was found that single wire CNT sensors show diverse behavior, whereas mat structures showed similar response for all these vapors. It was indicated that the sensing mechanism of random CNT networks cannot be simply interpreted as summation of the constituting individual CNT effects, but it is also associated with some another robust phenomenon, localized presumably at CNT–CNT junctions.

Brahim et al. [32] reported sensors containing metal-CNT hybrid materials as the active sensing layer for ethanol vapor detection at room temperature. Such metal-CNT hybrid materials were synthesized by infiltrating SWCNTs with the transition metals like Ti, Mn, Fe, Co, Ni, Pd or Pt. Each sensor was prepared by drop-casting dilute dispersions of a metal-CNT hybrid onto quartz substrate electrodes. They recorded impedimetric responses with different ethanol concentrations. AC impedance (Z') of the sensors was found to decrease on exposure to ethanol vapor. It was observed that sensor containing pristine CNT material was virtually non-responsive at low ethanol concentrations (< 50 ppm), but all metal-CNT hybrid sensors showed high sensitivity to trace ethanol levels with 100-fold or more gains in sensitivity. All hybrid sensors exhibited significantly larger sensor responses to ethanol vapor up to 250 ppm compared to the starting SWCNT sensor, except Ni filled CNT.

Penza et al. [33] demonstrated the integration of SWCNTs onto quartz crystal microbalance (QCM) and standard silica optical fiber (SOF) sensor for detection of alcohol at room temperature. Different transducing mechanisms have been used in order to explain sensing properties of these nanomaterials, but particular attention has been paid to two key parameters. These are: mass and refractive index changes due to gas absorption. Langmuir–Blodgett (LB) films consisting of tangled bundles of SWCNTs without surfactant molecules have been successfully transferred onto QCM and SOF. Mass-sensitive 10 MHz QCM SWCNTs sensor exhibited a resonant frequency decrease in presence of alcohols. It was reported that normalized optoelectronic signal ($\lambda = 1310$ nm) of the refractive index-sensitive SOF SWCNTs sensor was found to decrease. Highly sensitive, repeatable, and reversible responses of the QCM and SOF SWCNTs sensors indicated that the detection of alcohols, at room temperature, is possible in wide mm Hg vapor pressures

A dual sensitization gas sensor for selective detection of either H_2S or C_2H_5OH gases has been presented by Choi et al. [34] based on ZnO/CuO NP-decorated MWCNTs. They successfully deposited Cu-Zn layers of three different thicknesses (3, 6, and 9 nm) on the surfaces of CNTs by sputtering process and then converted to their corresponding oxides by thermal annealing. The gas sensing properties of metal oxide decorated-MWCNTs were tried for the presence of three gases, like NO_2, H_2S, and C_2H_5OH. It was observed that best sensing was obtained, when the Cu-Zn deposition layer was kept 6 nm thick. This sensor showed extraordinary responses to H_2S and C_2H_5OH both, at working temperatures of 100°C and 200°C, respectively. It is, therefore, possible to selectivity tune by selection of the working temperature. It will prove to be of great utility in fabricating dual sensitive gas sensors.

A single-use screen-printed carbon electrode strip was designed and fabricated by Chou et al. [35]. Nanohybrids were prepared by deposition of platinum NPs on MWCNT, and modified on the surface of screen-printed carbon electrode. A fast, sensitive, and cost-effective hydrogen peroxide detection amperometric sensor strip was developed. The current generated in response to H_2O_2 by the screen-printed carbon electrode strip with Pt-MWCNT nanohybrids was enhanced hundred-fold with an applied potential of 300 mV. As-prepared electrode strip quality was assured by the low coefficient of variation (CV) ($< 5\%$) of currents measured at 5 s. Three different linear detection ranges with sensitivity of 75.2, 120.7, and 142.8 $\mu A\ mM^{-1}\ cm^{-2}$ were observed for H_2O_2 concentration in the range of 1–15 mM, 0.1^{-1} mM, and 10–100 μM, respectively. The lowest H_2O_2 concentration that could be measured by this type of strip was 10 μM. H_2O_2 levels in green tea infusion and pressed Tofu could be rapidly detected by this sensor and results were comparable to that measured by ferrous oxidation-xylenol orange (FOX) assay and peroxidase colorimetric method.

Wang et al. [36] prepared monodispersed surfactant-free MoS_2 NPs with sizes of less than 2 nm from bulk MoS_2 through simple ultrasonication and gradient centrifugation. The ultrasmall MoS_2 NPs exposed a large fraction of edge sites, and have high surface area, leading to high electrocatalytic activity for reduction of H_2O_2. A real determination limit as low as 2.5 nM and wide linear range of five orders of magnitude was observed indicating that H_2O_2 sensor based on MoS_2 NPs is extremely sensitive. The trace amount of H_2O_2 released from Raw 264.7 cells was successfully detected by this sensor, and an efficient glucose sensor was also fabricated. As, H_2O_2 is a byproduct of many oxidative biological reactions, this sensor can be applied in the fields of electrochemical sensing and bioanalysis.

Thirumalai et al. [37] developed a new approach based on differential pulse voltammetry (DPV) for the simultaneous determination of ascorbic acid (AA), dopamine (DA), and uric acid (UA). They used a modified glassy carbon electrode (GCE). This sensor was constructed by a simple "one-step" by drop-casting de-bundled SWCNTs onto the GCE. It was observed that SWCNTs were poorly dispersed in aqueous solution and were ineffective for the one-step procedure without de-bundling. De-bundling was achieved using a small amount (0.1 wt%) of the synthesized polymer dispersant, sulfonated poly(ether sulfone) (SPES). De-bundled SWCNTs had a high aspect ratio (length = 2.5 ± 1.0 µm; height = 2 ± 1 nm) and enhanced electrocatalytic activity and selectivity of the modified sensor for the simultaneous determination of AA, DA, and UA in DPV measurements. It was reported that the peak-to-peak separation values were 221, 119, and 340 mV (vs. Ag/AgCl) for DA–AA, UA–DA, and AA–UA, respectively. The dynamic linear ranges for AA, DA, and UA were observed as 0.2–1.6 mM, 5.0–50 µM, and 5.0–60 µM, and the detection limits 10.6 µM, 15 nM, and 113 nM (S/N = 3), respectively. The developed sensor can be successfully used in the determination of AA and DA in commercial pharmaceutical samples (vitamin C tablets and DA injection).

Acid yellow 9 (4-amino–1–1′-azobenzene–3,4′-di-sulfonic acid, AY) is a good stabilizing agent for MWCNTs. A dispersion of MWCNTs in AY solution remained stable for about three months and also after centrifugation at 10,000 rpm for 30 min. Thin-films were prepared using this MWCNTs/AY dispersion on indium tin oxide coated glass electrode and GCEs by Kumar et al. [38]. These dried films were subjected to electropolymerization in 0.1 M H_2SO_4 solution, and as a result, adsorbed AY molecules on MWCNTs get polymerized yielding a polymer–MWCNTs nanocomposite film (PAY/

MWCNT) on electrode surface. This was found to be electrochemically active in wide pH range. These synthesized PAY–MWCNTs composite film showed excellent electrocatalytic activity towards oxidation of DA and AA with high sensitivity. The determination of DA in the presence of AA in the range of 2×10^{-7} to 1.4×10^{-6} M was carried out by linear sweep voltammetry. Amperometry was employed to determination of AA at 0.0 V in the range from 1×10^{-6} to 5.6×10^{-5} M, and it was observed that DA, and UA did not interfere on the steady-state current of AA. Satisfactory results were obtained, when PAY/MWCNTs composite modified electrode was used for analysis of real samples such as DA injection and AA spiked into human urine.

A highly sensitive and selective CNT-based vapor sensor was developed by Zhang et al. [39] toward the detection of methamphetamine vapor. It is one of the most widespread, harmful, and addictive illegal drugs. Poly[3-(6-carboxyhexyl)thiophene–2,5-diyl)] (P3CT) was selected for non-covalently functionalizing CNTs, which helps in formation of suspension of CNTs in the solvents and also in introducing a carboxylic acid group. This carboxylic acid group in the polymer acts as a binder of organic amines via acid-base interaction. It was reported that P3CT-functionalized CNT sensors had sensitivity to n-methylphenethylamine (NMPEA), a detection simulant of methamphetamine, at very low concentration to the level of 4 ppb. Proposed sensors were able to distinguish NMPEA from two other amine compounds, other volatile chemical compounds (VOCs), and water vapor, just by observing the recoverability of signal of the sensor after exposure. It showed higher sensitivity to amine vapors, even when the vapor of amines is present much lower than VOCs. This gives a simple but effective method for detecting trace amounts of methamphetamine in air.

SWCNT random networks can be easily fabricated on a wafer scale; thus, providing an attractive path to large-scale SWCNT-based thin-film transistor (TFT) manufacturing. The mixture of semiconducting SWCNTs and metallic SWCNTs (m-SWCNTs) in the networks limits the TFT performance as m-SWCNTs dominate the charge transport. Chen et al. [40] have achieved a uniform and high-density SWCNT network throughout a complete 3-in. Si/SiO$_2$ wafer using a solution-based assembly method. UV radiations were utilized to etch m-SWCNTs from the networks, and a remarkable increase in the channel current on/off ratio (Ion/Ioff) was achieved from 11 to 5.6×10^3. They used the SWCNT-TFTs as gas sensors to detect methyl methylphosphonate, which is a stimulant of benchmark threats. It was found that the SWCNT-TFT sensors treated with UV radiation had much higher sensitivity and faster response than without treatment.

Tarlekar et al. [41] proposed accurate monitoring of antiviral drug acyclovir (ACV) using GCE with SWCNTs and Nafion composite film. They employed square wave voltammetry. This sensor exhibited effective and sustained electron mediating behavior displaying higher peak currents at lower potential as compared to bare GCE. The oxidation current showed a wide linear response for this drug in the concentration range from 10 nM to 30 μM at optimal conditions. It has pronounced analytical performance with LOD corresponding to 1.8 nM and high sensitivity of 15.4 μA μM^{-1}. The modified sensor had high recognition selectivity, fair reproducibility and long term stability of signal response. This sensor was successfully implemented to quantify ACV in some commercially available pharmaceuticals. This method was also applied successfully in detecting ACV in real human urine sample of patient undergoing pharmacological treatment with ACV.

Uge et al. [42] deposited MWCNTs electrochemically on a GCE and, it was modified with nanoceria-poly(3,4-ethylene dioxythiophene) (CeO$_2$-PEDOT) composite. This composite was prepared in the presence of sodium dodecyl sulfate (SDS). The MWCNTs/CeO$_2$-PEDOT modified electrode was used with good sensitivity for the analysis of DA in the presence of UA and AA. The MWCNTs/CeO$_2$-PEDOT modified GCE exhibited effective electrocatalytic performance as compared with bare GCE for oxidation of DA. This sensor showed two wide linear ranges of 0.10–10 μmol L^{-1} and 40–400 μmol L^{-1} with a good detection limit of 0.03 μmol L^{-1}. No measurable interferences by K$^+$, Na$^+$, Ca^{2+}, urea, glucose, sucrose, citric acid, and cysteine were not observed on the differential pulse response of DA.

Chemoresistive multi-layer sensor was fabricated by Yuan and Chang [43] through drop-coating PANI solution on MWCNTs with chemical modification. Response of this sensor was investigated by exposing it to methanol, chloroform, dimethoxy methyl phosphonate (DMMP) and dichloromethane (DCM). They showed that 1wt% MWCNTs and 10, 5, and 1 wt% PANI sensor samples had high sensitivity, excellent selectivity and good reproducibility to DMMP and DCM vapors. Principal component analysis (PCA) was also applied to distinguish the performance of methanol, chloroform, DMMP, and DCM.

Kumar et al. [44] fabricated flexible SWCNT based sensor for detection of 2,4-dinitrotoluene (DNT), which is an explosive chemical. This flexible sensor was fabricated by vacuum filtration method for analyte sensing. This sensor was able to detect even low concentration traces of DNT (0.22 ppm) at room temperature and gives 0.28–0.32% repeatable response. It was observed that response of sensor increases with increase in the vapor concentration of the DNT vapors.

A facile synthesis method was developed by Kim et al. [45] for preparing hybrid 2-dimensional (2D) films. These films are based on large-scale molybdenum disulfide (MoS_2) nanosheets and SWCNTs, for flexible sensors. They combined 1-dimensional (1D) SWCNTs with MoS_2 nanosheets during their synthesis for improving the flexibility and stability of nanosheet. CVD method was used to synthesize uniform MoS_2 nanosheets using a porphyrin-type organic promoter. The high performance and enhanced sensitivity of the chemical gas sensors of fabricated hybrid MoS_2-SWCNT layers was due to sensitive gas adsorption by SWCNTs in the MoS_2 nanosheets. Hybrid MoS_2-SWCNT films, transferred on a flexible polyethylene terephthalate (PET) substrate, were used for the analysis of physical properties of chemical sensors as a function of the number of bending cycles. These hybrid MoS_2-SWCNT-based sensors showed quite a stable sensing performance after 105 bending cycles with almost 300% rise in the resistance of MoS_2-based sensors under the same bending process.

A threat to civilian populations and military personnel in operational areas all over the world is due to chemical warfare agents (CWA) as their measurements are quite critical in view of contamination detection, avoidance, and remediation. Although current deployed systems provide accurate detection of CWAs, but these are still limited by size, portability, and fabrication cost. A chemiresistive CWA sensor has been reported by Fennell et al. [46] using SWCNTs wrapped with poly(3,4-ethylene dioxythiophene) (PEDOT) derivatives. It was found that a pendant hexafluoroisopropanol group on the polymer enhances sensitivity to a nerve agent mimic, dimethyl methyl phosphonate, in both; nitrogen and air environments to concentrations as low as 5 and 11 ppm, respectively. It was interesting to note that these PEDOT/SWCNT derivative sensor systems experience negligible device performance in two weeks under ambient conditions.

Wearable electronics offer opportunities in sweat analysis using skin sensors, but there is a challenge to form sensitive and stable electrodes. The development of a wearable sensor has been reported by Roy et al. [47]

based on CNT electrode arrays for sweat sensing. They prepared solid-state ion-selective electrodes (ISEs), (sensitive to Na^+ ions) by drop coating plasticized poly(vinyl chloride) (PVC) doped with ionophore and ion exchange on CNT electrodes. Ion selective membrane (ISM) filled the intertubular spaces of the highly porous CNT film forming an attachment stronger than that is achieved with flat Au, Pt, or carbon electrodes. It was found that concentration of the ISM solution influenced the attachment to the CNT film, ISM surface morphology, and the overall performance of the sensor. A sensitivity of 56 ± 3 mV/decade to Na^+ ions was achieved. Solid-state reference electrodes (REs), suitable for this purpose, were prepared by coating CNT electrodes with colloidal dispersion of Ag/AgCl, agarose hydrogel with 0.5 M NaCl, and a passivation layer of PVC doped with NaCl. It was observed that CNT-based REs had low sensitivity (-1.7 ± 1.2 mV/decade) toward the NaCl solution and high repeatability and these were found to be superior to Ag/AgCl, metals, carbon, and CNT films. CNT-based ISEs were calibrated against CNT-based REs, and the short term stability of the system was also tested.

A simple, low cost, and highly sensitive chemical sensor was developed by Rahman et al. [48] with various nanocomposites (NCs) of Mn_3O_4-CNT. The performances were compared fabricated with Mn_3O_4, Mn_3O_4-CB (carbon black), and Mn_3O_4-GO (graphene oxide) prepared by simple wet-chemical approach in alkaline media (pH > 10). These NCs were embedded on a flat GCE; (surface area: 0.0316 cm^2 using binders; (Nafion with 5% ethanolic solution) for a selective hydrazine sensor. This fabricated GCE with Mn_3O_4-CNT NCs was used as chemical sensor to detect hydrazine in phosphate buffer solution using a simple and reliable electrochemical method within short response time. The fabricated chemical sensor showed higher sensitivity (1100.0 μA μM^{-1} cm^{-2}) with a very low LOD of 0.86 nM (signal-to-noise ratio at an SNR of 3). A calibration curve was plotted, which was found to be linear in concentration range of hydrazine (2.0 nM to 2.0 mM). It was opined that this selective chemical sensor has many advantages like reliability, reproducibility, ease of integration, higher stability, lower LOD, and higher selectivity.

Young and Lin [49] proposed a room-temperature CO_2 gas sensor with high sensitivity, superfast response, recovery, and good stability. They used CNTs on low-cost polyimide substrates. A response of this sensor under 50 ppm CO_2 vapor was measured. It was observed that time constants for the turn-on and turn-off transients were 12 and 56 s, respectively. Different bending states of the sensors showed good responses. It was revealed that this flexible sensor is reliable, even in passive recovery mode, without any significant sensitivity loss; thus, making it suitable for low-cost applications.

A carbon dioxide sensor was fabricated by Ong and Grimes [50] through depositing a thin layer of a MWCNT-silicon dioxide (SiO_2) composite on a planar inductor-capacitor resonant circuit. The complex permittivity of the coating material can be determined by just tracking the resonant frequency of the sensor. It was formed that the permittivity of MWCNTs changes linearly in response to CO_2 concentration, and hence, ambient CO_2 levels can be monitored. The passive sensor was remotely monitored with a loop antenna, so that measurements can be made within opaque, and sealed containers. It was shown that the response of the sensor is linear, reversible with no hysteresis between increasing and decreasing CO_2 concentrations, and a response time of about 45 s. An array of three such sensors was used to self-calibrate the measurement for operation in a variable humidity and temperature environment. These are comprised of uncoated, SiO_2 coated, and MWCNT-SiO_2 coated sensors. CO_2 levels can be measured in a variable humidity and temperature environment to ± 3% accuracy using such sensors.

Nanomechanical resonators have excellent capabilities in design of nanosensors for gas detection due to their unparalleled sensitivity resolution, but current challenge is to develop some new designs of the resonators so that they can differentiate distinct gas atoms with a high sensitivity. Arash and Wang [51] investigated the characteristics of impulse wave propagation in CNT-based sensors using molecular dynamics simulations to provide a new method for detection of noble gases. A sensitivity index based on wave velocity shifts in a SWCNT gives efficiency of the nanosensor; this index is induced by surrounding gas atoms. They indicated that the nanosensor can differentiate distinct noble gases at the same environmental temperature and pressure. The inertia and strengthening effects by the gases on wave characteristics of CNTs were discussed, and a continuum mechanics shell model was developed by them to interpret the effects.

An analytical method for the determination of three pesticides such as atrazine, methidathion, and propoxur in complex water samples has been proposed by Al-Degs et al. [52] without the need for chromatographic separation. It was observed that MWCNTs adsorbent showed a perfect extraction/preconcentration of pesticides present even at trace levels. The effect of different operating parameters on pesticides extraction by MWCNTs adsorbent were studied and optimized such as solution pH, sample & eluent volume, and extraction flow rate. The limits of detection 3, 2, and 3 µg L^{-1} and linear ranges 5–30, 3–60, and 5–40 µg L^{-1} were observed for atrazine, methidathion, and propoxur, respectively. A good precision was found for this method as R.S.D. values were always less than 5.0%. The percent recoveries for pesticides in tap water were extended from 95 to 104% and R.S.D. from 1 to 3%, while lower recoveries were observed in reservoir water: 84–93% (R.S.D.: 1–3). It was concluded that present method was found simple and can be operated at lower running costs.

A Tyr/Glu/Fe$_3$O$_4$/Nafion/CNT multilayer modified electrode was employed by Qu et al. [53] for the detection of carbofuran in samples. The configuration of the nanostructure on the electrode provided a favorable environment for the Tyr electrocatalytic characterization and fast electron transfer. Experimental conditions were also optimized by them for the detection of the pesticide. This developed biosensor showed high sensitivity, stability, and reproducibility with low detection limit of 2.0 × 10^{-9}molL^{-1} for carbofuran.

Deo et al. [54] reported an amperometric biosensor for organophosphorus (OP) pesticides, which is based on a CNT-modified transducer and an organophosphorus hydrolase (OPH) biocatalyst. A bilayer approach was used by them for preparing the CNT/OPH biosensor with the OPH layer atop of the CNT film. Here, CNT layer leads to an improved anodic detection of the enzymatically generated p-nitrophenol, with higher sensitivity and stability. The performance this sensor was optimized with respect to the surface modification and operating conditions. This biosensor may be used to detect as low as 0.15 µM paraoxon and 0.8 µM methyl parathion with sensitivities of 25 and 6 nA/µM, respectively at the optimal conditions.

A highly sensitive flow injection amperometric biosensor was developed by Liu and Lin [55] for organophosphate pesticides and nerve agents. This sensor was based on self-assembled acetylcholinesterase (AChE) on a CNT-modified glassy carbon (GC) electrode. AChE is immobilized on the negatively charged surface of CNT by assembling a cationic poly(diallyldimethylammonium chloride) (PDDA) layer and an AChE layer alternatively. TEM confirmed the formation of layer-by-layer nanostructures on carboxyl-functionalized CNTs. It was also indicated that AChE was immobilized successfully on the CNT/PDDA surface. Sandwich-like structure (PDDA/AChE/PDDA) on the CNT surface was unique in nature, and it was formed by self-assembling, which provides a favorable microenvironment to keep the bioactivity of AChE. Electrocatalytic activity of CNT leads to a greatly enhanced electrochemical detection of the enzymatically generated thiocholine product with low oxidation overvoltage (+150 mV), higher sensitivity, and stability. As developed biosensor integrated into a flow injection system was used by them to monitor organophosphate pesticides and nerve agents, like paraoxon. Performance of the sensor, inhibition time and regeneration conditions, were optimized. This biosensor was used to measure low concentration (0.4 pM) of paraoxon with inhibition time of 6 min. It had excellent operational lifetime stability with no appreciable decrease in the activity of enzymes, if the experiments are repeated for more than 20 times within a weak. The developed biosensor system can be used for online monitoring of organophosphate pesticides and nerve agents.

Younusov et al. [56] developed a simple and reliable technique for the construction of an amperometric AChE biosensor, which was based on screen-printed carbon electrodes. They proposed one-step modification using SWCNTs and Co phthalocyanine so as to decrease the working potential and to increase the signal of thiocholine oxidation. As-developed biosensor can detect 5–50 ppb and 2–50 ppb of paraoxon and malaoxon with detection limits of 3 and 2 ppb, respectively, when the incubation time was kept 15 min. It showed high reproducibility in measurements of substrate and inhibitor with R.S.D. about 1% and 2.5%, respectively. Spiked samples of sparkling and tape waters were used to test the reliability of the inhibition measurements.

Rahman et al. [57] prepared simple, new, low cost, and highly sensitive chemical sensing with various carbon materials of Ag_2O–CNT NCs, Ag_2O, Ag_2O-CB, and Ag_2O-GO. They used a simple wet-chemical method in alkaline medium with pH more than 10. These Ag_2O–CNT NCs were deposited on a flat GCE; (Surface area: 0.0316 cm^2) using 5% ethanolic solution of Nafion as binder to fabricate a selective p-nitrophenol (p-NP) sensor. This chemical sensor was used to detect p-NP with short response time in phosphate buffer phase. A simple and reliable electrochemical approach was attempted. As-fabricated Ag_2O–CNT NCs sensor exhibited higher sensitivity (103.89 μA μM^{-1} cm^{-2}) and has a very low detection limit of 0.091 ± 0.002 nM (signal-to-noise ratio of 3) as compared to other three composites Ag_2O, Ag_2O-CB, and Ag_2O-GO. A calibration plot was found to be linear ($r^2 = 0.983$) with the linear concentration range of 1.0 nM–0.01 mM. Proposed chemical sensor has many advantages like higher stability, reproducibility, reliability, selectivity, lower detection limit and ease of integration for toxic p-NP.

A novel potentiometric sensor was developed by Thayyath and Alexander [58] for the determination of lindane (γ-hexachlorocyclohexane; γ-HCCH) with high selectivity and sensitivity. It was based on the modification of γ-HCCH imprinted polymer film onto the surface of Cu electrode. A MWCNT was grafted using glycidyl methacrylate (GMA). The reaction of MWCNT with GMA produced MWCNT-g-GMA while epoxide ring of GMA produced vinylated MWCNT (MWCNT-CH=CH$_2$) on reaction with allylamine. MWCNT based imprinted polymer (MWCNT-MIP) was synthesized using methacrylic acid (MAA), ethylene glycol dimethacrylate (EGDMA), α,α'-azobisisobutyronitrile (AIBN) and γ-HCCH, as monomer, crosslinker, initiator, and template, respectively. The operational parameters were also optimized. The as-prepared sensor can detect lindane in the range 1×10^{-10}–1×10^{-3} M and the detection limit was 1.0×10^{-10} M.

Titanium nitride (TiN) NPs decorated MWCNTs nanocomposite was fabricated by Haldorai et al. [59] via a two-step process. These two steps are:

- Decoration of titanium dioxide NPs onto the MWCNTs surface; and
- Thermal nitridation.

TiN NPs with a mean diameter of ≤ 20 nm are homogeneously dispersed onto the MWCNTs surface as evident by TEM images. Direct electrochemistry and electrocatalysis of cytochrome c immobilized on the MWCNTs–TiN composite modified on a GCE was investigated for sensing of nitrite. The current response was found to be linear to its concentration from 1 μM to 2000 μM with a sensitivity of 121.5 μA μM^{-1} cm^{-2} and a low detection limit of 0.0014 μM under optimum conditions. This electrode shows good reproducibility as well as long-term stability. The applicability of the as-prepared sensor confirmed by using it for detection of nitrite in tap and seawater samples.

9.3 BIOSENSORS

A chemical sensor based on recognition of material of biological nature is called biosensor. A biosensor is an analytical device, which is used for detection of an analyte, that combines a biological component with a physicochemical detector.

Biosensors, in biomedicine and biotechnology, requires biological component, such as cells, protein, antibodies, cell receptors, enzymes, microorganisms, nucleic acids, organelles, tissue, etc., is a biologically derived material or biomimetic component that interacts, binds, or recognizes with the analyte under study.

Electrochemical bio-sensor (ECB) based on carbon material has received a great attention as an inexpensive and simple analytical method with remarkable detection sensitivity for neurochemicals detection. ECBs fabrication and structure have attracted efforts to obtain higher sensitivity and highly selectivity, simultaneously. Traditional film-based ECB have shown some disadvantages such as large sizes, random structure, time-consuming for film fabrication and low efficiency.

Oh et al. [60] fabricated CNT-based biosensor with a FET structure for detection of hepatitis B. A microfluidic channel was mounted on the device and the presence of hepatitis B antigen was detected by determining the electrical conductance as a function of time. It was observed that when hepatitis B antigen was exposed to the CNT biosensor with hepatitis B antibody immobilized, the conductance first increased and then it became almost constant within 10 min. The conductance increased further on increasing antigen concentration. It was demonstrated that hepatitis B may be possibly detected in real-time using such CNT biosensor.

Kim et al. [61] reported a method to build ultrasensitive CNT-based biosensors using immune binding reaction. They functionalized carbon nanotube–field effect transistors (CNT–FETs) with antibody binding fragments as a receptor. Binding event of target immunoglobulin G (IgG) onto the fragments was also detected by monitoring the gating effect, which is caused by the charges of the target IgG. As biosensors were used in buffer solution, it was difficult to use small-size receptors so that the charged target IgG could reach CNT surface within the Debye length distance to have a large gating effect. They showed that CNT–FET biosensors using whole antibody had very low sensitivity (detection limit $\gg 1000$ ngmL^{-1}), but those based on small Fab fragments could detect 1 pgmL^{-1}($\gg 7$ fM level). This

Fab-modified CNT–FET could successfully block the nontarget proteins and could detect the target protein selectively in the environment, which is very much similar to that of human serum electrolyte. This strategy can be applied to general antibody-based detection schemes also, and it may provide the production of label-free ultrasensitive electronic biosensors to detect clinically important biomarkers for disease diagnosis.

Badhulika and Mulchandani [62] reported the synthesis and fabrication of an enzyme-free sugar sensor based on molecularly imprinted polymer (MIP) on the surface of SWCNTs. Electropolymerization of 3-amino-phenyl boronic acid (3-APBA) in the presence of D-fructose and fluoride at neutral pH conditions resulted in the formation of a self-doped, molecularly imprinted conducting polymer (MICP). It formed a stable anionic boronic ester complex between poly(aniline boronic acid) and D-fructose. Template removal generated binding sites on the polymer matrix were complementary to D-fructose in structure, shape, size, and positioning of functional groups; thus, enabling sensing of D-fructose with high affinity and specificity over non-MIP based sensors. Using CNTs along with MICPs assisted in developing an efficient electrochemical sensor by increasing analyte recognition and signal generation. Such sensors could be regenerated and used number of times whereas conventional affinity-based biosensors suffer from physical as well as chemical stability.

SWCNTs were selectively wrapped with a water-soluble, eco-friendly, and biocompatible polymer chitosan (CHI) by Zhou et al. [63]. They employed it for the construction of a bioelectrochemical platform for the direct electron transfer (DET) of glucose oxidase (GOD) and biosensing purposes. It was observed that the wrapped small-diameter SWCNTs were dispersed within the CHI film and existed on the surface of the electrode as small bundles. The heterogeneous electron transfer rate constant and the surface coverage of GOD were determined as 3.0 s^{-1} and 1.3×10^{-10} mol cm^{-2}, respectively. Such immobilized GOD retained its catalytic activity towards the oxidation of glucose. It not only exhibited a rapid response time, wide linear range and low detection limits at a detection potential of −400 mV but also the effective anti-interference capability.

Biosensors having novel modified electrode structures for glucose determination was developed by Ghica and Brett [64] using different combinations of MWCNTs and polyazine redox polymer, poly(neutral red) or poly(brilliant cresyl blue) on GCEs. CNT films were first formed using functionalized CNTs, covalently immobilized by crosslinking in a CHI matrix, and then azine dyes were electropolymerized directly on GCE or on top of the GCE modified with CNTs. Glucose oxidase (GOx) was immobilized by crosslinking with glutaraldehyde on top of the GCE modified with CNT, poly(azines), or combinations of poly(azines) and CNT. They also studied an assembly with enzyme/nanotube mixture immobilized on top of polyazine films. The mechanism was investigated in the presence as well as absence of oxygen and also by using bienzymatic devices containing glucose oxidase and catalase. They obtained the best performance with a PBCB/CNT/GOx biosensor at a potential of −0.3V vs. SCE with a detection limit of 11 μmolL^{-1}.

Cella et al. [65] reported SWCNT-based chemiresistive affinity sensors. These sensors are useful in detection of small and/or weakly charged or uncharged molecules. The detection of glucose (a small and weakly charged molecule) by displacement of plant lectin (concavalin A) bound to a polysaccharide (dextran) immobilized on SWCNTs can be done with picomolar sensitivity and selectivity over other sugars and human serum proteins.

Bareket et al. [66] reported development of an electrochemical biosensor for the detection of formaldehyde in aqueous solution. This sensor is based on the coupling of the enzyme formaldehyde dehydroge-nase and CNTs-modified screen-printed electrode (SPE). Amperometric response to formaldehyde released from U251 human glioblastoma cells was monitored in response to treatment with different anticancer prodrugs of formaldehyde and butyric acid. This current response was relatively higher for prodrugs that release two formaldehyde molecules (AN–193) than those prodrugs that release only one formaldehyde molecule (AN–1, AN–7). However, homologous prodrugs that release one (AN–88) or two (AN–191) molecules of acetaldehyde showed no signal. The sensor was found to be rapid, sensitive, selective, inexpensive, and disposable, as well as simple to manufacture and operate for detecting formaldehyde.

Zhu et al. [67] designed a novel brush-like electrode based on CNT nanoyarn fiber for electrochemical biosensor applications. Its efficiently was tested as an enzymatic glucose biosensor. This fiber was spun directly from a chemical-vapor-deposition (CVD) gas flow reaction, where a mixture of ethanol and acetone was used as the carbon source and an iron nanocatalyst. The fiber was made of bundles of double-walled CNTs (DWNTs) (28 μm in diameter) concentrically compacted into multiple layers forming a nanoporous network structure. It showed a superior electrocatalytic activity for CNT fiber as compared to traditional Pt–Ir coil electrode. The electrode end tip of the CNT fiber was freeze-fractured, where a unique brush-like nano-structure was obtained resembling a scale-down electrical 'flex.' Glucose oxidase (GOx) enzyme was immobilized using glutaraldehyde crosslinking in the presence of bovine serum albumin (BSA), and an outer epoxy-polyurethane (EPU) layer was used as semipermeable membrane. The sensitivity, linear detection range, and linearity for detecting glucose using this miniature CNT fiber electrode were found to be better than Pt–Ir coil electrode. Thermal annealing of the CNT fiber at 250°C for 30 min prior to fabrication of the sensor resulted in about 7.5 fold increase in its sensitivity for glucose. As-spun CNT fiber based glucose biosensor was observed to be stable for 70 days. The glucose detection limit was evaluated to 25 µM on gold coating of the electrode connecting end of the CNT fiber.

Functionalized MWCNTs were casted on GC and carbon film electrodes (CFE) by Carla et al. [68]. These were characterized electrochemically and applied in a glucose-oxidase-based biosensor. Then, MWCNT-modified CFE were used to develop an alcohol oxidase (AlcOx) biosensor, in which AlcOx–BSA was cross-linked with glutaraldehyde and attached by drop-coating. They optimized applied potential and pH for ethanol monitoring. Ethanol was determined amperometrically at −0.3 V vs. SCE at pH 7.5. The sensitivity of sensor was 20 times higher with MWCNT than without MWCNT. Electrocatalytic effects of MWCNT were also observed and compared with unmodified CFE.

Disposable field effect transistors (FETs) biosensors (bio-FET) based on CNTs were fabricated by Ines et al. [69] for detection of domoic acid (DA). This acid belongs to the group of biotoxins associated with the amnesic shellfish poisoning. The analytical results obtained with this bio-FET were compared with results obtained with a traditional methodology (enzyme-linked immunosorbent assay (ELISA)). A calibration curve was constructed using standard solutions of DA with concentrations between 10 and 500 ng L^{-1}. Five bio-FET were tested for reproducibility while two measurements were performed for each bio-FET for repeatability estimation. Ten spiked artificial seawater samples were used to validate this bio-sensor. The results for reproducibility (0.52–1.43%), repeatability (0.57–1.27%), LOD (10 ng L^{-1}) and recovery range (92.3–100.3%) proved that the use of bio-FET for the detection of DA is adequate in environmental samples such as seawater samples.

Mandal et al. [70] reported the label-free, sensitive, and real-time electrical detection of whole viruses using carbon nanotube thin film (CNT-TF) field effect devices. Selective detection of about 550 model viruses, M13-bacteriophage was done using a simple two-terminal (no gate electrode) configuration. It was proposed that chemical gating through specific antibody-virus binding on CNT surface plays the role in sensing mechanism. CNT-TF sensors were found to exhibit sensitivity of about five times higher than electrical impedance sensors with identical microelectrode dimensions (without CNT). This approach could lead to a reproducible and cost-effective solution for rapid viral identification.

Deng et al. [71] developed CNT-based biosensors for fast detection of *Escherichia coli* and *Staphylococcus aureus* in water. They observed the adsorption kinetics and equilibrium of pure and mixed culture of *E. coli* and *S. aureus* on the CNT aggregates at ambient temperature and different culture concentrations. It was observed that CNT aggregates could absorb significant amounts of *E. coli* and *S. aureus* bacteria with different size and shape characteristics. The smaller size *S. aureus* has a five to ten times faster diffusion rate than *E. coli* and about hundred times higher adsorption affinity with the CNT aggregates. It may be due to the fact that CNT aggregates have separate adsorption sites for *E. coli* and *S. aureus*. High adsorption affinity and fast adsorption kinetics for *S. aureus* suggest that *S. aureus* and *E. coli* in water can be selectively differentiated even unmodified SWCNTs.

An impedimetric biosensor has been developed by Zhoua and Ramasamy [72] for rapid and selective detection of bacteria using T2 bacteriophage (virus) as the bio-recognition element on CNT modified SPE. Bacteriophage was immobilization on SPE via simultaneous covalent bonding and electrostatic interaction. The phage-modified electrodes were used for the detecting bacterial strain *E. coli B*. Electrochemical impedance spectroscopy was used for detecting variation of impedance of the electrodes resulting from the binding of *E. coli B* to T2 phage. It was interesting to note that detection was highly selective towards the B strain of *E. coli* as no signal was observed for its K strain. A detection limit of 50 CFU mL^{-1} was achieved using this biosensor.

Abdalhai et al. [73] used an electrochemical genosensor for detection of *E. coli* O157:H7. Capture and signalizing probes were modified by thiol (SH) and amine (NH$_2$), respectively, and it was connected using cadmium sulfide nanoparticles (CdSNPs). The genosensor was prepared by immobilizing complementary DNA on the gold electrode surface, which hybridizes with a specific fragment gene from pathogenic to form a sandwich-like structure. It was observed that conductivity and sensitivity of the sensor were increased by using MWCNTs modified with CHI deposited as a thin layer on the glass carbon electrode (GCE) surface, followed by a deposit of bismuth. The detection limit was observed as 1.97×10^{-14} M, and the correlation coefficient was 0.989. High sensitivity and selectivity of the electrochemical DNA biosensor to the pathogenic bacteria *E. coli* O157:H7 was also reported. This biosensor was used to detect pathogen in real beef samples, which are artificially contaminated.

CNT-based immunosensors for the detection of two types of microorganisms, bacteria, and viruses have been reported by García-Aljaroa et al. [74] *Escherichia coli* O157:H7 and the bacteriophage T7 were selected as model for bacteria and viruses, respectively. They used *E. coli* K12 and the bacteriophage MS2 to assess the selectivity of the biosensor. The transduction element consisting of SWCNTs aligned in parallel bridging two gold electrodes was used to function as a chemiresistive biosensor. SWCNTs were first functionalized with specific antibodies (Ab) for the different microorganisms by covalent immobilization to the non-covalently bound 1-pyrene butanoic acid succinimidyl ester. When this biosensor was exposed to *E. coli* O157:H7 whole cells or lysates, there was a significant increase in the resistance of the device. The detection of 105 and 103 CFU mL^{-1}corresponding to 103 and 101 CFU chip^{-1}, respectively were observed, but there was no response on exposing biosensor to *E. coli* K12. A significant resistance enhancement was observed due to interaction of the bacteriophages with the Abs, with a LOD of 103 PFU mL^{-1} corresponding to 101 PFU chip^{-1} and excellent selectivity against MS2 bacteriophage. It was observed that sensor exhibited a fast response time of ~5 min for bacteriophage detection, while it was 60 min for the detection of bacteria.

Villamizar et al. [75] reported a fast, sensitive, and label-free biosensor for the selective determination of *Salmonella Infantis*. This sensor is based on a FET, where a network of single-walled carbon nanotubes (SWCNTs) acts as the conductor channel. Anti-*Salmonella* antibodies were adsorbed onto the SWCNTs and these, SWCNTs were protected with tween 20 to prevent the non-specific binding of any other bacteria or proteins. This FET devices were able to detect *S. Infantis* at least 100 CFU/mL^{-1} in 1 h. *Streptococcus pyogenes* and *Shigella sonnei* were tested as potential competing bacteria for *Salmonella* just to evaluate the selectivity of FET devices. *Streptococcus*or *Shigella* did not interfere with the detection of

Salmonella at concentration of 500 CFU mL^{-1}. Thus, sensors can be used to detect *S. Infantis*.

A disposable electrochemical immunosensor for the simultaneous measurements of common food pathogenic bacteria namely *Escherichia coli* O157:H7 (*E. coli*), *campylobacter*, and *salmonella* were developed by Viswanathan et al. [76]. It was fabricated by immobilizing the mixture of anti-*E. coli*, anti-*campylobacter*, and anti-*salmonella* antibodies with a ratio of 1:1:1 on the surface of the MWCNT-polyallylamine modified screen-printed electrode (MWCNT-PAH/SPE). When the immunosensor was incubated in liquid samples, bacteria suspension became attached to the immobilized antibodies. The sandwich immunoassay was performed with three antibodies conjugated with specific nanocrystal (*E. coli*-CdS, -*campylobacter*-PbS, and -*salmonella*-CuS), which can release metal ions for electrochemical measurements. Square wave anodic stripping voltammetry (SWASV) technique was used to determine released metal ions from bound antibody nanocrystal conjugates. The calibration curves for their three selected bacteria were found with the LOD as 400, 400 and 800 cells mL^{-1} for *salmonella*, *campylobacter*, and *E. coli*, respectively. The precision and sensitivity of this method show that it is feasible to determine bacteria in milk samples.

High specific anti-*Salmonella* and anti-*Staphylococcus aureus* antibodies can be immobilized on hydrophobic and hydrophilic nanodiamond, and carbon-nanotube coated silicon substrates. The efficacy of antibody immobilization was evaluated by Huang et al. [77] through ELISA, while bacterial binding efficiency was analyzed by SEM. It was observed that immobilization efficacy of both; antibodies and bacterial binding efficiency on air plasma treated nanodiamond were found to be better than hydrogen plasma treated. Significant differences were obtained for antibody immobilization and *S. aureus* binding on the surfaces of hydrophobic and hydrophilic CNTs. It was revealed that bacteria binding ability on hydrophilic CNTs was much higher than hydrophobic CNTs.

Integration of SWCNTs and immobilized antibodies into a disposable bio-nano combinatorial junction sensor was fabricated by Kara et al. [78] for detection of *Escherichia coli* K–12. Gold tungsten wires (50 µm diameter) coated with polyethyleneimine (PEI) and SWCNTs were aligned to form a crossbar junction, which was functionalized with streptavidin and biotinylated antibodies to allow for enhanced specificity towards targeted microbes. The performance of sensor was evaluated by changes in electrical current after bioaffinity reactions between bacterial cells (*E. coli* K–12) and antibodies on the SWCNT surface. The average increased from 33.13 to 290.9 nA in the presence of SWCNTs in a 10^8 CFU mL^{-1} concentration of *E. coli*, which showed an improvement in magnitude of sensing. A linear relationship (R^2 = 0.973) was obtained between the changes in current and concentrations of bacterial suspension in range of 10^2–10^5 CFU mL^{-1}. It was found that current decreased on increased cell concentrations, which may be due to increased bacterial resistance on the bio-nanomodified surface. The detection limit of this sensor was 10^2 CFU mL^{-1} with a detection time of less than 5 min with nanotubes.

Choi et al. [79] developed a SWCNT based biosensor to detect *Staphylococcus aureus*. The specificity of eleven bacteria and polyclonal anti-*Staphylococcus aureus* antibodies (pAbs) was determined using an indirect ELISA. These pAbs were immobilized onto sensor platform after the hybridizing 1-pyrenebutanoic acid succinimidyl ester (PBASE). The optimum concentration of SWCNTs on this platform was determined to be 0.1 mg mL^{-1}. The binding of pAbs with *S. aureus* resulted in a significant increase in value of resistance of the biosensor ($p < 0.05$). The specific binding of *S. aureus* on the biosensor was confirmed by the SEM images. The SWCNT-based biosensor was able to detect *S. aureus* with a LOD of 4 log CFU mL^{-1}.

Nanoscale electronic devices can be used to detect any change in electrical properties, when receptor proteins bind to their corresponding antibodies functionalized on the surface of the device, in extracts from as few as ten lysed tumor cells. Protein arrays that measure multiple protein cancer biomarkers in clinical samples hold great promise for reliable early cancer detection. CNT-based drug delivery holds great promise for cancer therapy.

Titania NPs and CNTs are ordinarily used to modify the electrodes to enhance their detection sensitivity of biomolecular recognition. Shen et al. [80] prepared novel TiO_2/CNT NCs and doped on the carbon paper as the modified electrodes. Then, the redox behavior of the ferricyanide probe and the surface properties of the cancer cells coated on the modified electrodes were investigated using electrochemical and contact angle measurements. Enhanced electrochemical signals on the modified electrodes covered with cancer cells were observed as compared to bare carbon paper. Different leukemia cells (i.e., K562/ADM cells and K562/B.W. cells) could be also recognized by this sensor because of different electrochemical behavior and hydrophilic/hydrophobic features on the modified electrodes. It is all due to the specific components on the plasma membranes of the target cells. This new strategy may find promising in development of biocompatible and multi-signal responsive biosensors for the early diagnosis of cancers.

Shao et al. [81] reported a single nanotube FET array, functionalized with IGF1R-specific and Her2-specific antibodies. It exhibits highly sensitive and selective sensing of live, intact MCF7 and BT474 human breast cancer cells in blood. Single or small bundle of such nanotube devices functionalized with IGF1R-specific or Her2-specific antibodies showed 60% decrease in conductivity upon interaction with BT474 or MCF7 breast cancer cells in only two μL drops of blood. Control experiments were also carried out with non-specific antibodies or with MCF10A control breast cells, which showed less than 5% decrease in electrical conductivity. It was postulated that the free energy change due to multiple simultaneous cell–antibody binding events exerted stress along the nanotube surface, and as a result, its electrical conductivity is decreased due to an increase in band gap. Free energy change upon cell–antibody binding, stress exerted on the nanotube, and change in conductivity are quite specific to a specific antigen-antibody interaction, and therefore, these properties may be used as a fingerprint for the molecular sensing of circulating cancer cells. The binding of a single cell to a single nanotube FET produced changes in electrical conductivity. A nanoscale oncometer with single cell sensitivity has been reported with a thousand times smaller diameter than a cancer cell that functions in a drop of fresh blood.

A prototype 4-unit electrochemical immunoarray based on SWCNT forests has been reported by Chikkaveeraiah et al. [82] for the simultaneous detection of multiple protein biomarkers for prostate cancer. They designed immunoarray procedures to measure prostate-specific antigen (PSA), prostate specific membrane antigen (PSMA), platelet factor–4 (PF–4), and interleukin–6 (IL–6) simultaneously in a single serum sample. All of these proteins are known to be elevated in serum of patients having prostate cancer. Horseradish peroxidase (HRP) was used as label on detection (secondary) antibodies in a sandwich immunoassay scheme. Biotinylated secondary antibodies (Ab2) binding specifically to streptavidin−HRP conjugates provided 14−16 labels per antibody, which gave the necessary higher sensitivity required for PF–4 and IL–6 detection at physiological levels. Conventional singly labeled Ab2−HRP conjugates were enough for PSA and PSMA detection. Immunoarrays were used to measure four biomarkers in human serum samples of prostate cancer patients controls and showed excellent correlation to referee enzyme-linked immunosorbent (ELISA) assays.

Bhirde et al. [83] reported targeted, *in vivo* killing of cancer cells using a drug-SWCNT bioconjugate. They also demonstrated superior efficacy to nontargeted bioconjugates. Anticancer agent cisplatin and epidermal growth factor (EGF) were attached to SWCNTs to specifically target squamous cancer, while nontargeted control was SWCNT-cisplatin without EGF. Initial *in vitro* imaging studies with head and neck squamous carcinoma cells (HNSCC) overexpressing EGF receptors (EGFR) using Qdot luminescence and confocal microscopy indicated that SWCNT-Qdot-EGF bioconjugates internalized rapidly into the cancer cells. There was limited uptake for control cells without EGF as this uptake was blocked by siRNA knockdown of EGFR in cancer cells. It reveals the importance of EGF-EGFR binding. Three colors, two-photon intravital video imaging *in vivo* showed that SWCNT-Qdot-EGF injected into live mice was selectively taken up by HNSCC tumors, but SWCNT-Qdot controls with no EGF were cleared from the tumor region within less than 20 min. It was observed that HNSCC cells treated with SWCNT−cisplatin−EGF were killed selectively, but control systems not featuring EGF-EGFR binding did not influence cell proliferation. Regression of tumor growth was rapid in mice treated with targeted SWCNT−cisplatin−EGF as compared to nontargeted SWCNT-cisplatin.

A novel SWCNT-based tumor-targeted drug delivery system (DDS) was developed by Chen et al. [84] It consisted of a functionalized SWCNT linked to tumor-targeting modules as well as prodrug modules. Three key features are there of this nanoscale DDS:

- Use of functionalized SWCNTs as a biocompatible platform for the delivery of therapeutic drugs or diagnostics;
- Conjugation of prodrug modules of an anticancer agent (taxoid with a cleavable linker) that is activated to its cytotoxic form inside the tumor cells upon internalization and *in situ* drug release; and
- Attachment of tumor-recognition modules (biotin and a spacer) to the nanotube surface.

Three fluorescents and fluorogenic molecular probes were designed, synthesized, and characterized, to prove the efficacy of this DDS. These were subjected to the analysis of the receptor-mediated endocytosis and drug release inside the cancer cells (L1210FR leukemia cell line). The specificity and cytotoxicity of the conjugate have also been assessed and compared with L1210 and human noncancerous cell lines. It was observed that this shows high potency toward specific cancer cell lines.

A novel vertically aligned CNT based electrical cell impedance sensing biosensor (CNT-ECIS) was reported by Abdolahad et al. [85]. It was found to be more rapid, sensitive, and specific device for the detection of cancer cells. It is based on the fact that fast entrapment of cancer cells on vertically aligned CNT arrays leads to some mechanical and electrical interactions between CNT tips and entrapped cell membranes. As a result, the impedance of the biosensor changed. CNT-ECIS was fabricated through a photolithography process on $Ni/SiO_2/Si$ layers. CNTA grown on 9 nm thick patterned Ni microelectrodes by DC-PECVD. SW48 colon cancer cells were passed on to the surface of CNT covered electrodes. CNTA act as both; adhesive and conductive agents. Impedance changed as fast as 30 s (for whole entrapment and signaling processes). CNT-ECIS was able to detect the cancer cells with the concentration as low as 4000 cells cm^{-2} on its surface with a sensitivity of 1.7×10^{-3} Ω cm^2. Time and cell efficiency factor (TEF and CEF) of this sensor were much higher as compared to other biosensors.

An anticancer drug screening device for rapid electrochemical detection of cyclophosphamide was developed by Wang et al. [86] This deoxyribonucleic acid (DNA) biosensor was based on a CNT modified electrode. CNT-based biosensor could detect interaction of cyclophosphamide with double-stranded (ds) calf thymus DNA, and it was compared with the carbon paste based biosensor using DPV. It was indicated that the developed CNT-based DNA biosensor showed a faster response and higher detection reproducibility as compared to the carbon paste based biosensor.

Gomes et al. [87] investigated the influence of functionalized MWCNT in the presence of an ionic liquid 1-butyl–3-methylimidazolium hexafluorophosphate ($[BMIM]PF_6$) in different ratios on the acetaminophen (ACOP) electrochemical behavior. It exhibited a pair of well-defined redox peaks, which indicates an improvement in the reversibility of ACOP to irreversible oxidation peak on GCE. This type of redox process was controlled by adsorption, involving two electrons. The value of apparent rate constant (k_s) was determined as 14.7 ± 3.6 s^{-1}. The analytical curves were obtained for different concentrations of ACOP from 0.3 to 3.0 µmol L^{-1} and values of the detection limit were found to be 6.73×10^{-8} mol L^{-1}. This electrochemical sensor exhibited good stability and reproducibility, and it was applied for ACOP determination in some tablets giving satisfactory results.

CNT membranes were employed as the active element of a switchable transdermal drug delivery device that can facilitate more effective treatments of drug abuse and addiction. Due to the dramatically fast flow through CNT cores, high charge density, and small pore dimensions, highly efficient electrophoretic pumping through functionalized CNT membrane was achieved.

A simple and sensitive method for the detection of a prostate cancer marker (PSA-ACT complex) have been reported by Kim et al. [88] through label-free protein biosensors based on a CNT-FET. They functionalized CNT-FET with a solution containing various linker-to-spacer ratios and the binding of the target PSA-ACT complex onto the receptor was detected by monitoring the gating effect, which was caused by charges in the target complex. The results showed that CNT-FET biosensors modified with only linkers could not detect target proteins, till high concentration of the PSA-ACT complex solution ($\sim$500 ng mL^{-1}) was injected, but modified sensor with a 1:3 ratio of linker-to-spacer could detect even 1.0 ngmL^{-1} without any pretreatment. These linker and spacer-modified CNT-FET could successfully block non-target proteins and selectively detect the target protein in human serum.

Although CNTs are utilized in different fields, but in last few decades, a new feather has been added to their caps. Nowadays, CNTs are successfully utilized for detecting very little amount of some chemical vapors, like ammonia, alcohol, aldehyde, moisture, etc. Apart from this, it is easy to detect a minute amount of pesticides. Some biosensors have been developed to detect the presence of various pathogenic bacteria as well as early diagnosis of different types of cancers. CNTs will give a newer dimension in the technology in the field of chemical sensors and biosensors.

KEYWORDS

- biosensors
- cancer marker
- carbon nanotubes
- chemical sensors
- MWCNT
- SWCNT

REFERENCES

1. Parikh, K., Cattanach, K., Rao, R., Suh, D. S., Wu, A., & Manohar, S. K., (2006). Flexible vapor sensors using single walled carbon nanotubes. *Sens. Actuators B, Chem.*, *113*(1), 55–63.
2. Mitsubayashi, K., Wakabayashi, Y., Murotomi, D., Yamada, T., Kawase, T., Iwagaki, S., & Karube, I., (2003). Wearable and flexible oxygen sensor for transcutaneous oxygen monitoring. *Sens. Actuators B, Chem.*, *95*(1–3), 373–377.
3. Shih, W. S., Young, S. J., Ji, L. W., Water, W., Meen, T. H., & Shiu, H. W., (2011). Effect of oxygen plasma treatment on characteristics of TiO_2 photodetectors. *IEEE Sensors, J.*, *11*(11), 3031–3035.
4. Young, S. J., (2014). Photoconductive gain and noise properties of ZnO nanorods Schottky barrier photodiodes. *IEEE, J. Sel. Topics Quantum Electron*, *20*(6), 1–4, doi: 10.1109/JSTQE.2014.2316599.
5. Rout, C. S., Hegde, M., & Rao, C. N. R., (2008). H_2S sensors based on tungsten oxide nanostructures. *Sens. Actuators B, Chem.*, *128*(2), 488–493.
6. Kong, J., Franklin, N. R., Zhou, C., Chapline, M. G., Peng, S., Cho, K., & Dai, H., (2000). Nanotube molecular wires as chemical sensors. *Sci. Mag.*, *287*, 622–625.
7. Modi, A., Koratkar, N., Lass, E., Wei, B., & Ajayan, P. M., (2003). Miniaturized gas ionization sensors using carbon nanotubes. *Nature*, *424*, 171–174.
8. Joshi, K. A., Prouza, M., Kum, M., Wang, J., Tang, J., Haddon, R., Chen, W., & Mulchandani, A., (2006). V-type nerve agent detection using a carbon nanotube-based amperometric enzyme electrode. *Anal. Chem.*, *78b*(1), 331–336.
9. Ren, X., Chen, C., Nagatsu, M., & Wang, X., (2011). Carbon nanotubes as adsorbents in environmental pollution management: A review. *Chem. Eng, J.*, *170*(2 & 3), 395–410.
10. Sun, S. J., (2008). Gas adsorption on a single walled carbon nanotube-model simulation. *Phys. Lett, A.*, *372*(19), 3493–3495.
11. Gatica, S. M., Bojan, M. J., Stan, G., & Cole, M. W., (2001). Quasi-one- and two dimensional transitions of gases adsorbed on nanotube bundles. *J. Chem. Phys.*, *114*(8), 3765–3769.
12. Agnihotri, S., Mota, J. P. B., Rostamabadi, M., & Rood, M. J., (2005). Structural characterization of single-walled carbon nanotube bundles by experiment and molecular simulation. *Langmuir ACS, J. Surfaces Colloids*, *21*(3), 896–904.
13. Calbi, M. M., Cole, M. W., Gatica, S. M., Bojan, M. J., & Stan, G., (2001). Condensed phases of gases inside nanotube bundles. *Rev. Mod. Phys.*, *73*(4), 857–865.
14. Talapatra, S., Zambano, A. Z., Weber, S. E., & Migone, A. D., (2000). Gases do not adsorb on the interstitial channels of closed-ended single-walled carbon nanotube bundles. *Phys. Rev. Lett.*, *85*(1), 138–141.
15. Fujiwara, A., Ishii, K., Suematsu, H., Kataura, H., Maniwa, Y., Suzuki, S., & Achiba, Y., (2001). Gas adsorption in the inside and outside of single-walled carbon nanotubes. *Chem. Phys. Lett.*, *336*(3 & 4), 205–211.
16. Muris, M., Pavlovsky, N. D., Bienfait, M., & Zeppenfeld, P., (2001). Where are the molecules adsorbed on single-walled nanotubes? *Surface Science*, *492*(1 & 2), 67–74.
17. Byl, O., Kondratyuk, P., Forth, S. T., Fitz, G. S. A., Chen, L., Johnson, J. K., & Yates, J. T., (2003). Adsorption of CF4 on the internal and external surfaces of opened

single-walled carbon nanotubes: A vibrational spectroscopy study. *J. Am. Chem. Soc.*, *125*(19), 5889–5896.

18. Kondratyuk, P., Wang, Y., Johnson, J. K., & Yates, J. T., (2005). Observation of a one-dimensional adsorption site on carbon nanotubes: Adsorption of alkanes of different molecular lengths. *J. Phys. Chem. B.*, *109*(44), 20999–21005.

19. Snow, E. S., Perkins, F. K., Houser, E. J., Badescu, S. C., & Reinecke, T. L., (2005). Chemical detection with a single-walled carbon nanotube capacitor. *Science*, *307*(5717), 1942–1945.

20. Li, X., Chen, X., Ding, X., & Zhao, X., (2018). High-sensitive humidity sensor based on graphene oxide with evenly dispersed multiwalled carbon nanotubes. *Mater. Chem. Phys.*, *207*, 135–140.

21. Sheng, J., Zeng, X., Zhu, Q., Yang, Z., & Zhang, X., (2014). Facile fabrication of CNT-based chemical sensor operating at room temperature. *Mater. Res. Express.*, 12, doi: 10.1088/2053–1591/aa9ac7.

22. Lupan, O., Schütt, F., Postica, V., Smazna, D., Mishra, Y. K., & Adelung, R., (2017). Sensing performances of pure and hybridized carbon nanotubes-ZnO nanowire networks: A detailed study. *Scientific Rep.*, 7, doi: 10.1038/s41598–017–14544–0.

23. Liu, S. F., Petty, A. R., Sazama, D. G. T., & Swager, P. T. M., (2015). Single-walled carbon nanotube/metalloporphyrin composites for the chemiresistive detection of amines and meat spoilage. *Angew. Chemie. Int. Ed.*, *54*(22), 6554–6557.

24. Li, J., Lu, Y., Ye, Q., Cinke, M., Han, J., & Meyyappan, M., (2003). Carbon nanotube sensors for gas and organic vapor detection. *Nano Letters.*, *3*(7), 929–933.

25. Jeon, M., Choi, B., Yoon, J., Kim, D. M., Kim, D. H., Park, I., & Choi, S. J., (2017). Enhanced sensing of gas molecules by a 99.9% semiconducting carbon nanotube-based field-effect transistor sensor. *Appl. Phys. Lett.*, *111*, doi: 10.1063/1.4991970.

26. Kaniyoor, A., Jafri, R. I., Arockiadoss, T., & Ramaprabhu, S., (2009). Nanostructured Pt decorated graphene and multi walled carbon nanotube based room temperature hydrogen gas sensor. *Nanoscale*, *1*(3), 382–386.

27. Jung, H., Ahn, E., Le Hung, N., Oh, D., Kim, H., & Kim, D., (2009). Effects of CO doping on NO gas sensing characteristics of ZnO-carbon nanotube composites. *Korean, J. Mater. Res.*, *19*(11), 607–612.

28. Hashishin, T., & Tamaki, J., (2009). Chemical modification of carbon nanorubes for NO_2 detection. *Sensors Mater.*, *21*(5), 265–280.

29. Esser, D. B., Schnorr, J. M., & Swager, P. T. M., (2012). Selective detection of ethylene gas using carbon nanotube-based devices: Utility in determination of fruit ripeness. *Angewandte. Chemie.*, *51*(23), 5752–5756.

30. Ishihara, S., Labuta, J., Nakanishi, T., Tanaka, T., & Kataura, H., (2017). Amperometric detection of sub-ppm formaldehyde using single-walled carbon nanotubes and hydroxylamines: A referenced chemiresistive system. *ACS Sens.*, *2*(10), 1405–1409.

31. Neumann, P. L., Obreczán, V. I., Dobrik, G., Kertész, K., Horváth, E., Lukács, I. E., Biro, L. P., & Horvath, Z. E., (2014). Different sensing mechanisms in single wire and mat carbon nanotubes chemical sensors. *Appl. Physics, A.*, *117*(4), 2107–2113.

32. Brahim, S., Colbern, S., Gump, R., Moser, A., & Grigorian, L., (2009). Carbon nanotube-based ethanol sensors. *Nanotechnology*, *20*(23), doi: 10.1088/0957–4484/20/23/235502.

33. Penza, M., Cassano, G., Aversa, P., Antolini, F., Cusano, A., Cutolo, A., Giordano, M., & Nicolais, L., (2004). Alcohol detection using carbon nanotubes acoustic and optical sensors. *Appl. Phys. Lett.*, *85*, doi: 10.1063/1.1784872.

34. Choi, M. S., Bang, J. H., Mirzaei, A., Na, H. G., Kwon, Y. J., Kang, S. Y., et al., (2018). Dual sensitization of MWCNTs by co-decoration with p- and n-type metal oxide nanoparticles. *Sensors Actuators B: Chem.*, *264*, 150–163.

35. Chou, T. C., Wu, K. Y., Hsu, F. X., & Lee, C. K., (2018). Pt-MWCNT modified carbon electrode strip for rapid and quantitative detection of H_2O_2 in food. *J. Food Drug Anal.*, *26*(2), 662–669.

36. Wang, T., Zhu, H., Zhuo, J., Zhu, Z., Papakonstantinou, P., Lubarsky, G., Lin, J., & Li, M., (2013). Biosensor based on ultrasmall MoS_2 nanoparticles for electrochemical detection of H_2O_2 released by cells at the nanomolar level. *Anal. Chem.*, *85*(21), 10289–10295.

37. Thirumalai, D., Subramani, D., Yoon, J. H., Lee, J., Paik, H. J., & Chang, S. C., (2018). De-bundled single-walled carbon nanotube-modified sensors for simultaneous differential pulse voltammetric determination of ascorbic acid, dopamine, and uric acid. *New, J. Chem.*, *42*, 2432–2438.

38. Kumar, S. A., Wang, S. F., Yang, C. K., & Yeh, C. T., (2010). Acid yellow 9 as a dispersing agent for carbon nanotubes: Preparation of redox polymer-carbon nanotube composite film and its sensing application towards ascorbic acid and dopamine. *Biosensors Bioelectron.*, *15*(12), 2592–2597.

39. Zhang, Y., Bunes, B. R., Wu, N., Ansari, A., Rajabali, S., & Zang, L., (2018). Sensing methamphetamine with chemiresistive sensors based on polythiophene-blended single-walled carbon nanotubes. *Sensors Actuators B: Chem.*, *255*(2), 1814–1818.

40. Chen, T., Wei, L., Zhou, Z., Shi, D., Wang, J., Zhao, J., Yu, Y., Wang, T., & Zhang, Y., (2012). Highly enhanced gas sensing in single-walled carbon nanotube-based thin-film transistor sensors by ultraviolet light irradiation. *Nanoscale Res. Lett.*, *7*, doi: 10.1186/1556–276X–7–644.

41. Tarlekar, P., Khan, A., & Chatterjee, S., (2018). Nanoscale determination of antiviral drug acyclovir engaging bifunctionality of single-walled carbon nanotubes–Nafion film. *J. Pharma. Biomed. Anal.*, *151*, 1–9.

42. Üğe, A., Zeybek, D. K., & Zeybek, B., (2018). An electrochemical sensor for sensitive detection of dopamine based on MWCNTs/CeO_2-PEDOT composite. *J. Electroanal. Chem.*, *813*, 134–142.

43. Yuan, C. L., & Chang, C. P., (2009). MWNTs/polyaniline composite chemoresistive sensor array for chemical toxic agents detection. *J. Chung. Cheng. Ins. Technol.*, *38*(1), 147–156.

44. Kumar, D., Jha, P., Chouksey, A., Tandon, R. P., Chaudhury, P. K., & Rawat, J. S., (2018). Flexible single walled nanotube based chemical sensor for 2, 4-dinitrotoluene sensing. *J. Mater. Sci.: Mater. Electron.*, *29*(8), 6200–6205.

45. Kim, S., Han, J., Kang, M. A., Song, W., Myung, S., Kim, S. W., Lee, S. S., Lim, J., & An, K. S., (2018). Flexible chemical sensors based on hybrid layer consisting of molybdenum disulphide nanosheets and carbon nanotubes. *Carbon.*, *129*, 607–612.

46. Fennell, J. J. F., Hamaguchi, H., Yoon, B., & Swager, T. M., (2017). Chemiresistor devices for chemical warfare agent detection based on polymer wrapped single-walled carbon nanotubes. *Sensors*, *17*(5), doi: 10.3390/s17050982.

47. Roy, S., Pur, M. D., & Hanein, Y., (2017). Carbon nanotube-based ion selective sensors for wearable applications. *ACS Appl. Mater. Interfaces*, *9*(40), 35169–35177.
48. Rahman, M. M., Marwani, H. M., Algethami, F. K., & Asiri, A. M., (2017a). Comparative performance of hydrazine sensors developed with Mn_3O_4/carbon-nanotubes, Mn_3O_4/graphene-oxides and Mn_3O_4/carbon-black nanocomposites. *Mater. Express*, *7*(3), 169–179.
49. Young, S. J., & Lin, Z. D., (2017). Sensing performance of carbon dioxide gas sensors with carbon nanotubes on plastic substrate. *ECS, J. Solid State Sci. Technol.*, *6*(5), M72–M74.
50. Ong, K. G., & Grimes, C. A., (2001). A carbon nanotube-based sensor for CO_2 monitoring. *Sensors*, *1*(6), 193–205.
51. Arash, B., & Wang, Q., (2013). Detection of gas atoms with carbon nanotubes. *Scientific Reports.*, *3*, doi: 10.1038/srep01782.
52. Al-Degs, Y. S., Al-Ghouti, M. A., & El-Sheikh, A. H., (2009). Simultaneous determination of pesticides at trace levels in water using multiwalled carbon nanotubes as solid-phase extractant and multivariate calibration. *J. Hazard. Mater.*, *169*(1–3), 128–135.
53. Qu, Y., Xiao, F., Cheng, Y., Shi, G., & Jin, L., (2010). Construction of Tyr/Glu/Fe_3O_4/Nafion/CNT/GCE biosensor for detection of pesticide. *Acta. Chim. Sinica.*, *68*(6), 535–539.
54. Deo, R. P., Wang, J., Block, I., Mulchandani, A., Joshi, K. A., & Trojanowicz, M., (2005). Determination of organophosphate pesticides at a carbon nanotube/organophosphorus hydrolase electrochemical biosensor. *Anal. Chimi. Acta.*, *530*(2), 85–189.
55. Liu, G., & Lin, Y., (2006). Biosensor based on self-assembling acetylcholinesterase on carbon nanotubes for flow injection/amperometricdetection of organophosphate pesticides and nerve agents. *Anal. Chem.*, *78*(3), 835–843.
56. Younusov, R. R., Evtugyn, G. A., Arduini, F., Moscone, D., & Palleschi, G., (2011). Acetylcholinesterase biosensor based on single-walled carbon nanotubes-Co phtalocyanine for organophosphorus pesticides detection. *Talanta.*, *85*(1), 216–221.
57. Rahman, M. M., Marwani, H. M., Algethami, F. K., Asiri, A. M., Hameed, S. A., & Alhogbi, B., (2017b). Ultra-sensitive p-nitrophenol sensing performances based on various Ag_2O conjugated carbon material composites. *Environ. Nanotechnol. Monitor Manage*, *8*, 73–82.
58. Thayyath, S., & Alexander, A. S., (2015). Design and fabrication of molecularly imprinted polymer-based potentiometric sensor from the surface modified multiwalled carbon nanotube for the determination of lindane (γ-hexachlorocyclohexane), an organochlorine pesticide. *Biosensor. Bioelectron.*, *64*, 586–593.
59. Haldorai, Y., Hwang, S. K., Gopalan, A. I., Huh, Y. K., Han, Y. K., Walter, V. W., Gopalan, S. A., & Kwang, P., (2016). Direct electrochemistry of cytochrome c immobilized on titanium nitride/multi-walled carbon nanotube composite for amperometric nitrite biosensor. *Biosensors Bioelectron.*, *79*, 543–552.
60. Oh, J., Yoo, S., Chang, Y. W., Lim, K., & Yoo, K. H., (2009). Carbon nanotube-based biosensor for detection hepatitis, B. *Curr. Appl. Phys.*, *9*(4), e229–e231.
61. Kim, J. P., Lee, B. Y., Hong, S., & Sim, S. J., (2008). Ultrasensitive carbon nanotube-based biosensors using antibody-binding fragments. *Anal. Biochem.*, *381*, 193–198.

62. Badhulika, S., & Mulchandani, A., (2015). Molecular imprinted polymer functionalized carbon nanotube sensors for detection of saccharides. *Appl. Phys. Lett.*, *9*(107), doi: 10.1063/1.4930171.

63. Zhou, Y., Yang, H., & Chen, H. Y., (2008). Direct electrochemistry and reagentless biosensing of glucose oxidase immobilized on chitosan wrapped single-walled carbon nanotubes. *Talanta.*, *76*(2), 419–423.

64. Ghica, M. E., & Brett, C. M. A., (2010). The influence of carbon nanotubes and polyazine redox mediators on the performance of amperometric enzyme biosensors. *Microchimi. Acta.*, *170*(3), 257–265.

65. Cella, L. N., Chen, W., Myung, N. V., & Mulchandani, A., (2010). Single-walled carbon nanotube-based chemiresistive affinity biosensors for small molecules: Ultrasensitive glucose detection. *J. Am. Chem. Soc.*, *132*(14), 5024–5026.

66. Bareket, L., Rephaeli, A., Berkovitch, G., Nudelman, A., & Rishpon, J., (2010). Carbon nanotubes based electrochemical biosensor for detection of formaldehyde released from a cancer cell line treated with formaldehyde-releasing anticancer prodrugs. *Bioelectrochemistry*, *77*(2), 94–99.

67. Zhu, Z., Song, W. S., Burugapalli, K., Moussy, F., Li Y. L., & Zhong, X. H., (2010). Nano-yarn carbon nanotube fiber based enzymatic glucose biosensor. *Nanotechnol.*, *21*, 16. doi: 10.1088/0957–4484/21/16/165501.

68. Carla, G. C., Rasa, P. C., & Brett, M. A., (2008). Development of electrochemical oxidase biosensors based on carbon nanotube-modified carbon film electrodes for glucose and ethanol. *Electrochim. Acta.*, *53*(23), 6732–6739.

69. Ines, M., Joao, P. C., Celine, J., Patricia, S., Katia, D., Ana, F., Susana, C., Armando, D., & Teresa, R. S., (2017). Carbon nanotube field effect transistor biosensor for the detection of toxins in seawater. *Inter. J. Environ. Anal. Chem.*, *97*(7), 597–605.

70. Mandal, S. H., Su, Z., Ward, A., & Tang, X., (2012). (Shirley). Carbon nanotube thin film biosensors for sensitive and reproducible whole virus detection. *Theranostics.*, *2*(3), 251–257.

71. Deng, S., Upadhyayula, V. K. K., Smith, G. B., & Mitchell, M. C., (2008). Adsorption equilibrium and kinetics of microorganisms on single-wall carbon nanotubes. *IEEE Sensors, J.*, *8*(6), 954–962.

72. Zhoua, Y., & Ramasamy, R. P., (2015). Phage-based electrochemical biosensors for detection of pathogenic bacteria. *ECS Trans.*, *69*(38), 1–8.

73. Abdalhai, M. H., Fernandes, A., Xia, X., Musa, A., Ji, J., & Sun, X., (2015). Electrochemical genosensor to detect pathogenic bacteria (*E.coli* O157:H7) as applied in real food samples (fresh beef) to improve food safety and quality control. *J. Agric. Food Chem.*, *63*(20), 5017–5025.

74. García-Aljaroa, C., Cella, L. N., Shirale, D. J., Park, M., Munoz, F. J., & Yates, M. V., (2010). Carbon nanotubes-based chemiresistive biosensors for detection of microorganisms. *Biosensors Bioelectron.*, *26*, 1437–1441.

75. Villamizar, R. A., Marotoa, A., Rius, F. X., Inza, I., & Figueras, M. J., (2008). Fast detection of *Salmonella Infantis* with carbon nanotube field effect transistors. *Biosensors and Bioelectron.*, *24*, 279–283.

76. Viswanathana, S., Rani, C., & Ho, J. A., (2012). Electrochemical immunosensor for multiplexed detection of food-borne pathogens using nanocrystal bioconjugates and MWCNT screen-printed electrode. *Talanta.*, *94*, 315–319.

77. Huang, T. S., Tzeng, Y., Liu, Y. K., Chen, Y. C., Walker, K. R., & Guntupalli, R., (2004). Immobilization of antibodies and bacterial binding on nanodiamond and carbon nanotubes for biosensor applications. *Diamond Rel. Mater.*, *13*, 1098–1102.

78. Kara, Y., Kim, C. T., Kim, J. H., Chung, J. H., Lee, H. G., & Jun, S., (2014). Single-walled carbon nanotube-based junction biosensor for detection of *Escherichia coli*. *J. Pone.Org.*, *9*(9), doi: org/10.1371/journal.pone.0105767.

79. Choi, H. K., Lee, J., Park, M. K., & Oh, J. H., (2017). Development of single-walled carbon nanotube-based biosensor for the detection of *Staphylococcus aureus*. *J. Food Quality*, doi: 10.1155/2017/5239487.

80. Shen, Q., You, S. K., Park, S. G., Jiang, H., Guo, D., & Chen, B., (2008). Electrochemical biosensing for cancer cells based on TiO_2/CNT nanocomposites modified electrodes. *Electroanalysis*, *20*(23), 2526–2530.

81. Shao, N., Wickstrom, E., & Panchapakesan, B., (2008). Nanotube-antibody biosensor arrays for the detection of circulating breast cancer cells. *Nanotechnol.*, *19*(46), doi: 10.1088/0957–4484/19/46/465101.

82. Chikkaveeraiah, B. V., Bhirde, A., Malhotra, R., Patel, V., Gutkind, J. S., & Rusling, J. F., (2009). Single-wall carbon nanotube forest arrays for immunoelectrochemical measurement of four protein biomarkers for prostate cancer. *Anal. Chem.*, *81*(21), 9129–9134.

83. Bhirde, A. A., Patel, V., Gavard, J., Zhang, G., Sousa, A. A., & Masedunskas, A., (2009). Targeted killing of cancer cells *in vivo* and *in vitro* with EGF-directed carbon nanotube-based drug delivery. *ACS Nano.*, *3*(2), 307–316.

84. Chen, J., Chen, S., Zhao, X., Kuznetsova, L. V., Wong, S. S., & Ojima, I., (2008). functionalized single-walled carbon nanotubes as rationally designed vehicles for tumor-targeted drug delivery. *J. Am. Chem. Soc.*, *130*(49), 16778–16785.

85. Abdolahad, M., Taghinejad, M., Taghinejad, H., Janmaleki, M., & Mohajerzadeh, S., (2012). A vertically aligned carbon nanotube-based impedance sensing biosensor for rapid and high sensitive detection of cancer cells. *Lab Chip.*, *12*(6), 1183–1190.

86. Wang, S., Wang, R., Sellin, P. J., & Chang, S. X., (2009). Carbon nanotube-based DNA biosensor for rapid detection of anti-cancer drug of cyclophosphamide. *Current Nanoscience*, *5*(3), 312–317.

87. Gomes, R. N., Sousa, C. P., Casciano, P. N. S., Ribeiro, F. W. P., Morais, S., De Lima-Neto, P., & Correia, A. N., (2018). Dispersion of multi-walled carbon nanotubes in [BMIM]PF$_6$ for electrochemical sensing of acetaminophen. *Mater. Sci. Eng. C.*, *88*, 148–156.

88. Kim, J. P., Lee, B. Y., Lee, J., Hong, S., & Sim, S. J., (2009). Enhancement of sensitivity and specificity by surface modification of carbon nanotubes in diagnosis of prostate cancer based on carbon nanotube field effect transistors. *Biosensors Bioelectron.*, *24*, 3372–3378.

CHAPTER 10

APPLICATIONS OF MICELLEAR PHASE OF PLURONICS AND TETRONICS AS NANOREACTORS IN THE SYNTHESIS OF NOBEL METAL NANOPARTICLES

RAJPREET KAUR, NAVDEEP KAUR, DIVYA MANDIAL, LAVANYA TANDON, and POONAM KHULLAR

Department of Chemistry, B.B.K. D.A.V. College for Women, Amritsar–143005, Punjab, India, E-mail: virgo16sep2005@gmail.com

ABSTRACT

Poloxamer and poloxamines belong to the unique category of environmental friendly polymers that have innumerous academic and industrial applications. Their micellar form plays an important role and it is must for the synthesis of gold nanoparticles of various morphologies. The thermoresponsive nature makes them suitable candidate in biomedical field.

10.1 INTRODUCTION

Nanoscience is the emerging science of nanoscale materials. The word "nano," derived from the Greek word nano, meaning dwarf, is used to describe any material or property which occurs with dimensions on the nanometer scale (1–100 nm). Metal nanoparticles (NPs) have long been considered to exhibit unique physical and chemical properties differing from those of the bulk state or atoms, due to the quantum size effect resulting in specific electronic structures [1–6].

Among all the nanostructured materials, gold nanoparticles (Au NPs) have attracted considerable interest and find promising applications in various fields such as catalysis, sensing, bioimaging, photo thermal therapy, drug delivery, nano electronics and in the fabrication of photonic and plasmonic devices [7–17]. The first scientific description of the properties of nanoparticles was provided in 1857 by Michael Faraday in his famous paper "Experimental Relations of gold to light" [18]. In 1959, Richard Feynman gave a talk describing molecular machine, built with atomic precision. This was considered the first talk on nanotechnology. This was entitled "There's Plenty of Space at the Bottom."

Au NPs can be prepared using two approaches. First one is "top down" approach in which bulk material is broken down to generate nanoparticles of desired dimensions. But this approach suffers from the limitations of controlling the size and shape of particle as well as further functionalization [19]. Another approach called "bottom up" approach involves the formation of Au NPs from individual molecules using either a chemical or a biological reduction [20]. Chemical reduction further involves two steps i.e. nucleation and successive growth. However when both the steps are completed in the same process, it is termed as *in situ* synthesis while the other one is called seed growth method. Nonionic amphiphilic copolymer micelles composed of PPO and PEO units and surfactants are extensively used in the synthesis of nanostructured materials.

This chapter accounts for the recent advance in the synthesis of Au NPs using environmental friendly, non toxic linear chained poloxamers and branched poloxamines and further scope of the block copolymers coated gold NPs in various biomedical fields.

10.2 POLOXAMERS AND POLOXAMINES

Recently, water soluble block copolymers have attracted significant interest due to the environmental concerns [21, 22] and their water based food and pharmaceutical formulations [23, 24]. Amphiphilic block copolymers (BCPs) based on the polyethylene oxide (PEO) and polypropylene oxide (PPO) are an important class of biocompatible polymers in both industry as well as academics. The linear block copolymers with the structure PEO-b-PPO-b-PEO are known as poloxamers (Figure 10.1) and tetra branched copolymers with four PEO-PPO blocks central ethylene diamine bridge are called poloxamines (Figure 10.2). These surfactants were first introduced in the 1950s by BASF, NJ, USA.

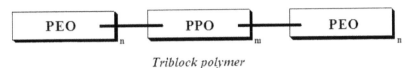

Triblock polymer

FIGURE 10.1   Structure of Poloxamers Composed of Two PEO Blocks and a Central PPO Block.

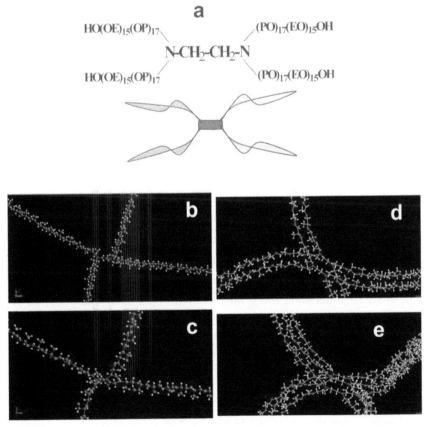

FIGURE 10.2   (See color insert.) (a) Structural formula of star shaped T904 and its schematic representation. H (white), C (grey), O (red), and N (blue). (b) A graphical model of the structure and (c) its relaxed structure. (d) A graphical model of an aggregate of three molecules and (e) their relaxed structure.

Poloxamers are also known as Pluronics® (BASF) or Synperonics (ICI) or Genapol are water soluble polymers which are commercially available in a variety of molecular weights and PEO/PPO ratios. The micellization process of these polymers in aqueous solutions is governed by the interplay between the hydrophobicity and hydrophilicity of the building blocks and their interactions with the solvent. In block copolymer, each segment exhibit particular function such as the reducing and anchoring function. It is because of the hydrophilic nature of PEO blocks and hydrophobic nature of PPO blocks that these block copolymers are amphiphilic in nature. An appropriate variation in the number of PEO and PPO repeating units cause a shift in the overall nature from predominant hydrophilic to predominant hydrophobic. Such a versatile nature also make them widely useful in various applications such as dispersion stabilizer [25, 26], pharmaceutical ingredients, [27–29], biomedical materials [30, 31], and template for the synthesis of mesoporous materials and NPs [32–37].

Pluronics with different molecular composition can be categorized using an easy description of structure and properties of a given copolymer, In all the trade names, i.e., Pluronic (BASF), Lutrol (BASF) or Synperonics (Croda), the prefix letters describe the physical appearance of the pure copolymer, i.e., 'L' refers to liquid, 'P' refers to paste and 'F' refers to flakes. In addition, the name contains information about the block length ratio in the respective polymer. The approximate molecular weight of PPO can be obtained by multiplying the first one or two digits by 300 and that of PEO can be obtained by multiplying the last digit with 10. It is also possible to convert the nomenclature for Pluronics to the generic name for poloxamers by multiplying the first one or two digit by a factor of 3 [38].

In order to understand the synthesis of Au NPs using the micelles of block copolymers, we need to ensure the presence of micellar phase along with the gold salt. It is to be mentioned that the monomeric form of block copolymer does not effectively involve in the reduction reaction due to the lack of surface cavities and hence, reduction is caused by the micelles only (Figure 10.3).

The reducing ability of the cavity which is formed by the PPO and PEO block is directly connected to its size. Since one polymer is contributing only one surface cavity because it accepts only one guest ion, i.e., oxidizing agent as accepted by the partially hydrated or dehydrated cavities with high aggregation number at high temperature [39]. Such type of reducing agents is termed as *structured reducing agents* where structural factors are the deciding factors for reduction. Reduction of gold salt is carried out by the surface cavities produced by the compact arrangement of the TBP monomers in the corona layer of TBP micelles (Figure 10.4).

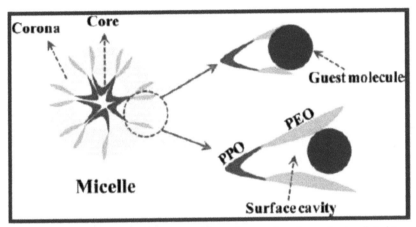

ₐ Red dotted circle shows a possible surface cavity whose size is related to the number of PEO and PPO units. A larger cavity can easily accommodate a guest molecule in comparison to a smaller cavity.

FIGURE 10.3   (See color insert.) A TBP Micelle with the Core Occupied by PPO Units and the Corona Constituted by PEO Unitsa

The extent of hydration of these surface cavities is the deciding factor as the greater hydration reduces the nucleation process because it hinders the approach of gold ions. But with the increase of dehydration, compact micelles are formed that bring the nucleation center in close proximity to produce larger gold NPs. Since soft micelles are unable to hold the larger NPs, and hence such NPs find their way into the bulk phase. On the contrary, larger micelles are fully capable to hold these NPs. The overall shape, structure as well as the temperature are the key parameters for the production of Au NPs.

10.3   MICELLE ARCHITECTURE AND MIXED MICELLES

Triblock polymers TBPs with a greater number of PEO units rather than PPO units form compound micelles but predominantly hydrophobic TBPs usually produce well defined micelles. Micelles undergo several structure transitions (i.e., micelles to thread like micelles to vessels, etc.) with concentration and temperature variations. It is observed that whenever a transition in the structure of the micelle occurs, it alters the interfacial arrangement of the surface cavities and hence it affects the overall mechanism that is actually occurring through the LMCT complex. TBP with larger number of cavities with sufficient size is able to produce ordered morphologies of gold NPs whereas a TBP with a fewer surface cavities and small micelle will lead to unstable LMCT complex and hence the nucleation depends on the extent of inter micelle collisions due to diffusion.

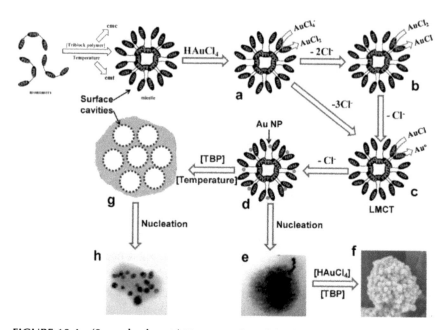

FIGURE 10.4   (See color insert.) Demonstration of the Overall Redox Process Taking Place in the Surface Cavities at the Micelle-Solution Interface of TBP Micelles.

In order to achieve well-defined morphologies of Au NPs, predominantly hydrophilic TBP with greater number of PEO units is desired because of the presence of larger number of PEO units, the instant reduction can be achieved. The other deciding parameter is the PPO/PEO ratio, TBPs with high PPO/PEO ratio produce stable micelle w.r.t. temperature (Figure 10.5).

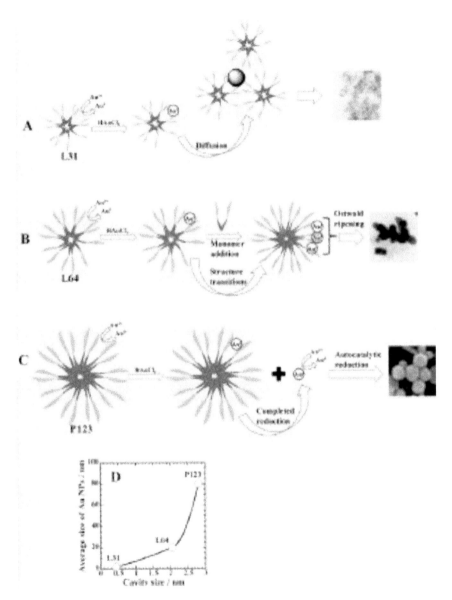

FIGURE 10.5 (See color insert.) (A–C) Schematic Representation of the Proposed Mechanism for the Synthesis of Au NPs by Using Micelles of L31, L64, and P123, Respectively (see details in the text). (D) Plot of the Average Size of NPs Estimated from TEM Images versus Cavity Size for L31, L64, and P123.

TBPs which are predominantly hydrophobic e.g. L121 are used as templates for the synthesis of Au NPs. However, the presence of convectional surfactant is must to solubilize L121 in aqueous phase. By the comparison of different zwitter ionic surfactant like DPS, TPS, HPS, it is observed that the order of hydrophobicity of these surfactant follows the DPS<TPS<HPS (C12<C14<C16). Less hydrophobic nature of DPS produces well defined micelles whereas the more hydrophobic TPS and HPS dismantle the micelle template because of their enhanced solubilizing effects. Low concentration and low hydrophobicity of neutral surfactant causes minimal disturbance to the arrangement and hence produces well defined micelles loaded with tiny NPs (Figure 10.6).

10.4 SYNTHESIS OF VARIOUS MORPHOLOGIES OF GOLD NANOPARTICLES (AU NPS)

Over the past decade, various methods such as seed-mediated growth processes [40], template-directed patterning [41, 42], biomineralization [43], two-phase reactions [44], and inverse micelles [45], have been used to synthesize nanostructures including rods [46] (Keul 2007), plates [43], spheres [41], and cages [47]. Template-based approach (either hard or soft) is the most commonly used approach to prepare uniform gold nanorods. The hard templates usually are alumina or polycarbonate membranes for 1D gold nanostructure whereas the soft templates are usually surfactants that form rod-shaped micelles. The block copolymer mediated synthesis method offers many advantages and need only environmental and economic way but also it is a very simple procedure that simply requires the mixing of metal salt with a block copolymer ratio. This type of methodology offers many possibilities like simply varying the block copolymer type, amphiphilic character, concentration, temperature, and solvent.

It is observed that when the nucleation and growth is kinetically controlled or when it deviates from a thermodynamically controlled to a kinetically controlled pathway as a result of slowing down of precursor decomposition or using a weak reducing agent or by Ostwald ripening [6].

Seed growth method can be employed for anisotropic nanoparticle synthesis such as nano prisms [48, 49]. Another mechanism that is normally employed to describe the growth of NPs is Lamer Mechanism [50, 51]. In this mechanism, after initial nucleation, the nuclei grow into particles by the molecular addition of nutrient species on the surface of the particles known as Ostwald ripening. In this method, the particles will be mostly mono disperse in nature. 'Aggregative nanocrystal model' is used to describe the Au nanocrystal growth [52, 53]. This model explains that initial nucleation and growth results from a number of critically sized aggregates of smaller nanocrystallites in a non-classical aggregative nucleation step.

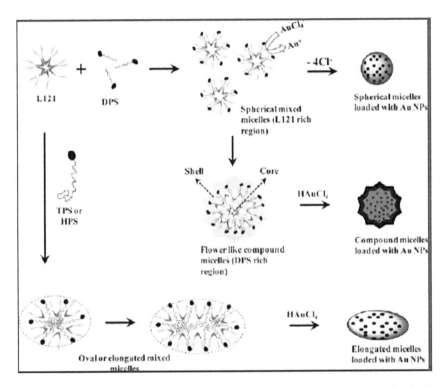

FIGURE 10.6 (See color insert.) Schematic representation of the mixed micelle formation between L121 and zwitterionic surfactants and their subsequent use as micelle templates for the self-assembled Au NPs. Top reaction shows the formation of spherical L121 + DPS mixed micelles in the L121-rich region of the mixture by incorporating the hydrocarbon chains of DPS molecules in the L121 micelles. In the DPS-rich region, predominantly hydrophobic L121 is solubilized by the hydrocarbon tails of DPS in the form of a typical compound micelle where DPS molecules occupy the shell while L121 resides in the core, resulting in the formation of flower-like morphologies with Au NPs mainly accommodated in the core due to the presence of L121 surface cavities. Lower reaction shows that using TPS or HPS instead of DPS induces their longer hydrocarbon tails in the L121 micelles, thus causing structure transitions with the formation of oval or elongated morphologies that are visualized by the self-assembled Au NPs.

Seed growth method is a soft template approach, and it is one of the most popular techniques for the synthesis of Au NPs. Initially, gold seeds of ~3–4 nm size are prepared by the borohydride reduction of the gold salt in the presence of citrate or CTAB as the capping agent. Then these performed seeds are added into the growth solution, which contains a reducing agent such as Ascorbic Acid. The addition of silver nitrate in the growth stage helps improve and increase the yield of nanorods. The nanostructures thus obtained, have potential applications in medicines due to tunable plasmonic properties, but it is restrained due to the toxicity of CTAB [9]. Also, CTAB binds strongly to the surface and restricts subsequent functionalization.

This limitation can be improved by using a mixture of CTAB and Pluronic (F-127) in the seed growth method (Figure 10.7). The nanorods thus prepared in the presence of Pluronic (F-127) are quite stable and also the presence of pluronic results in the higher yield of nanorods. The reducing nature of PEO blocks of pluronics is responsible for the enhanced yield of nanorods (Figure 10.8). The formation of stable complex between CTAB and block copolymer due to hydrophobic interactions between the PPO blocks of pluronics and hydrophobic surfactant tail is the cause of stabilization of gold nanorods. The presence of ascorbic acid is required for the synthesis of gold nanorods as the pluronic alone is not sufficient to reduce the metal salt in the presence of CTAB. However, at high concentration, PEO can reduce the gold ions, but the PPO blocks get adsorbed onto the gold cluster and lead to the stabilization of gold nanoparticles that hinder the growth of NPs into nanorods.

Generally, the size and shape of the metal nanoparticles depend on the competition between the nucleation (metal ion reduction in bulk) and the growth process (metal ion reduction on nuclei). If the metal ion reduction on nuclei is more dominant than in bulk solution, then particle growth is more significant. However, if the metal ion reduction in bulk is more dominant than on nuclei, then new particle formation is more significant than particle growth.

10.4.1 NANOPLATES

Pluronics like F-127 has also been used in single step seed mediated synthesis to prepare biocompatible IR responsive gold nanoplates. The

FIGURE 10.7 SEM images of gold nanorods prepared from 25 mL of seed in the presence of AgNO$_3$ (a) without Pluronic F-127 and (b) with Pluronic F-127 (4.75 6 1024 M). Scale bars: 500 nm.

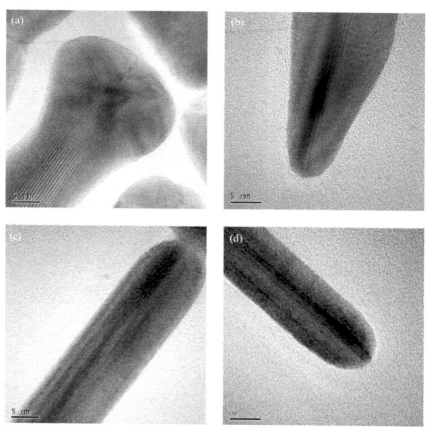

FIGURE 10.8 TEM images of gold nanorods prepared from 25 mL of seed (5.5 6 1027 M): (a) in the presence of AgNO$_3$ (without Pluronic F-127), after storage for one week in aqueous environment; (b) in the presence of AgNO$_3$ and Pluronic F-127 (c) in the absence of Pluronic F-127 and AgNO$_3$; (d) in the presence of Pluronic F-127 and in the absence of AgNO$_3$. Scale bars: 5 nm.

pluronics performs multiple functions i.e. it acts as a reducing agent, shape directing agent as well as its stabilizes the nanoparticles obtained. F-127 is approved by FDA (Food and Drug Administration of USA) for in vivo biomedical applications. The PPO blocks of F-127 binds to the surface of NPs through hydrophobic interacts and it controls the growth of Au NPs by specific crystallographic directions. In this case, the growth rate of nanoparticles is kinetically controlled rather than thermodynamically by different sticking probabilities by atoms on a particular crystallographic face [55]. The order of sticking probabilities of different faces is

$\gamma(110){>}\gamma(100){>}\gamma(111)$. Therefore, FCC metal tends to grow and nucleate along the [111] facets. Because of the fact that [111] facet of FCC metals has the lowest surface energy as compared to other facets, i.e., [110] > [100] > [111]. The FCC metal confers its tendency to nucleate and grows into NPs with their surface bounded by [111] facet [56]. At a particular F127/HAuCl$_4$ molar ratio, the PPO blocks of the pluronics get adsorbed on the [111] planes of FCC Au nuclei therefore inhibiting the growth on [111] crystallographic planes and promote anisotropic growth along [110] facets [6, 55, 57–68] (Figures 10.9–10.11).

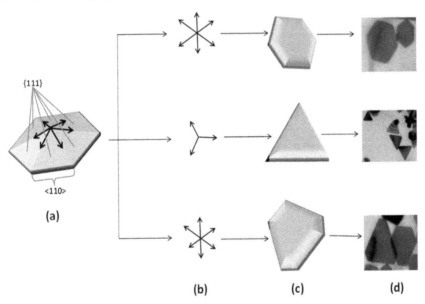

FIGURE 10.9 Schematic illustration of proposed growth mechanism of Au-NPs: (a) crystallographic facets with basal {111} plane and h110i side facet, (b) the formation of hexagonal plate, triangular plate and truncated triangular plate like shapes along the different h110i directions, (c) the expected particles shapes, (d) representative TEM images of hexagonal NPs, triangular NPs, and truncated triangular NPs.

Under ideal conditions, the nuclei grow uniformly along six [110] directions if the growth results in the formation of hexagonal plate. But deviation of the crystal growth forms uniform growth rate along [110] directions due to the local fluctuations, results in the formation of triangular and truncated triangular plate. Such nanostructures are found to be quite stable when preserved in pluronic F-127 (Chudasama 2014) and

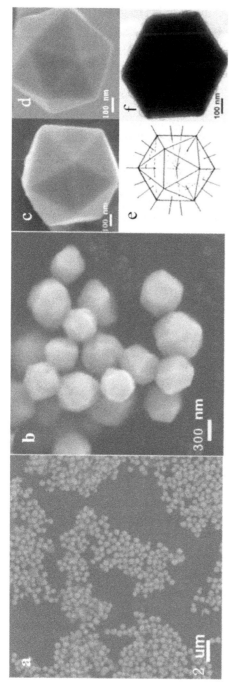

FIGURE 10.10 Characterization of gold icosahedrons synthesized at 40 °C with a reaction time of one day, and the concentrations of F88 and HAuCl₄ are 0.84 mM and 5.8 mM, respectively: (a) SEM image; (b) higher magnification SEM image; (c) and (d) SEM image of a single gold particle observed from different angles of view; (e) geometrical model of the obtained icosahedron particles; (f) representative TEM image.

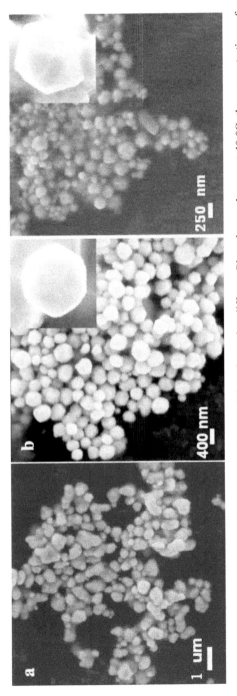

FIGURE 10.11 SEM images of the as-synthesized particles when using different Pluronic copolymers at 40 °C, the concentrations of HAuCl₄ and copolymers were 0.84 mM and 5.8 mM, respectively. (a) P85, (b) P104, (c) P105. The concentration of the Pluronics was 8.4 mM. The insets of (b) and (c) show the magnified SEM images of the icosahedrons.

hence they have potential to be used for in vivo applications of cancer diagnosis and therapy.

10.4.2   ICOSAHEDRAL GOLD NANOPARTICLES (AU NPs)

F88, P85, P104, P105 pluronics has also been used to prepare icosahedral Au NPs. But a regular icosahedron shapes are formed only when a high molecular weight copolymer i.e. P85 is used. It is due to the fact that larger molecular weight copolymer is more flexible and it can adsorb on the surface of gold cluster more efficiently which is required for the formation of regular icosahedral Au NPs. Also the size of these NPs can be controlled in the range of 100 nm to 1 micrometer by simply varying the experimental conditions such as temperature, concentration, etc.

10.5   BIMETALLIC NANOPARTICLES

Pluronic F-127 has also been employed for the synthesis of Ag-Au bimetallic NPs in which maltose coated Ag NPs are added to the aqueous solution of F-127 followed by the addition of $HAuCL_4$. These bimetallic NPs have enhanced catalytic activities for the reduction of 4-nitrophenol by $NaBH_4$ as compared to individual Ag and Au NPs. The enhanced efficiency is attributed to the greater surface areas containing higher energy facets and Ag/Au interfacial regions where homogeneous surface aggregation (alloying) occurs that have excess electron density compared to monometallic NPs.

10.6   COMPARISON OF POLOXAMER AND POLOXAMINES

In comparison to the linear pluronic copolymers, tetronics consist of X-shaped structure which is made up of ethylenediamine central group bonded to four chains of PPO-PEO. It is because of this unique structure that tetronics are widely used in various biomedical and pharmaceutical applications [69]. The presence of tertiary amine group plays a very important role for response to pH variation as well as in conferring thermodynamic stability [70–72] that distinguishes their properties from poloxamers.

The amphiphilic nature of the tetronics is due to the presence of both PPO and PEO block, whereas PEO block is responsible for the reduction of the gold salt, the PPO block causes the adsorption of the copolymer on the nanoparticle surface. This creates a competition between the reduction (in bulk) and growth (on the surface). The Au(III) ion bind both to the diamine groups and PEO groups present in the copolymer via ion-dipole interactions. As a result of which the metal ion Au(III) gets entrapped and reduced in the micelle to Au(s). The reduction process is directly proportional to the number of micelles. Greater the number of micelles with such a type of loosely packed structures the more easily the metal ion gets penetrated inside the micelle and hence more the number of NPs will be produced.

Tetronics which gives rise to micelle with loose structure acts as a better stabilizer for the synthesis of Au NPs than their linear counterparts pluronics because of better coverage of NP surface. A comparison of tetronic T904 and pluronic P105, both having same percentage of PEO and PPO units, demonstrate the good reducing power of T904 even at as much as low concentration of 0.07M with 0.1 mM $HAuCl_4$. It is due to presence of peculiar X-shape structure (Figure 10.12) that eases the formation of pseudocrown cavities, secondly presence of reducing amino groups in the copolymer molecule [73, 74]. Both these factors favor reduction (Taboada 2010).

10.6.1 EFFECT OF pH

pH plays a very crucial role during the synthesis of AuNPs using BCPs. At low pH, due to addition of protons, the hydrogen bond between BCPs and water molecule gets strengthened. It is due to the attachment of protonated water molecules with the PEO block and hence PEO corona holds more positive charge. However, when AuNPs are produced, H^+ are also produced and the presence of H^+ ion interrupts the binding of metal ion to the PEO block and the reduction will get slow down, it increases the size of the micelle [75], and hence less number of AuNPs will be produced. Also, at low pH, because of hydration of micelle, the AuNPs can't be entrapped by the micelle and hence they find their way into the bulk. But as the pH is increased, according to Le-Chatelier's principle, more and more number of H^+ ions will be neutralized and hence the reduction of the gold salt is increased resulting in a large number of AuNPs.

Low pH, i.e., acidic conditions results in poor yield of (less stable) Au NPs and high pH, i.e., basic conditions results in good yield of stabilized Au NPs. This stability at high pH can be attributed to the hydrophobic interactions because of the coating of Au NPs withe the block copolymer. Therefore, the coating makes the NPs hydrophilic and easily dispersible in water without further surface modification that is desirable for biomedical applications. Increasing pH value enhances both the reaction rate [76] and the coordination ability of the block copolymers as well as results in a change in the morphology of Au NPs that results in Ostwald Ripening process. Pluronics get adsorbed on the gold surface to form a core shell structure via hydrophobic interactions.

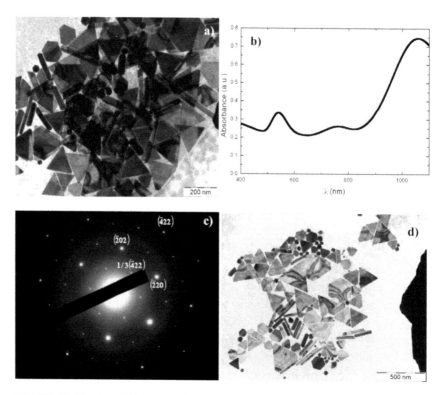

FIGURE 10.12 (a) TEM image of gold nanoplates formed by reduction of HAuCl₄ in the presence of T904 at a copolymer/metal salt molar ratio (MR) of 1.5 (0.5 mM HAuCl₄) at 25 °C. (b) Absorption spectrum of the nanoplates. (c) SAED pattern taken from an individual nanoplate and its assigned reflection indexes. (d) TEM image of gold nanoplates displaying bending contours.

Poloxamine, i.e., Tetronic T904 contains two nitrogen atoms and its titration shows two inflection points pK_a^1 and pK_a^2 at 4.0 and 8.8, respectively. Below pH < 4, T904 exists in diprotonated form that results in coulombic repulsions among the positively charged amine groups and hence prevent self-aggregation. Tetronics are highly pH sensitive, and size and shape of gold nanoparticles could be controlled by varying the pH of the reaction mixture. At low pH 3.0, triangular nanoparticles are formed and as the pH is increased more spherical NPs are formed. With the further increase of pH, the size of the spherical particles further decrease. It is also observed that acid used to decrease the pH itself exerts a strong effect on the shape and size of NPs because of the specific adsorption of Cl⁻ ions (HCl/NaCl) on the NP surface. As the pH decreases, the micellization becomes more difficult which causes increase in the CMC and decrease in micellar size.

At low pH, i.e., under acidic conditions, H⁺ ion disturbs the binding of metal ions to the PEO block and therefore, reduction rate decreases. Also increased population of plates is observed at low pH (Figure 10.13) due to the adsorption of Cl⁻ ion on the specific planes whereas high pH, spherical NPs are produced due to accumulation of T904 on different crystallographic planes, that promotes the growth in all directions and hence produces spherical particles. Spherical particles are also produced at high temperature which indicates that reduction rate is under thermodynamic control (Figure 10.14). In tetronics, the presence of amine groups raises the pH of the solution up to 8.9–9.3.

It is easier to predict the morphology of micelles in pluronics rather than in tetronics. As in pluronics, the hydrophobic PPO blocks form the core and hydrophilic PEO blocks form the corona/shell. Significant steric constrains in the four arms of tetronics lead to the formation of highly hydrated micelles.

In pluronics, PEO blocks act as mild reducing agent and the redox reaction is mainly carried out in the micelle surface cavities that act as active site for the entrapment of gold ions [39, 77]. It is also observed that tetronic coated NPs, produced at pH ~ 2 can prove to be excellent models for pH-responsive drug delivery vehicles [78, 79] in the colon at pH~ 10, where they change from ionic to the nonionic form for controlled drug delivery. In the case of tetronics, the presence of micellar phase is must for the synthesis of AuNPs as it is true for the pluronics [39, 77, 80]. The formation of surface cavity by PPO and PEO blocks in the aggregated state is the most important requirement for the reduction process.

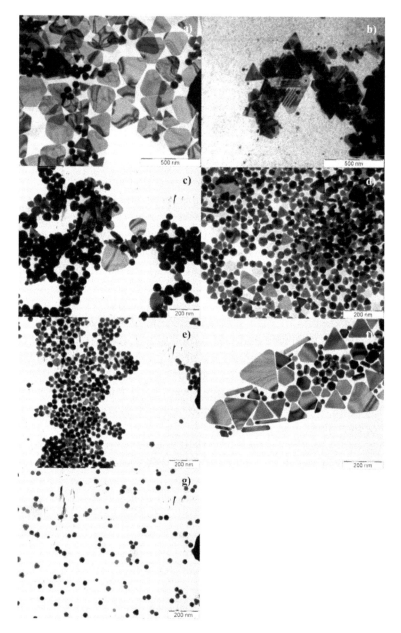

FIGURE 10.13   TEM images of gold particles formed by reduction of HAuCl$_4$ at MR) 4 (0.5 mM HAuCl$_4$) and pH (a) 3.0, (b) 4.0, (c) 5.5, (d) 6.5, and (e) 7.5 at 25 °C and at MR) 16 and pH (f) 3.0 and (g) 7.5 at 25 °C.

FIGURE 10.14   TEM images of gold particles formed by reduction of HAuCl$_4$ at MR)
1.5 (0.5 mM HAuCl$_4$) and (a) 35, (b) 45, (c) 60, and (d) 75 °C and at 75 °C for (e) MR) 4
and (f) MR)

In case of tetronic T904, it is observed that at low pH, diamine groups
get protonated which results in the formation of highly hydrated vesicles/
micelles that electrostatically attract negatively charged AuCl$_4^-$ ion. It
results in the formation of ligand and metal Charge transfer (LMCT)
complex that requires high temperature for the formation of compact

micelles and facilitate the reduction reaction. On the other hand, at high pH, no protonation of amine group occurs, and hence the reduction of Au^{3+} to $Au(0)$ occurs instantaneously at low temperature. Further growth process occurs at the surface of the micelles, that is why gold nanoparticles get decorated at the micellar assemblies. Tetronics are less surface active than cationic or nonionic surfactants, and hence they do not show any shape control effects and hence results in the formation of only spherical or polyhedral morphology (Figure 10.15).

However, the presence of any other surfactant, e.g., C14E8 which is a nonionic surfactant lead to the formation of well-defined geometries of AuNPs (Figure 10.16). It is because of the active participation of the surfactants in the shape control effects on the crystal growth of NPs. The micelles of tetronics due to their pH and thermoresponsive behavior acts as excellent nanoreactor for the nonmaterial synthesis.

10.7   BIOMEDICAL APPLICATIONS

Two important factors that control targeting the attachment of cell-specific ligands required for the increased selectivity are the size and shape of the particles. In medical field, it is desired to achieve the selective delivery of drugs to specific areas in the body to increase the efficiency of drugs and decrease their side effects. Each drug, in addition to benefits also offers some side effects. This kind of problem is commonly encountered in drugs used for the treatment of cancer chemotherapy. Similarly, cytotoxic compounds can kill not only target cells but also normal cells in the body.

Stimuli-responsive polymers or smart or intelligent polymers are those that respond sharply to small changes in physical or chemical conditions with relatively large phase. Below the cloud point (C_p), the copolymer dissolves in water. This polymeric delivery system respond to even small changes in temperature, and it is due to their tendency to undergo sol-gel transitions near body temperature and therefore, controlling the release rate of incorporated drugs while maintaining their physiochemical stability and biological activity.

Poloxamers and poloxamine nonionic surfactants are approved by FDA to be used as drug carriers in parenteral system, food additives and pharmaceutical ingredients have diverse applications in various biomedical fields ranging from drug delivery and medical imaging to management of vesicular diseases and disorders. The PPO block of the copolymers undergoes hydrophobic interactions with the hydrophobic surfaces of the nanospheres. This

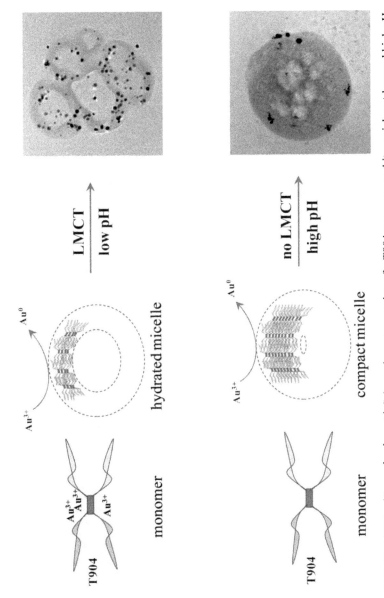

FIGURE 10.15 (See color insert.) Schematic representation of a T904 monomer and its vesicles at low and high pH. Hydrated vesicle with core–shell type morphology is formed at low pH while compact compound micelle with no clear core–shell regions is produced at high pH.

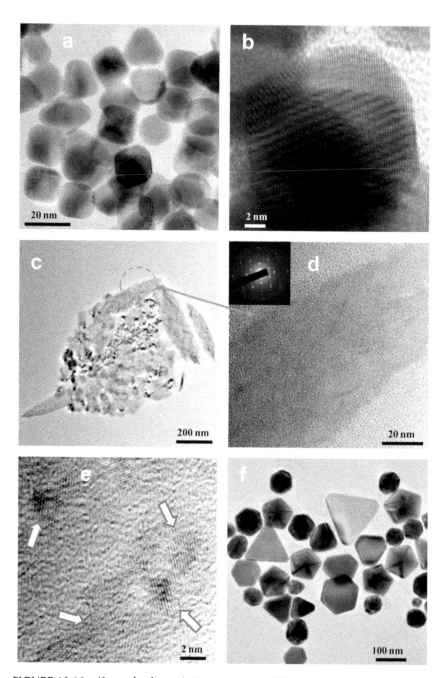

FIGURE 10.16   **(See color insert.)** Geometries of AuNPs.

kind of adsorption results in the free movement of PEO side arms which also causes steric repulsions. The extent of adsorption depends on both size of PEO and PPO block as well as type of interactions present such as nanoparticle surface charge, hydrogen bond between PEO unit and the constituent groups on the particle surface. These kind of engineered nanoparticles exhibit reduced adsorption of proteins and blood as compared to the uncoated nanoparticles and hence resist ingestion by phagocytic scavenger cells [81, 82].

One of the remarkable features of these copolymers is their thermo-responsive behavior. The micellization process is extremely temperature dependent. With increase in temperature, CMC values decrease dramatically. Due to this property, such polymers are called "smart and intelligent polymers" and find applications in controlled drug delivery, thermal printing, biomedical processing and sensor development [83–85].

Poloxamer micelles are composed of PPO core and PEO corona. Hence, it permits encapsulation of a large number of hydrophobic compounds. As a result of which they have full potential to be used as drug and gene delivery systems [29]. Because of the self-assembled nature, emulsifying properties as well as their biocompatible nature, pluronics have been used for biomedical and pharmaceutical applications and as well as for the development of the drug delivery systems [29, 86]. Pluronic formulations have shown enhancement in cytotoxic activity of chemotherapeutic drugs like dexorubin towards multi-drug resistant (MDR) cancer cells [87–89]. Also, these formulations have potential to modify the sensitivity of MDR cancer cells and led to increase in the drug transport across the cell membrane.

This tendency is directly proportional to the number of PEO and PPO units. It is observed that these BCPs which have 40 PPO units and adjacent PEO segment consisting of at least 70 PEO units (e.g., Poloxamine-908, poloxamine-1508, poloxamer-407). Such engineered NPs are observed to remain in systemic circulation for prolonged periods when injected intravenously into mice, rats, and rabbits. These nanovehicles are the promising candidates for various applications such as medical imaging and drug delivery for controlled release of therapeutic materials.

Because of the tendency of some poloxamers, e.g., poloxamer-407 to get converted to gel form, they have potential to be used for the slow release of peptide and therapeutic proteins, which include interleukin-2, urease, and human growth hormone [90–92]. When injected into the body, the gel slowly dissolves and slowly releases the entrapped protein molecules over a period of 1–2 days and a substantial fraction is removed from the body in the form of renal excretion.

Poloxamer-407 has also been used as an artificial skin for the treatment of third-degree burns because of its bactericidal properties [93]. Owing to its surfactant nature, it also cleans the wound of tissue detritus.

Another major problem in chemotherapy is the development of drug resistance [94, 95]. BCPs can be used for the treatment of such problems. But this activity depends on HLB (Hydrophilic Lipophilic Balance) properties and the size of poloxamer molecules. It is suggested that greater the size of PPO segment, more will be the effectiveness of block copolymer. It is due to the interaction of PPO block with the cell membrane [96]. Therefore, the use of these BCPs in pharmaceutical formulations results in increased drug bioavailability as well as drug accumulation in selected organs (e.g. brain) and might overcome the problem of drug resistance which limits the effectiveness of many therapeutic reagents.

Nearly 70% of the drug molecules coming from the synthesis have solubility problems9 while nearly 40% of newly developed active pharmaceutical ingredient (API) are rejected in early phase development. The surface active nature, low toxicity, minimal immune response make pluronics as promising candidate for controlled drug delivery. Due to the improved bioavailability, low drug degradation make them the topic of research. Due to presence of PPO blocks in the micellar core, these micelles serve as the reservoirs for hydrophobic drugs. Due to low CMC values, these micelles have greater thermodynamic and kinetic stability. Two types of routes, i.e., physical, and chemical can be used to keep the drugs into the hydrophobic core of the micelle. Nanocrystallization is the process to enhance the bioavailability of poorly soluble drugs and drug association rate. Drugs are available in submicron size range with a crystalline API core covered by a stabilizer layer. It offers the following advantages:

- improved bioavailability;
- high drug loading; and
- potential for targeted drug delivery.

The selection of the stabilizer is a key factor. A limited number of stabilizers in nanorange are being used, i.e., polymeric stabilizers [97–99], pluronics [100], surfactant stabilizer, etc. Due to the difference in the physiochemical properties, each API needs a specific stabilizer.

Indomethacin is a water-insoluble hydrophobic drug. It is being used as a model drug to study the interactions of poloxamers (L64, F68), reverse poloxamers (17R4), and poloxamines (T908, T1107). Surface plasmon resonance and contact angle techniques are used to study the interaction of polymer with the drug using wet ball milling technique.

These studies concluded that a good stabilizer must fulfill the following requirements:

- it should firmly attach to the surface; and
- the polymer should properly coat the particle.

Compared to their linear counterparts, i.e., pluronics, poloxamines have been neglected for a long time. But these are the potential candidates for drug delivery and tissue engineering tetronics can stabilize DNA® (Pitard metal) and hence negatively charged DNA-poloxamine supramolecular assemblies were used in gene delivery system for the therapy of skeletal and heart muscle related diseases.

Poloxamine has also been employed in the form of coating over the hydrophobic NPs in order to prolong the particles blood circulation time and render them potential drug delivery systems for medical and pharmaceutical purpose. The molecular weight and composition of the PEO-PPO are the deciding factors for the coating efficiency. It is observed that with increase of the hydrophilicity and molecular weight, the shielding effect becomes more important.

Tetronic, T304, is a promising drug carrier and an effective transport inhibitor for the treatment of MDR tumors. Similarly, aqueous micellar solution of Tetronic T904 micelles, have been employed to solubilize Quercetin (QN), a hypolipidemic drug, at different salt concentration, pH, and temperature. It is the special architecture of tetronics that pH, temperature, and ionic strength of the medium strongly influences their solution behavior [101]. The solubility of drug in the micellar system was examined by UV-visible spectrophotometer. The micellar size was determined using dynamic light scattering (DLS), and the possible locus of the drug molecule in the micellar aggregate was estimated from two dimension nuclear overhauser effect spectroscopy (2D-NOESY).

Theoretical studies suggest that as the molecular weight of the copolymer increases, micellar size also increases. As the concentration of T904 increases, the number of micelles also increases and also the solubility of QN increases. Most of the drugs are usually salts for pH buffering and ionic strength balance, QN tends to influence the behavior of T904 in aqueous medium.

It is due to the presence of 2° amine group in tetronics that they show pH and thermoresponsive behavior. When tetronics T904 is compared with P84 which have the same percentage of PEO block, i.e., 40%, it is observed that QN is very less soluble in T904 at low pH; but as the pH is increased up to 6, due to its ionization, a drastic increase in solubility is observed while in P84, up to 3 mM of QN is soluble at low pH but with increase in pH, no drastic increase in solubility is observed. It is due to the absence of pH-sensitive moieties in pluronics. That is why pluronics P85, shows considerable solubility even at low pH and with increase in pH, solubility increases due to dehydration. But this increase is not drastic as compared to that in T904.

The enhanced solubility of QN in T904 at high pH is due to both hydrophobic and electrostatic interactions. At low pH, T904 gets protonated that results in columbic repulsions and hence micellization does not occur. At low pH, QN molecules interact with T904 unimers. But at high pH, micellization occurs. The QN gets solubilized in these micelles. Above pH 10, the phenolic –OH group of QN gets dissociated (at position 17, 19) and results in the formation of mixture of neutral, anionic species which indicates the existence of both hydrophobic and electrostatic interactions that results in dramatic increase in the QN solubility.

But the drug-loaded micelles suffer from thermodynamic stability, i.e., with dilution, these drug carrier micellar systems become unstable in nature. This situation holds for very hydrophilic BCPs having high cms/cmt values. Several strategies have been employed to overcome the shortcoming; e.g., formation of interpenetrating network via light initiated cross-linking of a tetra-functional acrylate monomer. Another approach involves conversion of –OH group into aldehyde and introducing mine linkages via the addition of diamine [102].

Pluronics are also used for the surface coating of the drug-loaded hydrophobic NPs to enhance the blood circulation time of the drug carriers. The size and the properties of the coated surfaces are deciding factors for the site of deposition within the body is observed. Effective enhancement of serum lifetime have been achieved for polymeric NPs in a size ranging from 70–200 nm.

Several investigations have been done on the solubilization and delivery of hydrophobic drugs using block copolymer micelles [103–109].

e.g., carbamazepine and hydrochlorothiazide have been solubilized using plutonic micelles [103, 104]. Another study involves the solubilization of Nimesulide using star block four-armed PEO-PPO-BCP, and it concluded that the solubility increases with temperature and pH while it decrease in the presence of added salt. Pluronics such as P103, P104, and P105 have also been used for the solubilization of the Nimesulide drug. It is observed that CMC is the major thermodynamic parameter for deciding the micellar stability. Micelles below the CMC values in the body fluids disintegrate, and the drug is released into the external media [110]. CMC and CMT are vital parameters for the application of block copolymer micelle in controlled drug delivery system.

The CMC and CMT values of pluronics can be determined by UV-visible measurement, and it indicates that they form stable micelles that remain intact on dilution. It is observed that the entrapment of drugs in these nano-sized core-shell micelles results in increased bioavailability as well as improvement of the membrane transportation. The greater the hydrophobicity of the pluronics, more will be the solubility of the drug.

Due to the biodegradable and biocompatible nature of NPs from plouronics, they are employed in various biomedical applications. Also, they undergo phase separation in the concentration above CMC resulting in micelles with cargo space in the core for lyophilic drugs and hydrated exterior for stabilizing the micelles.

There are several advantages of employing drug incorporated micellar systems are traditional drug formulation such as: reduction of multidrug resistance, increase of bioavailability, targeted drug delivery

There are several factors that actually control the pharmaceutical properties of the NPs:

- the miscibility of drug and core-forming block;
- the physical state of the micellar core;
- the amount of the loaded drug;
- the molecular volume of the drug;
- the length of the core-forming block; and
- the localization of the drug within the micelle [111].

To improve the low encapsulation capacity, high PDI and large size of pure F–68 NPs, mixed micelles of F68 with polycaprolactone (PCL) derivative have been synthesized. The PCL derivative including PCL homopolymers, triblock, pentablock, and spherical NPs are synthesized.

Out of the all given PCL derivatives, the pentablock based particles in both pure ratio (2:1 MR of PCL to PPO in the core) and mixed ratio (1:1 MR of PCL to PPO in the core) forms displaying minimum crystallinity, gave the best results. They were found to have best drug encapsulation, most optimized particle size, better PDI's and finally faster mono-mechanism releases. These investigations also indicated the role of core-compatibilizer molecule for the miscibilization of the PCL and PPO in the hydrophobic part of the mixed micelles.

Currently, micelles of Pluronic L61 and F127 loaded with dexorubin are tested for treatment of cancer cells. Similar investigations are done using paclitaxel as anticancer agent. These formulations have shown improved pharmokinetic properties and enhanced blood circulation time. In addition to drug carrier system, these are also known to penetrate the blood-brain barrier [112] as well as potential candidate for gene therapy to increase the efficiency of the gene transfer technologies.

Compared to pluronics, at fixed PEO/PPO ratio, the tetronics show unique properties such as better penetrability power, more soluble in water and higher interfacial activity. The aggregation behavior of tetronics can be altered by changing the block sequence or temperature. In case of tetronics T1107, the hydrophilic PEO blocks are present at both the ends and results in "brush" like adsorption model whereas that for T904, we get "umbrella." With the increase in temperature, the PPO blocks progressively lose their hydration layer and hence results in the compact packing arrangement. Such studies help in the development of pharmaceutical formulation as well as for controlled drug delivery.

Tetronics are nowadays also used as the matrix to produce nanocomposite hydrogels. Such materials are emerging as an attractive concept to craft materials with tailored properties such as optical, electronic, mechanical as well as promoting a specific biological function. Like tetronic T1107 has been used to study the phase behavior using various techniques such as SANS, DLS, and FTIR- IR spectrum.

At low concentration of T1107, spherical micelles are formed, but at high concentration, it forms gels. As this process initiates, the shell of the micelles gets dehydrated, and long-range bcc order is revealed. However, in the presence of BT NPs (Barium titanate), that are modified with CD (cyclodextrin), the sol-gel transition (temperature decrease from 25°C to 12°C) as well as broadening of the gel phase region is observed. The presence of NPs do not disturb the BCC arrangement of the micelles in the gel. The low pH hinders the formation of gel. Hence, the gel phase can be readily used for biomedical applications.

Polymeric micelles are appealing trojan horses to make water-soluble hydrophobic drugs formulated for oral administration [113]. Mixed micelles formed by the co-micellization of two amphiphiles displaying different hydrophilic-lipophilic balance (HLB) have become an effective approach to optimize the encapsulation performance of poorly water-soluble drugs and the physical stability of the system [114, 115].

Nowadays, developing novel poorly water-soluble anti-HIV drugs and preparing highly concentrated aqueous formulations of EFV is the main goal to improve the pharmacotherapy of the pediatric population [116]. It is observed that the co-micellization of poloxamine/poloxamer polymeric micelles, i.e., T904/F127 enhance the encapsulation capacity as well as the physical stability. Such drug-loaded nanocarriers are sufficient in size so as to ensure the appropriate absorption in the intestine after oral administration. In such systems, F127 micelles play the role of micellar templates and then the incorporation of T904, makes the system more hydrophobic. These mixed micelles show synergistic behavior. It was observed that the mixed micelles are more stable than pure poloxamer/poloxamine micelles. Such mixed micelle behavior offers following advantages:

1. the greater encapsulation capacity of tetronic T904.
2. the greater physical stability of poloxamer F127, which helps to prepare more concentrated and stable aqueous solution.

Such systems help to prepare scalable and cost-effective PEO-PPO polymeric micelles to fine-tune not only the encapsulation performance but also the size of the drug-loaded aggregate.

10.8 FUTURE PERSPECTIVES

This chapter introduces the applications and uses of generally non-toxic and environmentally friendly pluronics and tetronics for the synthesis, characterization, and applications of nanomaterials. Their micellar assemblies play an important role, and by changing the micellar environment, one can easily control the shape and size of the nanomaterials. The thermoresponsive nature makes them suitable for various biomedical applications.

KEYWORDS

- **biomedical field**
- **gold nanoparticles**
- **pluronics**
- **tetronics**

REFERENCES

1. Rotello, V., (2004). *Nanoparticles, Building Block for Nanotechnology*. Kluwer Academic Publishers, New York.
2. Schmid, G., (2006). *Nanoparticles: From Theory to Application*, Wiley-VCH, Weinheim.
3. Caruso, F., (2004). *Colloids and Colloid Assemblies*, Wiley-VCH, Weinheim.
4. Burda, C., Chen, X., Narayanan, R., & El-Sayed, M. A., (2005). Chemistry and properties of nanocrystals of different shapes. *Chem. Rev., 105*(1025).
5. Chen, X., & Mao, S. S., (2007). Titanium dioxide nanomaterials: Synthesis, properties, modifications, and applications. *Chem. Rev., 107*, 2891–2959.
6. Xia, Y., Xiong, Y., Lim, B., & Skrabalak, S. E., (2009). Shape-controlled synthesis of metal nanocrystals: Simple chemistry meets complex physics? *Angew. Chem. Int. Ed., 48*, 60–103.
7. Eustis, S., & El-Sayed, M. A., (2006). Why gold nanoparticles are more precious than pretty gold: Noble metal surface Plasmon resonance and its enhancement of the radiative and nonradiative properties of nanocrystals of different shapes. *Chem. Soc. Rev., 35*, 209.
8. Jain, P. K., Huang, X., El-Sayed, I. H., & El-Sayed, M. A., (2008). Noble metals on the nanoscale: Optical and photothermal properties and some applications in imaging, sensing, biology, and medicine. *Acc. Chem. Res., 41*, 1578.
9. Murphy, C. J., Gole, A. M., Stone, J. W., Sisco, P. N., Alkilany, A. M., Goldsmith, E. C., & Baxter S. C., (2008). Gold nanoparticles in biology: Beyond toxicity to cellular imaging. *Acc. Chem. Res., 41*, 1721–1730.
10. Hu, M., Chen, J., Li, Z. Y., Au, L., Hartland, G. V., Li, X., Marquez, M., & Xia, Y., (2006). Gold nanostructures: Engineering their plasmonic properties for biomedical applications. *Chem. Soc. Rev., 35*, 1084.
11. Cobley, C. M., Chen, J., Cho, E. C., Wang, L. V., & Xia, Y., (2011). Gold nanostructures: A class of multifunctional materials for biomedical applications. *Chem. Soc. Rev., 40*, 44.

12. Sardar, R., Funston, A. M., Mulvaney, P., & Murray, R. W., (2009). Gold nanoparticles: Past, present, and future. *Langmuir, 25*, 13840.

13. Ghosh, S. K., & Pal. T., (2007). Interparticle coupling effect on the surface Plasmon resonance of gold nanoparticles: From theory to applications. *Chem. Rev., 107*, 4797.

14. Hashmi, A. S. K., & Hutchings, G. J., (2006). Gold catalysis. *Angew. Chem. Int. Ed., 45*, 7896.

15. Sperling, R. A., Gil, P. R., Zhang, F., Zanella, M., & Parak, W. J., (2008). Biological applications of gold nanoparticles. *Chem. Soc. Rev., 37*, 1896.

16. Boisselier, E., & Astruc, D., (2009). Gold nanoparticles in nanomedicine: Preparations, imaging, diagnostics, therapies, and toxicity. *Chem. Soc. Rev., 38*, 1759–1782.

17. Giljohann, D. A., Seferos, D. S., Daniel, W. L., Massich, M. D., Patel, P. C., & Mirkin, C. A., Angew., (2010). Gold nanoparticles for biology and medicine. *Chem. Int. Ed., 49*, 3280.

18. Faraday, M., (1857). *The Bakerian Lecture: Experimental Relations of Gold (and Other Metals) to Light* (Vol. 147, pp. 145–181). Philosophical Transactions of the Royal Society, London.

19. Nguyen, D. T., Kim, D. J., & Kim, K. S., (2011). Controlled synthesis and biomolecular probe application of gold nanoparticles. *Micron, 42*, 207.

20. Parab, H., Jung, C., Woo, M. A., & Park, H. G., (2011). An anisotropic snowflake-like structural assembly of polymer-capped gold nanoparticles. *J. Nanopart. Res., 13*, 2173.

21. Krishna, S., Ayothi, R., Hexemer, A., Finlay, J. A., Sohn, K. E., Perry, R., Ober, C. K., Kramer, E. J., Callow, M. E., Callow, J. A., & Fischer, D. A., (2006). *Anti-Biofouling Properties of Comblike Block Copolymers with Amphiphilic Side Chains* (Vol. 22, pp. 5075–5086). Langmuir.

22. Hobbs, R. G., Farrell, R. A., Bolger, C. T., Kelly, R. A., Morris, M. A., Petkov, N., & Holmes, J. D., (2012). Selective sidewall wetting of polymer blocks in hydrogen silsesquioxane directed self-assembly of PS-b-PDMS. *ACS Appl. Mater. Interfaces, 4*, 4737–4642.

23. Naohika, S., & Atsushi, M., (2013). Thermoresponsive polymers with functional groups selected for pharmaceutical and biomedical applications. *Tailored Polymer Architectures for Pharmaceutical and Biomedical Applications*. American Chemical Society.

24. Yusa, S. I., Fukuda, K., Yamamoto, T., Ishihara, K., & Morishima, Y., (2005). Synthesis of well-defined amphiphilic block copolymers having phospholipid polymer sequences as a novel biocompatible polymer micelle reagent. *Biomacromolecules, 6*, 663–670.

25. Lin, Y., & Alexandridis, P., (2002). Temperature-dependent adsorption of pluronic F127 block copolymers onto carbon black particles dispersed in aqueous media. *J. Phys. Chem. B., 106*(42), 10834–10844.

26. Barnes, T. J., & Prestidge, C. A., (2000). PEO-PPO-PEO block copolymers at the emulsion droplet-water interface. *Langmuir, 16*(9), 4116–4121.

27. Yang, L., & Alexandridis, P., (2000). Physicochemical aspects of drug delivery and release from polymer-based colloids. *Curt. Opin. Colloid Interface Sci., 5*, 132–143.

28. Ivanovo, R., Lindman, B., & Alexandridis, P., (2002). Effect of pharmaceutically acceptable glycols on the stability of the liquid crystalline gels formed by poloxamer 407 in water. *J. Colloid Interface Sci., 252*(1), 226–235.

29. Kabanov, A. V., Batrakova, E. V., & Alakhov, V. Y., (2002). Pluronic block copolymers as novel polymer therapeutics for drug and gene delivery. *J. Controlled Release, 82,* 182–212.

30. Ahmed, F., Alexandridis, P., Shankaran, H., & Neelmegham, S., (2001). The ability of poloxamers to inhibit platelet aggregation depends on their physicochemical properties. *Thromb. Hemostasis, 86*(6), 1532–1539.

31. Cohn, D., Sosnik, A., & Levy, A., (2003). Improved reverse thermo-responsive polymeric systems. *Biomaterials, 24*(21), 3707–3714.

32. Solee-Illia, G. J., De, A. A., Crepaldi, E. L., Grosso, D., & Sanchez, C., (2003). Block copolymer-templated mesoporous oxides. *Curt. Opin. Cooled Interface Sci., 8*(1), 109–126.

33. Zhao, D., Feng, J., Huo, Q., Melosh, N., Fredrickson, G. H., Chmelka, B. F., & Stucky, G. D., (1998). Triblock copolymer synthesis of mesoporous silica with periodic 50 to 300-angstrom pores. *Science, 279*(23), 548–552.

34. Han, Y. J., Kim, J. M., & Stucky, G. D., (2000). Preparation of noble metal nanowires using hexagonal mesoporous silica SBA–15. *Chem. Mater., 12*(8), 2068–2069.

35. Karanikolos, G. N., Alexandrines, P., Itskos, G., Petrou, A., & Mountziaris, T. J., (2004). Synthesis and size control of luminescent ZnSe nanocrystals by a microemulsion-gas contacting technique. *Langmuir, 20*(3), 550–553.

36. Kim, J. U., Cha, S. H., Shin, K., Jho, J. Y., & Lee, J. C., (2004). Preparation of gold nanowires and nanosheets in bulk block copolymer phases under mild conditions. *Adv. Mater., 16*(5), 459–464.

37. Wang, L., Chen, X., Zhan, J., Sui, Z., Zhao, J., & Sun, Z., (2004). Controllable morphology formation of gold nano- and micro-plates in amphiphilic block copolymer-based liquid crystalline phase. *Chem. Lett., 33*(6), 720–721.

38. Torcello-Gomez, A., Wulff-Perez, M., Galvez-Ruiz, M. J., Martín-Rodríguez, A., Cabrerizo-Vílchez, M., & Maldonado-Valderrama, J., (2014). Block copolymers at interfaces: Interactions with physiological media. *Adv. Colloid Interface Sci., 206,* 414–427.

39. Khullar, P., Singh, V., Mahal, A., Kaur, H., Singh, V., Banipal, T. S., Kaur, G., & Bakshi, M. S., (2011). Tuning the shape and size of gold nanoparticles with triblock polymer micelle structure transitions and environments. *J. Phys. Chem. C, 115,* 10442–10454.

40. Iqbal, M., Chung Y. I., & Tae, G., (2007). An enhanced synthesis of gold nanorods by the addition of Pluronic (F-127) *via*seed-mediatedted growth process *J. Mater. Chem., 17,* 335–342.

41. Liang, H. P., Wan, L. J., Bai, C. L., & Jiang, L., (2005). Gold hollow nanospheres: Tunable surface Plasmon resonance controlled by interior-cavity sizes. *J. Phys. Chem. B, 109,* 7795–7800.

42. Johnson, C. J., Dujardin, E., Davis, S. A., Murphy, C. J., & Mann, S., (2002). Growth and form of gold nanorods prepared by seed-mediated, surfactant-directed synthesis. *J. Mater. Chem., 12,* 1765–1770.

43. Xie, J. P., Lee, J. Y., & Wang, D. I. C., (2007). Synthesis of single-crystalline gold nanoplates in aqueous solutions through biomineralization by serum albumin protein. *J. Phys. Chem. C., 111*, 10226–10232.

44. Daniel, M. C., & Astruc, D., (2004). Gold nanoparticles: Assembly, supramolecular chemistry, quantum-size-related properties, and applications toward biology, catalysis, and nanotechnology. *Chem. Rev., 104*, 293–346.

45. Pileni, M. P., (2003). The role of soft colloidal templates in controlling the size and shape of inorganic nanocrystals. *Nat. Mater., 2*, 145–150.

46. Niidome, Y., Honda, K., Higashimoto, K., Kawazumi, H., Yamada, S., Nakashima, N., et al., (2007). Surface modification of gold nanorods with synthetic cationic lipids. *Chem. Commun., 36*, 3777–3779.

47. Chen, J. Y., Willey, B., Li, Z. Y., Campbell, D., Saeki, F., Cang, H., et al., (2005). Gold nanocages: Engineering their structure for biomedical applications. *Adv. Mater., 17*, 2255–2261.

48. Millstone, J., Métraux, G., & Mirkin, C., (2006). Controlling the edge length of gold nanoprisms via a seed-mediated approach. *Adv. Funct. Mater., 16*, 1209.

49. Millstone, J, E., Park, S., Shuford, K. L., Qin, L., Schatz, G. C., & Mirkin, C. A., (2005). Observation of a quadrupole plasmon mode for a colloidal solution of gold nanoprisms. *J. Am. Chem. Soc., 127*, 5312.

50. LaMer, V. K., & Dinegar, R. H., (1950). Theory, product, on and mechanism of formation of monodispersed hydrosols. *J. Am. Chem. Soc., 72*, 4847–4854.

51. Mer, V. K. L., (1952). Nucleation in phase transitions. *Ind. Eng. Chem., 44*, 1270.

52. Njoki, P. N., Luo, J., Kamundi, M. M., Lim, S., & Zhong, C., (2010). Aggregative growth in the size controlled of monodispersed gold nanoparticles. *J. Langmuir, 26*, 13622.

53. Shields, S. P., Richards, V. N., & Buhro, W. E., (2010). Nucleation Control of size and dispersity in aggregative nanoparticle growth. A study of the coarsening kinetics of thiolate-capped gold nanocrystals. *J. Chem. Mater., 22*, 3212.

54. Alexandridis, P., & Hatton, T. A., (1995). Poly (ethylene oxide) poly (propylene oxide) poly (ethylene oxide) block copolymer surfactants in aqueous solutions and at interfaces: Thermodynamics, structure, dynamics, and modeling. *Colloids and Surfaces A: Physicochemical and Engineering Aspects, 96*, 1–46.

55. Nam, H. S., Hwang, N. M., Yu, B. D., & Yoon, J. K., (2002). Formation of an icosahedral structure during the freezing of gold nanoclusters: Surface-induced mechanism. *Phys. Rev. Lett., 89*, 275502.

56. Xu, J., Li, S., Weng, J., Wang, X., Zhou, Z., Yang, K., Liu, M., et al., 2008. *Adv. Funct. Mater., 18*, 277–284.

57. Khan, Z., Al Thabaiti, S., Obaid, A., Khan, Z., & Al Youbi, A. J., (2012). Shape-directing role of cetyltrimethylammonium bromide in the preparation of silver nanoparticles. *Colloid Interface Sci., 367*, 101–108.

58. Jin, R., Cao, Y. W., Mirkin, C. A., Kelly, K. L., Schatz, G. C., & Zheng, J. G., (2001). Photoinduced conversion of silver nanospheres to nanoprisms. *Science, 294*, 1901–1903.

59. Smith, D. J., Long, A. K. P., Wallenberg, L. R., & Bovin, J. O., (1986). Dynamic atomic-level rearrangements in small gold particles. *Science, 233*, 872–875.

60. Jana, N. R., Gearheart, L., & Murphy, C. J., (2001). Seed-mediated growth approach for shape-controlled synthesis of spheroidal and rod-like gold nanoparticles using a surfactant template. *Adv. Mater., 13*, 1389–1393.

61. Sau, T. K., Pal, A., Jana, N. R., Wang, Z. L., & Pal, T., 200 Size-controlledled synthesis of gold nanoparticles using photochemically prepared seed particles. *J. Nanopart. Res., 3*, 257–261.

62. Sau, T. K., & Murphy, C. J., (2004). Room temperature, high-yield synthesis of multiple shapes of gold nanoparticles in aqueous solution. *J. Am. Chem. Soc., 126*, 8648–8649.

63. Zou, X., Ying, E., & Dong, S., (2006). Seed-mediated synthesis of branched gold nanoparticles with the assistance of citrate and their surface-enhanced Raman scattering properties. *Nanotechnology, 17*, 4758–4764.

64. Zhang, J. M., Ma, F., & Xu, K. W., (2004). Calculation of the surface energy of FCC metals with modified embedded-atom method. *Appl. Surf. Sci., 229*, 34–42.

65. Wiley, B., Sun, Y., Chen, J., Cang, H., Li, Z. Y., Li, X., & Xia, Y., (2005). Silver and gold nanostructures with well-controlled shapes. *MRS Bull., 30*, 356–361.

66. Heinz, H. B. L., Farmer, R. S., Pandey, J. M., Sloeik, S. S., Patnaik, R. P., & Naik, R. R., (2009). Nature of molecular interactions of peptides with gold, palladium, and Pd–Au bimetal surfaces in aqueous solution. *J. Am, Chem. Soc., 131*, 9704–9714.

67. Feng, J., Pandey, R. B., Berry, R. J., Farmer, B. L., Naik, R. R., & Heinz, H., (2011). Adsorption mechanism of single amino acid and surfactant molecules to Au {111} surfaces in aqueous solution: Design rules for metal-binding molecules. *Soft Matter., 7*, 2113–2120.

68. Kan, C., Zhu, X., & Wang, G., (2006). Single-crystalline gold microplates: Synthesis, characterization, and thermal stability. *J. Phys. Chem. B., 110*, 4651–4656.

69. Dong, J., Chowdhry, B. Z., & Leharne, S. A., (2004). Solubilization of polyaromatic hydrocarbons in aqueous solutions of poloxamine T803. *Colloids Surf A Physiochem. Eng. Aspects, 246*, 91–98.

70. Alvarez-Lorenzo, C., Gonzalez-Lopez, J., Fernandez-Tarrio, M., Sandez-Macho, I., & Concheiro, A., (2007). *Eur. J. Pharm. Biopharma., 66*, 244–252.

71. Gonzalez-Lopez, J., Alvarez-Lorenzo, C., Taboada, P., Sosnik, A., Sandez-Macho, I.,& Concheiro, A., (2008). Self-associative behavior and drug-solubilizing ability of poloxamine (tetronic) block copolymers. *Langmuir, 24*, 10688–10697.

72. Longenberger, L., & Mills, G., (1995). Formation of metal particles in aqueous solutions by reactions of metal complexes with polymers. *J. Phys. Chem., 99*, 475–478.

73. Sakai, T., & Alexandridis, P., (2005). Size shape-controlled synthesis of colloidal gold through autoreduction of the auric cation by poly (ethylene oxide)-poly (propylene oxide) block copolymers in aqueous solution at ambient conditions. *Nanotechnology, 16*, S344–S353.

74. Newman, J. D. S., & Blanchard, G., (2006). Formation of gold nanoparticles using amine reducing agents. *J. Langmuir, 22*, 5882–5887.

75. Yang, B., Guo, C., Chen, S., Ma, J. H., Wang, J., Liang, X. F., Zheng, L., & Liu, H. Z., (2006). Effect of acid on the aggregation of poly (ethylene xide)-poly (propylene oxide)-poly (ethylene oxide) block copolymers. *J. Phys. Chem. B., 110*(23068).

76. Piao, Y. Z., Jang, Y. J., Shokouhimehr, M., Lee, I. S., & Hyeon, T., (2007). *Small, 3*, 255.
77. Khullar, P., Singh, V., Mahal, A., Kaur, H., Singh, V., Banipal, T. S., Kaur, G., & Bakshi, M. S., (2013). Block copolymer micelles as nanoreactors for self-assembled morphologies of gold nanoparticles. *J. Phys. Chem. B, 117*, 3028–3039.
78. Mishra, S., Peddada, L. Y., Devore, D. I., & Roth, C. M., (2012). Poly (Alkylene Oxide) copolymers for nucleic acid delivery. *Acc. Chen. Res., 45*, 1057–1066.
79. Cuestas, M. Ü. L., Sosnik, A., & Mathet, V. Ü. L., (2011). Poloxamines display a multiple inhibitory activity of ATP-binding cassette (ABC) transporters in cancer cell lines. *Mol. Pharmaceutics, 8*, 1152–1164.
80. Sakai, T., & Alexandridis, P., (2005). Mechanism of gold metal ion reduction, nanoparticle growth and size control in aqueous amphiphilic block copolymer solutions at ambient conditions, *J. Phys. Chem. B, 109*, 7766–7777.
81. Moghimi, S. M., et al., (1993). Coating particles with blocopolymermer (poloxamine–908) suppresses opsonization but permits the activity of dysopsonins in the serum. *Biochim. Biophys. Acta., 1179*, 157–165.
82. Li, J. T., & Caldwell, K. D., (1996). Plasma protein interactions with Pluronic treated colloids. *Colloids Surfaces B-Biointerfaces, 7*, 9–22.
83. Parmar, A., Parekh, P., & Bahadur, P., (2013). Solubilization and release of a model drug nimesulide from PEO-PPO-PEO block copolymcore-shellell micelles: Effect of size of PEO blocks. *J. Solut. Chem., 42*, 80–101.
84. Zou, P., Suo, J. P., Nie, L., & Feng, S. B., (2012). Temperature-sensitive biogradable mixed star-shaped block copolymers hydrogels for an injection application. *Polymer, 53*, 1245–1257.
85. Huynh, C. T., Nguyen, Q. V., Kang, S. W., & Lee, D. S., (2012). Synthesis and characterization of poly (amiurea-urethaneane)-based block copolymer and its potential application as injectable pH/temperature-sensitive hydrogel for protein carrier. *Polymer, 53*, 4069–4075.
86. Fusco, S., Borzacchiello, A., & Netti, P. A., (2006). Perspectives on: PEO-PPO-PEO triblock copolymers and their biomedical applications. *J. Bioact. Compat. Polym., 21*, 149–164.
87. Batrakova, E., Lee, S., Li, S., Venne, A., Alakhov, V., & Kabanov, A., (1999). Fundamental relationships between the composition of pluronic block copolymers and their hypersensitization effect in MDR cancer cells. *Pharm. Res., 16*, 1373–1379.
88. Alakhova, D. Y., & Kabanov, A. V., (2014). Pluronics and MDR reversal: An update. *Mol. Pharmaceutics, 11*, 2566–2578.
89. Alakhov, V. Y., Moskaleva, E. Y., Batrakova, E. V., & Kabanov, A. V., (1996). Hypersensitization multidrug-resistant human ovarian carcinoma cells by plutonic P85 block copolymer. *Bioconjugate Chem., 7*, 209–216.
90. Morikawa, K., et al., (1987). Enhancement of therapeutic effects of recombinant interleukin–2 on a transplantable rat fibrosarcoma by the use of a sustained release vehicle. *Pluronic Gel Cancer, 47*, 37–41.
91. Fults, K. A., & Johnson T. P., (1990). Sustained-release of urease from a poloxamer gel matrix. *J. Parenter. Sci. Technol., 44*, 58–65.

92. Katakana, M., et al., (1997). Controlled release of human growth hormone in rats following parenteral administration of poloxamer gels. *J. Control. Release, 49,* 21–26.

93. Nalbandian, R. M., et al., (1987). Pluronic F-127 gel preparation as an artificial skin in the treatment third-degree burns in pigs. *J. Biomed. Mater. Res., 21,* 1135–1148.

94. Van Even, H. W., & Konings, W. N., (1997). Multidrug transport from bacteria to man: Similarities in structure and function. *Semin. Cancer Biol., 8,* 183–191.

95. Germany, U. A., et al., (1993). p-Glycoproteins: Mediators of multidrug resistance. *Semi. Cell Biol., 4,* 63–76.

96. Batrakova, E. V., et al., (1999). Fundamental relationships between the composition of Pluronic block copolymers and the hyper sensitisation effect in MDR cancer cells, *Pharm. Res., 16,* 1373–1379.

97. Lindfors, L., Foreseen, S., Westergren, J., & Olsson, U., (2008). Nucleation and crystal growth in supersaturated solutions of a model drug. *J. Colloid Interf Sci., 325*(2), 404–413.

98. Lee, J., Choi, J. Y., & Park, C. H., (2008). Characteristics of polymers enabling nano communication of water-insoluble drugs. *Int. J. Pharm., 355*(1–2), 328–36.

99. Lai, F., Sinico, C., Ennas, G., Marongiu, F., Marongiu, G., & Fadda, A. M., (2009). Diclofenacnano suspensions: Influence of preparation procedure and crystal form on drug dissolution behavior. *Int. J. Pharm., 373*(1 & 2), 124–132.

100. Xiong, R., Lu, W., Li, J., Wang, P., Xu, R., & Chen, T., (2008). Preparation and characterization of intravenously injectable nimodipine nanosuspension. *Int. J. Pharm., 350*(1 & 2), 338–43.

101. Nivaggioli, T., Tsao, B., Alexandridis, P., & Hatton, T. A., (1995). Microviscosity in pluronic and tetronicpoly (ethylene oxide)-poly (propylene oxide) block copolymer micelles. *Langmuir, 11,* 119–26.

102. Yang, T. F., Chen, C. N., Chen, M. C., Lai, C. H., Liang, H. F., & Sung, H. W., (2007). Shell-crosslinked pluronic L121 micelles as a drug delivery vehicle. *Biomaterials, 28,* 725–734.

103. Kadam, Y., Yerramilli, U., Bahadur, A., & Bahadur, P., (2009Solubilizationion of poorly water-soluble drcarbamazepineine in Pluronic® micelles: Effect of molecular characteristics, temperature and added salt on solubilizing capacity. *Colloids Surf. B., 72,* 141–147.

104. Kadam, Y., Yerramilli, U., Bahadur, A., & Bahadur, P., (2011). Micelles from PEO-PPO-PEO block copolymers as nanocontainers solubilization of a poorly water-soluble drug hydrochlorothiazide. *Colloids Surf. B. Biointerfaces, 83,* 49–57.

105. Parekh, P., Singh, K., Marangoni, D. G., & Bahadur, P., (2011). Micellization solubilization of a model hydrophobic drug nimesulide in aqueous salt solutions of Tetronic® T904. *Colloids Surf. B., 83,* 69–77.

106. Barreiro-Iglesias, R., Bromberg, L., Temchenko, M., Hatton, T. A., Alvarez-Lorenzo, C., & Concheiro, A., (2005). Pluronic-g-poly (acrylic acid) copolymers as novel excipients site-specific sustained-release tablets. *Eur. J. Pharm. Sci., 26,* 374–385.

107. Bae, K. H., Lee, Y., & Park, T. G., (2007). Oil-encapsulating PEO-PPO-PEO/PEG shell cross-linked nanocapsules for target-specific delivery of paclitaxel. *Biomacromolecules, 8,* 650–656.

108. Oh, K. S. K., Lee, E., Han, S. S., Cho, S. H., Kim, D., & Yuk, S. H., (2005). Formation of core/shell nanoparticles with a lipid core and their application as a drug delivery system. *Biomolecules, 6,* 1062–1067.

109. Bhattacharya, A., Kankanala, K., Pal, S., & Mukherjee, A. K., (2010). A nimesulide derivative with potential anti-inflammatory activity: Synthesis, x-ray powder structure and DFT study. *J. Mol. Struct., 975,* 40–46.

110. Jones, M. C., & Leroux, J. C., (1999). Polymeric micelles a new generation of colloidal drug carriers. *Eur. J. Pharm. Biopharm., 48,* 101–111.

111. Allen, C., Maysinger, D., & Eisenberg, A., (1999). Nano-engineering block copolymer aggregates for drug delivery. *Colloids and Surf B Biointerfaces, 16*(1–4), 3–27.

112. Kabanov, A. V., Batrakova, E. V., & Miller, D. W., (2003). Pluronic® block copolymers as modulators of drug efflux transporter activity in the blood-brain barrier. *Adv. Drug Delivery Rev., 55,* 151–164.

113. Sosnik, A., Carcaboso, A. M., & Chiappetta, D. A., (2008). Polymer nanocarriers: New endeavors for the optimization of the technological aspects of drugs. *Recent Pat. Biomed. Eng., 1,* 43–59.

114. Oh, K. T., Bronich, T. K., & Kabanov, A. V., (2004). Micellar formulations for drug delivery based on mixtures of hydrophobic and hydrophilic Pluronic® block copolymers. *J. Control Release, 94* 411–422.

115. Wei, Z., Hao, J., Yuan, S., Li, Y., Juan, W., Sha, X., et al., (2009). Paclitaxel-loaded pluronic P123/F127 mixed polymeric micelles: Formulation, optimization and in vitro characterization. *Int. J. Pharm., 376,* 176–185.

116. Sosnik, A., (2010). Nanotechnology contributions to the pharmacotherapy pediatric HIV: A dual scientific and ethical challenge and a still pending agenda. *Nanomedicine, 5,* 833–837.

117. Alkilany, A. M., Nagaria, P. K., Hexel, C. R., Shaw, T. J., Murphy, C. J., & Wyatt, M. D., (2009). Cellular uptake and cytotoxicity of gold nanorods: Molecular origin of cytotoxicity and surface effects. *Small, 5*(6), 701–708.

118. Bromberg, L., (2005). Intelligent hydrogels for the oral delivery of chemotherapeutics. *Expert Opin. Drug Delivery, 2,* 1003–1013.

119. Goy-Lopez, S., Taboada, P., Cambon, A., Juarez, J., Lorenzo, A. C., Concheiro, A., & Mosquera, V., (2010). Modulation of size and shape of Au nanoparticles using amino-x-shaped poly (ethylene oxide)-poly (propylene oxide) block copolymers. *J. Phys. Chem. B, 114,* 66–76.

120. Iqbal, M., Chung, Y., & Tae, G., (2007). An enhanced synthesis of gold nanorods by the addition of Pluronic (F-127) viaseed-mediatedted growth process. *J. Mater. Chem., 17,* 335–342.

121. Kaur, P., & Chudasama, B., (2014). Single step synthesis of pluronic stabilized IR responsive gold nanoplates. *RSC Adv. 4,* 36006–36011.

122. Khi, S., Hassanzadeh, S., & Goliaie, B., (2007). Effects of hydrophobic drug-polyesteric core interactions on drug loading and release properties of poly (ethylene glycol)-polyester-poly (ethylene glycol) trip block core-shell nanoparticles. *Nanotechnology, 18*(17). doi: 10.1088/0957–4484/18/17/175602.

123. Khullar, P., Singh, V., Mahal, A., Kaur, H., Singh, V., Banipal, T. S., Kaur, G., & Bakshi, M. S., (2010). How PEO-PPO-PEO triblock polymer micelles control the synthesis of gold nanoparticles: Temperature and hydrophobic effects, *Langmuir, 26,* 11363–11371.

124. Lim, B., Jiang, M., Tao, J., Camargo, P. H. C., Zhu, Y., & Xia, Y., (2009). Shape-controlled synthesis of Pd nanocrystals in aqueous solutions. *Adv. Funct. Mater., 19,* 189–200.

125. Moschwitzer, J. P., (2013). Drug nanocrystals in the commercial pharmaceutical development process. *Int. J. Pharm., 453*(1), 142–56.

126. Muller, R. H., Gohla, S., & Keek, C. M., (2011). State of the art of nanocrystals-special features, production, nanotoxicology aspects and intracellular delivery. *Eur. J. Pharm. Biopharm., 78*(1), 1–9.

127. Rabinow, B. E., (2004). Nanosuspensions in drug delivery. *Nat. Rev. Drug Discov., 3,* 785–96.

128. Zia, Y., Xiong, Y., Lim, B., & Skrabalak, S. E., (2009). Shape-controlled synthesis of metal nanocrystals: Simple chemistry meets complex physics? *Angew. Chem., Int. Ed., 48,* 60.

129. Zhang, C., Zhang, J., Han, B., Zhao, Y., & Li, W., (2008). Synthesis of icosahedral gold particles by a simple and mild route. *Green Chem., 10,* 1094–1098.

130. Murphy, C. J., Gole, A. M., Hunyadi, S. E., Stone, J. W., Sisco, P. N., Alkilany, A., Kinard, B. E., & Hankins, P., (2008). Chemical sensing and imaging with metallic nanorods. *Chem. Commun.,* 544–557.

131. Lliz-marfan, L. M., & Kamat, P. V., (2003). *Nanoscale Materials.* Kluwer Academic Publishers, New York.

132. Singh, V., Khullar, P., Dave, P. N., Kaura, A., Bakshi, M. S., & Kaur, G. (2014). pH and thermo-responsive tetronic micelles for the synthesis of gold nanoparticles: effect of physiochemical aspects of tetronics. *Phys. Chem. Chem. Phys., 16,* 4728–4739.

CHAPTER 11

METHODS FOR CALCULATING THE THERMOELECTRIC CHARACTERIZATIONS OF NANOMATERIALS

A. V. SEVERYUKHIN, O. YU. SEVERYUKHINA, and
A. V. VAKHRUSHEV

Udmurt Federal Research Center of the Ural Branch of the Russian Academy of Sciences, Institute of Mechanics, Izhevsk, Russia, E-mail: vakhrushev-a@yandex.ru

ABSTRACT

The review of main methods definitions of thermoelectric parameters of nanomaterials is considered. Possible ways of increasing the efficiency of thermoelectric conversion in low-dimensional systems are given. The influence of such parameters as the Seebeck coefficient, electrical conductivity, thermal conductivity and temperature on the value of the thermoelectric figure of merit is determined. Relations for determining the density of the number of states of carriers for systems of different dimensions are described.

11.1 THEORETICAL BASIS

Thermoelectric nanomaterials have great prospects for use in modern electronics, cooling systems, and other technical devices. Therefore, the study of thermoelectric properties of nanomaterials is one of the urgent problems of modern nanotechnology. A lot of papers [1–20] devoted to the general ideas of thermoelectricity have been published. In this chapter, we consider thermoelectric phenomena taking place in nanoscale and low-dimensional systems and their characteristics.

What is thermoelectricity? Thermoelectricity occurs when there is a different temperature on each side of a metal or semiconductor sample. Electrons having large thermal energy at the heated edge of the sample diffuse to the region with a lower temperature, as a result of which their concentration in this zone increases and a negative charge accumulates, while a positive charge remains at the opposite side (Figure 11.1) [21]. This process leads to the formation of an internal electric field. This effect is called the Seebeck effect and can be described as:

$$E = -S\nabla T, \tag{1}$$

where E is the electromotive field, S is the Seebeck coefficient, ∇T is the temperature gradient.

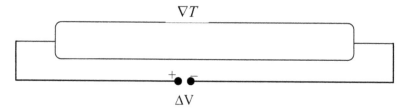

FIGURE 11.1   Scheme of the Seebeck effect.

In addition to the Seebeck effect, the Peltier effect takes place. This effect is due to the appearance of an electric current in the circuit when creating the supply and removal of heat at different ends of the sample. In this case, the relationship between the heat flux and the electric current is determined by the Peltier coefficient.

$$\Pi = \frac{Q}{I}, \tag{2}$$

where Q is the heat generated per unit time, I is the electric current. The Peltier effect is reversed by the Seebeck effect, thus:

$$\Pi = TS. \tag{3}$$

These effects are the basis of the work of electric power generators and refrigerators and other technical devices. The efficiency of a generator or a refrigerator can be estimated by a dimensionless figure of merit determined by the formula:

$$ZT = \frac{\sigma S^2 T}{\kappa},$$ (4)

where σ is the electrical conductivity, κ is the thermal conductivity.

Thermoelectric refrigerators have several advantages, for example, they do not have moving parts, but their efficiency is several times less than the efficiency of mechanical refrigerators. In 1851, William Thomson (Lord Kelvin) discovered the relation between Seebeck and Peltier coefficients. The spatial gradient of temperature can lead to a gradient of the Seebeck coefficient in different materials. The Thomson effect describes the heating or cooling of a simple homogeneous conductor when the current flows through the material at a constant temperature gradient.

$$q = -KJ \cdot \nabla T,$$ (5)

where K is the Thomson coefficient. The relation between the Thomson and Seebeck coefficients is given by:

$$K = T \frac{dS}{dT}.$$ (6)

In macro-samples there are definite dependences of the Seebeck coefficient, electrical conductivity and thermal conductivity on the carrier concentration. The thermoelectric figure of merit, ZT, depends on these values, and hence on the carrier concentration. The absolute values of all thermoelectric coefficients increase with decreasing carrier concentration; therefore, in semiconductors, they are tens and hundreds of times greater than in metals and alloys [22].

The Seebeck coefficient S, electrical conductivity σ, thermal conductivity κ, as well as temperatures of hot T_h and cold T_c sides affect the efficiency of the process of converting thermal energy into electricity.

11.2 THERMOELECTRICITY IN NANOSCALE

From the point of view of increasing the efficiency of thermoelectric materials, the creation of thermoelectric nanostructures is of particular interest, since new effects such as tunneling of charge carriers between nanoparticles, as well as additional phonon scattering at the boundary of polycrystals and the associated decrease in thermal conductivity, can appear in nanostructured thermoelectrics. These factors cause an improvement in thermoelectric properties in bulk thermoelectric nanomaterials.

Great progress in the field of creating and investigating the properties of nanostructured thermoelectric materials has been achieved in the last few decades. Thermoelectric nanostructures have a number of features. First, nanoscale objects have a special structure, which can lead to both a decrease in electric propulsion, and an increase in the value of the thermo-emf coefficient and figure of merit in comparison with the crystalline macro sample. The reason for this is the fact that in nanosystems, with increasing density of states near the Fermi level, the width of the band gap of a semiconductor increases. Second, there is a scattering of phonons by defects and interfaces. This leads to the fact that the decrease in heat conductivity occurs faster than the decrease in electrical conductivity.

Theoretical and experimental studies have shown that the use of nanoscale systems contributes to a significant increase in the thermoelectric figure of merit, ZT [23–35]. It should be noted that the thermoelectric figure of merit of nanostructured materials, for example, based on Bi_2Te_3, PbTe, SiGe, is 30–40% higher in comparison with crystalline analogs [36].

A feature of thermoelectric phenomena in nanoscale systems is the influence of scale factors on the behavior of conduction electrons. The electrical conductivity, electron and phonon thermal conductivities of materials depend on the corresponding mean free paths of electrons and phonons, l_e, l_{ph}. By mean free path l is understood the distance that the particle (electron or phonon) passes between two successive scattering events. To increase the thermoelectric figure of merit, it is necessary that the electrical conductivity σ grows, and the phonon thermal conductivity κ_{ph} decreases. This can be ensured if the characteristic size of the structural element of nanosystem will limit the phonon free path l_{ph}, but not the electrical conductivity σ. This leads to the relation $l_e < l_{ph}$. When the mean free path of electrons decreases, the electrical conductivity of the material must decrease. Thus, the influence of the boundaries of the structural elements of a nanosystem is significant.

If the wavelength of the electrons is much less than the electron mean free path $\lambda \ll l_e$, then the classical band picture remains valid. In three-dimensional nanosystems, the following dispersion relation for electrons is valid [21]:

$$\varepsilon = \frac{\hbar^2 k_x^2}{2m_x^*} + \frac{\hbar^2 k_y^2}{2m_y^*} + \frac{\hbar^2 k_z^2}{2m_z^*}, \tag{7}$$

where x, y, z are the indices defining the direction of motion, and m^* is the effective mass of the carriers.

For two-dimensional systems with a characteristic size $\lambda_e \ll l_e$, the preceding equation takes the form:

$$\varepsilon = \frac{\hbar^2 k_x^2}{2m_x^*} + \frac{\hbar^2 k_y^2}{2m_y^*} + \frac{\hbar^2 \pi^2 n^2}{2m_z^* d^2}, \tag{8}$$

where $n = 1, 2, 3,...$

For one-dimensional systems, such as nanowires or quantum wires, the motion of an electron is possible in one direction. In such systems, the energy is quantized along the y and z-axes, so Eq. (7) takes the form:

$$\varepsilon = \frac{\hbar^2 k_x^2}{2m_x^*} + \frac{\hbar^2 \pi^2 j^2}{2m_y^* d^2} + \frac{\hbar^2 \pi^2 n^2}{2m_z^* d^2}, \tag{9}$$

where d is the diameter of the nanowire.

For quantum dots of diameter d, the expression takes the form:

$$\varepsilon = \frac{\hbar^2 \pi^2 i^2}{2m_x^* d^2} + \frac{\hbar^2 \pi^2 j^2}{2m_y^* d^2} + \frac{\hbar^2 \pi^2 n^2}{2m_z^* d^2}, \tag{10}$$

Thus, using the dispersion relations for energy considered earlier, the density of the number of states of carriers can be defined as:

$$g(\varepsilon) = \frac{\sqrt{m_x^* m_y^* m_z^*}}{\hbar^3 \pi^2} \sqrt{2\varepsilon} \tag{11}$$

For two-dimensional system:

$$g(\varepsilon) = \sum_n g_n(\varepsilon),$$

$$\begin{cases} g_n(\varepsilon) = \dfrac{\sqrt{m_x^* m_y^*}}{\hbar^2 \pi d}, & \text{if } \varepsilon > \varepsilon_n \\ g_n(\varepsilon) = 0, & \text{if } \varepsilon \leq \varepsilon_n \end{cases} \tag{12}$$

For one-dimensional system:

$$g(\varepsilon) = \sum_{n,j} g_{n,j}(\varepsilon),$$

$$\begin{cases} g_{n,j}(\varepsilon) = \dfrac{\sqrt{m_x^*}}{\hbar \pi d^2} \dfrac{1}{\sqrt{\varepsilon - \varepsilon_{n,j}}}, & \text{if } \varepsilon > \varepsilon_{n,j} \\ g_{n,j}(\varepsilon) = 0, & \text{if } \varepsilon \le \varepsilon_{n,j} \end{cases} \qquad (13)$$

And for quantum dots:

$$g(\varepsilon) = \sum_{n,j,i} g_{n,j,i}(\varepsilon),$$

$$\begin{cases} g_{n,j,i}(\varepsilon) = \dfrac{1}{d^3}, & \text{if } \varepsilon = \varepsilon_{n,j,i} \\ g_{n,j,i}(\varepsilon) = 0, & \text{if } \varepsilon \ne \varepsilon_{n,j,i} \end{cases} \qquad (14)$$

As noted above, the magnitude of the electrical conductivity σ affects the value of the thermoelectric figure of merit ZT. The electrical conductivity, in turn, is determined by the mobility of the carriers μ. From the Boltzmann kinetic equation, the mobility of the carriers, when carrying over in the x-direction, is

$$\mu = |e| \frac{\tau}{m_x^*} = |e| \frac{l_e}{v m_x^*}, \qquad (15)$$

where τ is the relaxation time.

The thermal conductivity κ consists of the electron and phonon components [37, 38]:

$$\kappa = \kappa_e + \kappa_{ph}. \qquad (16)$$

The relationship between electrical conductivity and the electronic component of thermal conductivity is determined by the Wiedemann–Franz law law:

$$\kappa_e = LT\sigma, \qquad (17)$$

where L is the Lorentz number. It should be noted that in the general case the concentration of carriers influences the value of the Lorentz number, so for materials, with a low carrier concentration, the Lorentz number L decreases.

There is another approach to determining the Seebeck coefficient. So, in accordance with the Mott approach, the Seebeck coefficient can be expressed in terms of the density of states and the distribution function. In this approach, the electrical conductivity is an integral over the entire internal energy region is given by the Fermi-Dirac distribution:

$$\sigma = \int \sigma(\varepsilon)\left(-\frac{\partial f_0}{\partial \varepsilon}\right) d\varepsilon \, . \tag{18}$$

Thus, the Seebeck coefficient can be written as [39]:

$$S = \frac{k_B}{e\sigma} \int_0^\infty \sigma(\varepsilon)\left(\frac{\varepsilon - \varepsilon_F}{k_B T}\right)\left(-\frac{\partial f_0}{\partial \varepsilon}\right) d\varepsilon \, , \tag{19}$$

where k_B is the Boltzmann constant.

For metals and heavily doped semiconductors, the Mott ratio can be written:

$$S = \frac{\pi^2 k_B^2 T}{3e}\left(\frac{d\{\ln(\sigma(\varepsilon))\}}{d\varepsilon}\right)_{\varepsilon = \varepsilon_F} . \tag{20}$$

From the relation (20), one can see that the increase in electrical conductivity $\sigma(\varepsilon)$ leads to an increase in the Seebeck coefficient S and, as a consequence, to an increase in the thermoelectric figure of merit ZT. It should be noted that in nanosystems in comparison with volumetric systems, the value of the thermoelectric figure of merit ZT is optimal in a narrower range of Fermi energies.

Now consider the phonon transport in nanoscale systems. For the further development of nanostructured thermoelectric materials, it is necessary to understand the physics of the processes that describe phonon transport. Before the appearance of nanostructured materials, it was believed that the limit of the lowest values of thermal conductivity was achieved by the creation of solid solutions. This theory has dominated for 50 years and only recently was refuted. This happened because similar nanostructures were obtained for existing solid solutions. The thermal conductivity of the resulting structures was lower [40, 41]. What is the reason for this phenomenon?

As is known, phonons in the material have a wave spectrum [42]. Phonons with different wavelengths contribute differently to the total thermal conductivity of the material. In alloys and solid solutions, short-wave phonons undergo scattering by impurity atoms, however, in such a material, the transport of phonons from a longer wavelength is not difficult. By introducing structural elements commensurate with the phonon wavelength, it is possible to achieve a further reduction in the lattice thermal conductivity.

There are a number of theoretical models created to describe phonon interactions. A review of theoretical methods for describing phonon transport in nanostructures is given in Ref. [42].

In [43, 44], the model of the Callaway thermal conductivity is presented. This model is based on the Debye model with the addition of the effect of the phenomena that arise at the interfaces. On the whole, the Callaway model quite accurately describes the behavior of the integral value of the thermal conductivity of the material, but, more importantly, the accuracy of calculating the individual quantities included in its composition is low. If we analyze this model, we can conclude that it cannot accurately describe phonon interactions in the nanostructured material. Thus, even for crystalline silicon, the Callaway model demonstrates significant deviations from a more accurate molecular dynamics model when calculating the mean free path and the phonon wavelength.

On the basis of the assumption that a large number of interfaces with characteristic dimensions less than the mean free path of phonons $d < l_{ph}$ creates boundary barriers between regions of the nanostructure, theoretical calculations of the thermal conductivity of materials have been carried out [45]. The boundary resistance was calculated using the Boltzmann kinetic equation and the Monte Carlo method. A new approach to the analytical calculation of the thermal conductivity of nanostructured materials based on this was presented by Minik and Chen [46]. In [47] Mingo and his colleagues succeeded in performing a simulation of phonon transport in nanotubes with defects using the nonequilibrium Green's function without using adjustable parameters. In ref. [48], the authors' group simulated thermal transport for various phonon spectra using a spatiotemporal model that takes into account the reflection coefficient of thermal radiation.

The above works bring us closer to an understanding of the physical processes accompanying phonon transport in the nanostructured material.

Let us consider the processes of heat transfer in thermoelectric nanomaterials. It should be noted that in the general case the phonon heat capacity can be written in the form:

$$\begin{cases} C_{ph} \sim T^d v_{ph}^d, \text{ if } T < \theta_D \\ C_{ph} = const, \text{ if } T \ll \theta_D \end{cases}, \tag{21}$$

where v_{ph} is the group velocity of phonons, θ_D is the Debye temperature.

The dependence of the thermal conductivity on the temperature is given in the works [37, 38]. The use of layers of materials with a low thermal conductivity to the formation of multilayer nanostructures can lead to a decrease in the thermal conductivity of the nanosystem. There are two approaches for describing the behavior of superlattices. The first one assumes that in the superlattices there are scattering mechanisms, a change in the group velocity, and the density of the number of states. However, it is time to consider phonons as waves [49]. The second approach regards phonons as particles, which have a simple mechanism of reflection from the boundary [50]. Simulation confirms the scattering of phonons at the boundaries. Therefore, phonons can be considered as particles and the Boltzmann equation can be used:

$$\frac{\partial f}{\partial t} + \frac{\partial f}{\partial \mathbf{r}} \cdot \frac{\mathbf{p}}{m^*} + \frac{\partial f}{\partial \mathbf{p}} \cdot \mathbf{F} = \frac{df}{dt}\Big|_{coll}, \tag{22}$$

where $f(\mathbf{r}, \mathbf{p}, t)$ is the probability density function, $\mathbf{r}=(x, y, z)$, $\mathbf{p}=(p_x, p_y, p_z)$, m^* is the mass of particles, $\mathbf{F}(\mathbf{r}, t)$ is the field of forces acting on particles, and the term on the right-hand side of the equation takes into account collisions between the particles.

It should be noted that there are a number of experimental data that indicate that the Fourier law:

$$\vec{q} = -\kappa \nabla T \tag{23}$$

shows a lower thermal conductivity drop in superlattices compared to the real one in comparison with bulk materials.

The origin of thermoelectricity is due to the difference between the Fermi energy and the average energy of conducting electrons or holes. For most homogeneous materials, an increase in electrical conductivity is a consequence of an increase in carrier concentration. The growth of the effective mass of carriers leads to an increase in S, reduces mobility μ and electrical conductivity σ. In accordance with the Wiedemann-Franz law, the increase in electrical conductivity is accompanied by an increase in the electronic component of the thermal conductivity κ_e. A decrease in the thermal conductivity of phonons κ_{ph} with addition of defects reduces mobility μ and electrical conductivity σ

It should be noted that the value of the thermoelectric figure of merit ZT can be increased by using degenerate semiconductors. Reduction of the phonon component of thermal conductivity is possible due to the use of heterogeneous materials.

As noted earlier, an increase in the thermoelectric figure of merit will allow finding the most effective thermoelectric materials. The effectiveness of thermoelectric conversion is

$$\eta = \frac{P}{Q_h} = \eta^* \frac{T_h \sqrt{1+Z\overline{T}}}{T_h \sqrt{1+Z\overline{T}}+T_c},$$

$$\eta^* = \frac{T_h - T_c}{T_h} \tag{24}$$

where η^* is the thermodynamic efficiency of Carnot; P is the output electric power; T_c, T_h are temperatures of cold and hot receivers, respectively; $\overline{T}$ is the average temperature; Q_h is the heat flow.

The method of increasing the thermoelectric figure-of-merit of a bulk nanostructured material in comparison with a single crystal by scattering of phonons and charge carriers at the interfaces is given in Ref. [51]. Two approximations are considered in [51]: the mean free path of phonons and electrons is assumed to be constant; the dependence of the phonon relaxation time on their frequency is taken into account. Then the ratio of the thermoelectric figure of merit of a bulk nanostructured material to the thermoelectric figure of merit of a single crystal in accordance with the summation rule for the reciprocal mean free paths is:

$$\frac{ZT_n}{ZT_m} = \left[\int_0^1 \int_{-1}^1 \frac{\left(\frac{r}{l_e}\right)\sqrt{y^2+2zy+1}y^2 \, dz \, dy}{\left(\frac{r}{l_e}\right)\sqrt{y^2+2zy+1}+1} \right] \left[\int_0^1 \int_{-1}^1 \frac{\left(\frac{r}{l_{ph}}\right)\sqrt{y^2+2zy+1}y^2 \, dz \, dy}{\left(\frac{r}{l_{ph}}\right)\sqrt{y^2+2zy+1}+1} \right]^{-1}, \tag{25}$$

where r is the particle radius; l_e, l_{ph} are the mean free path of electrons and phonons, respectively.

For a second approximation of the above, the ratio of the thermal conductivity of a nanoparticle to the thermal conductivity of a single crystal is expressed as:

$$\frac{\kappa_n}{\kappa_m} = \frac{\frac{3}{2}\int\limits_0^1\int\limits_0^1\int\limits_{-1}^1\frac{z^2x^4\exp\left(\frac{x}{\theta}\right)}{\left[\exp\left(\frac{x}{\theta}\right)-1\right]^2}\left(\frac{\frac{r}{L^*}\sqrt{z^2-2yz+1}}{1+\frac{r}{L^*}Q_{l\parallel}(x)\sqrt{z^2-2yz+1}}+\frac{2\frac{r}{L^*}\sqrt{z^2-2yz+1}}{1+\frac{r}{L^*}Q_{r\parallel}(x)\sqrt{z^2-2yz+1}}\right)dydzdx}{\left(\int\limits_0^1\frac{x^4\exp\left(\frac{x}{\theta}\right)}{\left[\exp\left(\frac{x}{\theta}\right)-1\right]^2}\left[\frac{1}{Q_{l\parallel}(x)}+\frac{2}{Q_{r\parallel}(x)}\right]dx\right)}, \quad (26)$$

where index $\parallel$ refers to the phonon component of the thermal conductivity in the direction parallel to the layers, $\theta = T / \theta D$, θ_D is the Debye temperature; $L^* = \rho\hbar^4 v_\parallel^6 / \gamma^2 (k_B\theta_D)^5$.

Then:

$$\frac{ZT_n}{ZT_m} = \frac{3}{2}\left(\int\limits_0^1\int\limits_{-1}^1\frac{\frac{r}{l_e}y^2\sqrt{y^2+2zy+1}}{\frac{r}{l_e}\sqrt{y^2+2zy+1}+1}dzdy\right)\left(\frac{\kappa_m}{\kappa_n}\right). \quad (27)$$

Let us consider the electrical resistivity, which is another characteristic of nanomaterials associated with thermoelectric processes. The scattering of electrons by phonons, structural defects, and impurities causes the electrical resistivity of metallic solids. For many metal-like nanomaterials (Cu, Pd, Fe, Ni, Ni-P, Fe-Cu-Si-B, NiAl, nitrides, and borides of transition metals, etc.), a decrease in the size of the structural element leads to a significant increase in the resistivity ρ. This phenomenon is due both to the features of the phonon spectrum and to the increased role of defects in nanomaterials. For isotropic materials, the electrical resistivity is inversely proportional to the specific conductivity. However, in the case of anisotropic materials, the relationship between the specific electrical resistance and the conductivity takes the form:

$$J_i(\mathbf{r}) = \sum_{j=1}^3 \sigma_{ij}(\mathbf{r})E_j(\mathbf{r}),$$
$$E_i(\mathbf{r}) = \sum_{j=1}^3 \rho_{ij}(\mathbf{r})J_j(\mathbf{r}) \quad (28)$$

Almost all metal-like nanomaterials are characterized by a large residual electrical resistivity at $T \approx 1-10°K$ and a small value of the temperature coefficient of electrical resistivity. At a nanosystem size of less than

100 nm, there is a significant change in the electrical resistivity. It was shown in [52] that for nanocrystalline materials the resistivity at the grain boundary is practically equal to $\rho^* \sim 3 \times 10-12$ $\Omega \cdot cm$, which corresponds to the value for bulk crystalline materials. Thus, the electrical resistance of a nanomaterial can be calculated by the formula:

$$\rho = \rho_0 + \rho^* \left(\frac{S^*}{V} \right), \tag{29}$$

where ρ_0 is the electrical resistivity of a single-crystal material with a specified content of impurities and defects, V is the volume, S^* is the area of intergranular interfaces.

The electrical resistance of thin films depends on the scattering of electrons by external surfaces, topography, and structural features. The size of the structural element and the thickness of the film normalized to the mean free path exert a large influence on the magnitude of the electrical resistivity.

In semiconductors, a decrease in particle size leads to an increase in the width of the forbidden band to the level of dielectrics (for example, for GaAs). The dependence of the electrical resistivity and the dielectric constant on the size of the structural element can be ambiguous, since many factors influence the properties of semiconductors.

For a non-conducting matrix with metallic nanoparticles, there is a sharp increase in conductivity at a certain percentage of the conductive component, which is due either to a barrier transition or, mainly, to a tunneling (hopping transition).

Considering the growing influence of quantum effects on the conductivity of nanostructures is especially important in the design of such devices as: nanodiodes, nanotransistors, nanoswitches, etc.

11.3   CONCLUSIONS

Despite the known advantages of thermoelectric energy conversion, it has a relatively low efficiency. The efficiency of this transformation can be improved by using thermoelectric nanomaterials. An increase in the value of the Seebeck coefficient and the electrical conductivity of the material along with a decrease in its thermal conductivity will lead to an increase in the value of ZT.

The increase in thermoelectric efficiency in nanostructured thermoelectrics is mainly due to a decrease in the phonon component of the thermal conductivity because of an increase in phonon scattering at the nanoscreen boundaries and structural defects inside the grains.

There are ways to increase the thermoelectric figure of merit of nanomaterials, for example, through tunneling of charge carriers.

Obtaining high-efficiency thermoelectric nanomaterials will significantly expand the prospects for their application in many fields of science and technology, such as telecommunications, space, high-precision weapons, medicine, and cooling systems.

ACKNOWLEDGMENTS

The works was carried out with financial support from the Research Program of the Ural Branch of the Russian Academy of Sciences (project 18–10–1–29).

KEYWORDS

- density of states
- electrical conductivity
- electrical resistivity
- mathematical modeling
- nanostructures
- the Seebeck coefficient
- thermal conductivity
- thermoelectric figure of merit
- thermoelectric properties

REFERENCES

1. Zaiman, J., (1975). *Principles of Solid State Theory*. Moscow: Mir, 472 p.
2. Zaiman, J., (1962). *Electrons and Phonons*. Moscow: IL, 488 p.

3. Landau, L. D. & Lifshitz, E. M. (1982). Electrodynamics of continuous media. *Theoretical Physics* (Vol. 8 of Landau, L. D. & Lifshitz, E. M.). Moscow: Nauka, 621 p.

4. Lifshitz, E. M., Pitaevskii, L. P., (1979). Physical kinetics. Theoretical Physics (Vol. 10 of Landau, L. D. & Lifshitz, E. M.). Moscow: Nauka, 528 p

5. Ashcroft, N., & Mermin, N., (1979). *Solid State Physics*. Moscow: Mir, Vol. 2, 422 p.

6. Nolas, G., Sharp, J., Goldsmid, H., et al., (2001). *Principles of Thermoelectrics: Basics and New Materials Development*. Springer Verlag, Vol. 8, 293 p.

7. *Thermoelectrics Handbook*, (2005). In: Rowe, D., (ed.), *Macro to Nano*. CRC Press. Boca Raton.

8. Goldsmid, H., (1964). *Thermoelectric Refrigeration*. Plenum. New York.

9. MacDonald, D., (1962). *Thermoelectricity: An Introduction to the Principles*. Wiley. New York.

10. Mahan, G., Sales, B., & Sharp, J., (1997). Thermoelectric materials: New approaches to an old problem. *Physica Today*, *50*, 42–47.

11. Abelson, R., (2006). In: Rowe, D. (ed.), *Thermoelectrics Handbook: Macro to Nano* (Vol. 56, pp. 1–29). CRC Press. Boca Raton. FL.

12. Vining, C., Rowe, D., et al., (2005). In: Rowe, D., (ed.), *History of the International Thermoelectric Society, Thermoelectrics Handbook: Macro to Nano* (pp. A1–1, 7). FL USA: CRC Press.

13. Sharp, J., Goldsmid, H., et al., (2001). *Principles of Thermoelectrics: Basics and New Materials Development*, Springer Verlag, 293 p.

14. Vining, C. Z. T., (2007). 3.5: fifteen years of progress and things to come. *Proc. European Conf. on Thermoelectrics, ECT 2007* (pp. 1–6). Odessa. Ukraine.

15. Bell, L., (2008). Cooling, heating, generating power and recovering waste heat with thermoelectric systems. *Science*, *321*, 1457–1461.

16. Bahk, J., & Shakouri, A., (2014). Electron transport engineering by nanostructures for efficient thermoelectric. In: Wang, X., & Wang, Z., (eds.), *Nanoscale Thermoelectrics* (pp. 41–92). Springer. Int. Publ. Switzerland.

17. Chen, G., (2005). *Nanoscale Energy Transport and Conversion*. Oxford. University Press. New York.

18. Ioffe, F. A., (1957). *Semiconductor Thermoelements and Thermoelectric Cooling*. Info Search Limited. London, 192 p.

19. Tritt, T., Bottner, H., et al., (2008). Thermoelectrics: Direct solar thermal energy conversion. *MRS Bulletin.*, *33*(4), 366–368.

20. Mahan, G., Sales, B., & Sharp, J., (1997). Thermoelectric materials: New approaches to an old problem. *Physica Today.*, *50*, 42–47.

21. Dmitriev, A. S., (2015). *Introduction to Nanotempophysics*. Moscow: BINOM. Laboratoriya znaniy, 790 p.

22. Sivukhin, D. V., (2004). *General Course of Physics Stereotyped* (4th edn., Vol. III). Electricity Moscow: Fizmatlit, Publishing house MIPT.

23. Shakouri, A., & Zebarjadi, M., (2009). Nanoengineered materials for thermoelectric energy conversion. Ch. 9. In: Volz, S., (ed.), *Thermal Nanosystems and Nanomaterials, Topics in Applied Physics* (Vol. 118, pp. 225–299).

24. Snyder, G., & Toberer, E., (2008). Complex thermoelectric materials. *Nature Materials*, *7*, 105–114.

25. Majumdar, A., (2004). Thermoelectricity in semiconductor nanostructures. *Science*, *303*, 777–778.
26. Minnich, A., Dresselhau, M., Ren, Z., et al., (2009). Bulk nanostructured thermoelectric materials: current research and future prospects. *Energy Environ. Science*, *2*, 466–479.
27. Harman, T., Walsh, M., et al., (2005). Nanostructured thermoelectric materials. *Journal Electronic Materials*, *34*, L19–L22.
28. Vining, C., Rowe, D., et al., (2005). In: Rowe, D., (ed.), *History of the International Thermoelectric Society, Thermoelectrics Handbook: Macro to Nano* (pp. A1–1:7). Boca Raton. FL USA. CRC Press.
29. Dresselhaus, M. S., Chen, G., Tang, M. Y. et al., (2007). *New Directions for Thermoelectric Materials. Advanced Materials, 19*, 1–12.
30. Kajikawa, T., (2006). Present status and prospect of the development for advanced thermoelectric conversion systems. *Proc. 25ᵗʰ Intern. Conf. Thermoelectrics. ICT2006* (pp. 93–140).
31. Bottner, H., Nurnus, J., Gavrikov, A., et al., (2004). New thermoelectric components using microsystem technologies. *Journal Microelectromechanic Systems*, *13*(3), 414–420.
32. Shakouri, A., (2006). Nanoscale thermal transport and micro refrigerators on a chip. *Proc. of the IEEE. 94*, pp. 1613–1638.
33. Heremans, J., (2007). Nanometer-scale thermoelectric materials. In: Bharat, B., (ed.), *Ch. 12. Handbook Springer of Nanotechnology*. Springer Science: Business Media. Inc., pp. 345–374.
34. Chen, G., (2006). Nanoscale heat transfer and nanostructured thermoelectrics. *IEEE Transactions Comp. & Pack. Technology*, *29*(2), 238–246.
35. Yang, R., & Chen, G. (2006). Nanostructured thermoelectric materials: From superlattices to nanocomposites. *Material Integration*, *18*, 31–36.
36. Minnich, A. J., et al., (2009). Bulk nanostructured thermoelectric materials: Current research and future prospects. *Energy & Environmental Science*, *2*(5), 466–479.
37. Vakhrushev, A. V., Severyukhin, A. V., & Severyukhina, O. Yu., (2017). Calculation of macro-characteristics of nanosystems. Part 1. Coefficient of thermal conductivity of homogeneous nanosystems. *Chemical Physics and Mesoscopy*, *19*(2), 167–181.
38. Vakhrushev, A. V., Severyukhin, A. V., & Severyukhina, O. Yu., (2017). Calculation of macro-characteristics of nanosystems. Part 2. Coefficient of thermal conductivity of multicomponent nanosystems. *Chemical Physics and Mesoscopy*, *19*(4), 538–546.
39. Cutler, M., & Mott, N. F., (1969). Observation of Anderson localization in an electron gas. *Physical Review*, *181*.
40. Wang, X. W., et al., (2008). Enhanced thermoelectric figure of merit in nanostructured n-type silicon germanium bulk alloy. *Applied Physics Letters*, *93*(19), 193121.
41. Joshi, G., Lee, H., Lan, Y., et al., (2008). Enhanced thermoelectric figure-of-merit in nanostructured p-type silicon germanium bulk alloys. *Nanoletters*, *8*(12), 4670–4674.
42. Usenko, A. A., (2016). Investigation of nanostructured thermoelectric materials on the basis of n-type and p-type silicon germanium solid solutions. *Thesis for the Degree of Candidate of Physical and Mathematical Sciences*. Moscow, 153 p.
43. Callaway, J., (1959). Model for lattice thermal conductivity at low temperatures. *Physical Review*, *113*(4), 1046–1051.

44. Steigmeier, E. F., (1964). Scattering of phonons by electrons in germanium-silicon alloys. *Physical Review, 136*(4A), A1149.

45. Swartz, E. T., (1989). Thermal boundary resistance. *Reviews of Modern Physics, 61*(3), 605.

46. Hochbaum, A. I., Chen, R., & Delgado, R. D., (2008). Rough silicon nanowires as high-performance thermoelectric materials. *Nature, 451*, 163–168.

47. Mingo, N., Stewart, D. A., Broido, D. A., & Srivastava, D., (2008). Phonon transmission through defects in carbon nanotubes from first principles. *Physical Review B., 77*(3), 033418.

48. Koh, Y. K., & Cahill, D. G., (2007). Frequency dependence of the thermal conductivity of semiconductor alloys. *Physical Review B., 76*(7), 075207.

49. Liu, W., Ren, Z., & Chen, G., (2013). Nanostructured thermoelectric materials. In: Koumoto, K., & Mori, T., (eds.), *Ch. 11. Thermoelectric Nanomaterials, Materials Design, and Applications.* (pp. 255–285). Springer-Verlag Berlin Heidelberg.

50. Chen, G., (2001). Phonon heat conduction in low-dimensional structures. *Semiconduct. Semimetals, 71*.

51. Gorsky, P. V., & Mikhalchenko, V. P., (2013). To the question of the mechanism for increasing the thermoelectric figure of merit of bulk nanostructured materials. *Thermoelectricity International Scientific Journal, 5*, 5–10.

52. Gleiter, H., et al., (2001). Nanocrystalline materials: A way to solid with tunable electronic structure and properties. *Acta Materialia., 48*, 737–745.

INDEX

A

Absorption spectrum, 258
Acceleration of gravity, 173
Acetaminophen, 232
Acetylcholinesterase (AChE), 219, 220
Acid-base interaction, 214
Acidic conditions, 258, 259
Active pharmaceutical ingredient (API),
  266
Adsorption mechanism, 205
Adsorption sites, 205, 210, 226
Advanced
  manufacturing, 101, 102
  oxidation
    processes (AOPs), 4, 5, 7–13, 15,
      22–24, 28, 33, 38, 42, 43, 54–56,
      76, 79, 80, 83
    techniques, 8, 22
Aerosol gas generator, 166, 185, 188, 199
Agglomerates, 187, 189, 198
Aggregative nanocrystal model, 249
Alcohol, 203, 211, 233
  oxidase, 225
Aldehyde, 203, 233, 268
Algorithms, 167, 170, 176, 186, 198
Alumina
  beads, 155
  substrate, 196
  templates, 196
Aluminum oxide, 166, 193, 199
Ammonia, 138, 203, 208, 233
Amperometric detection, 210
Amperometry, 214
Amphiphilic character, 248
Analyte, 204, 206–208, 216, 222, 223
Analytical device, 206, 222
Anisotropic
  geometry, 155
  materials, 291

Anodic porous alumina, 200
Anticancer agent, 231, 270
Anti-HIV drugs, 271
Antimony, 122, 192, 193
Antithyroid drug methimazole, 139
Antiviral drug acyclovir (ACV), 215
Application of block copolymer micelle, 269
Aqueous medium, 268
Arginine, 139
Ascorbate, 98, 108
Ascorbic acid (AA), 213–215, 250
Atomic
  force microscope, 180
  velocities, 171
Atrazine, 218, 219

B

Barium titanate, 270
Barostats, 170, 177
Barrier resistance, 155
BCC arrangement, 271
Bimetallic nanoparticles, 67, 256
Bioanalysis, 213
Biocompatible polymer chitosan (CHI), 223
Bioimaging, 136, 137, 144, 242
Biological
  activity, 137, 262
  applications, 135, 140
Biomedical
  applications, 137, 252, 258, 262, 269,
    271, 272
  field, 241, 272
  materials, 244
  processing, 265
Biomineralization, 248
Biorestoration, 21
Biosensors, 120, 135, 138, 154, 203, 205,
  219, 220, 222–227, 229, 232, 233
Biosorption treatment technologies, 21

Block copolymers, 242, 244, 258
micelles, 268
ratio, 248
type, 248
Blood circulation time, 267, 268, 270
Blood-brain barrier, 270
Blow molding, 160
Blue phase liquid crystal (BPLC), 131
Boltzmann's
constant, 173, 287
equation, 289
kinetic equation, 286, 288
Borides, 291
Born-Oppenheimer approximation, 169
Borohydride reduction, 250
Bovine serum albumin (BSA), 225
Butadiene rubbers, 160

C

Cadaverine, 208
Cadmium sulfide nanoparticles (CdSNPs), 226
Calcium oxides, 187, 198
Calendaring, 160
Calibration curve, 139, 217, 225
Callaway model, 288
Campylobacter, 228
Cancer
cells, 229–232, 265
chemotherapy, 262
marker, 233
Candida albicans, 137
Carbamazepine, 269
Carbofuran, 219
Carbon
black, 186, 217
dioxide, 12, 42, 178, 187, 188, 198, 199, 218
film electrodes, 225
methyl cellulose (CMC), 259, 265, 266, 269
nanomaterials, 102
nanotubes (CNTs), 103, 152, 154, 155, 160, 161, 203–205, 207, 209–211, 214, 217–220, 222–227, 229, 232, 233
arrays (CNTA), 207, 208, 232

membranes, 233
thin film (CNT-TF), 226
Carrier concentration, 283, 286, 289
Casting, 135, 154, 160, 208, 209, 211, 213
Catalysis, 72, 99, 160, 242
Cationic or nonionic surfactants, 262
Cell
efficiency factor, 232
membrane, 265, 266
Cellulose, 103
Ceramic
materials, 154
matrix nanocomposites, 153
Chemical
engineering science, 27, 28, 30, 32, 38, 39, 51
mechanical
method, 157
techniques, 158
process engineering, 6, 7, 9, 10, 14, 15, 20, 24, 28–32, 39, 48–51, 56, 72, 80, 81
sensors, 154, 203, 206–208, 216, 217, 220–222, 233
synthesis, 107
vapor-deposition (CVD), 207, 216, 224
warfare agents, 216
Chemiresistive detectors, 208
Chemiresistors, 208
Chionanthus retusus, 137
Chitosan, 103, 223
Chloride pentahydrate, 141
Chloroform, 215
Chromatographic separation, 218
Ciprofloxacin, 142
Cisplatin (anticancer agent), 231
Clay minerals, 155
Climate change, 3, 8, 21, 23, 50, 106
Cloud point, 262
Clusters, 189
Cobalt meso-aryl porphyrin complexes, 208
Coefficient of
friction, 173
variation (CV), 212
Colloidal quantum dot (CQDs), 121–123, 125, 126, 132, 139, 141–144
CQDs solar cells (CQDSCs), 121, 122, 125

Columbic attraction, 121
Compression molding, 156, 158, 160
Conductivity, 205, 209, 210, 227, 230,
    284, 286–292
Controlled
    drug delivery, 259, 265, 266, 269, 270
    NOT (CNOT) gate, 127
Conventional manufacturing methods, 159
Cooling systems, 281, 293
Copolymer, 130, 155, 242, 244, 248, 250,
    256–258, 262, 266–269
Copper, 141, 181–183, 197, 198
Coreactant, 138
Coulomb peaks, 127
Coulombic repulsions, 259
Cryptococcus neoformans, 137
Crystalline, 131, 192, 266, 284, 288, 292
    analogs, 284
Crystallinity, 153, 270
Crystallographic facets, 253
Cyclodextrin, 270
Cytotoxicity, 137–140, 144, 232

D

Debye
    model, 288
    temperature, 289
Defects, 156, 210, 284, 288, 290–292
Dehydration, 245, 268
Dehydrogenation, 209
Density
    functional theory (DFT), 141
    states, 284, 287, 293
Deoxyribonucleic acid, 232
Deposition time, 195, 196
Desalination, 8, 14, 15, 22, 23, 36, 41,
    57–59
Dexorubin, 265, 270
Dextran, 103, 224
D-fructose, 223
Diamine, 242, 257, 261, 268
Dichloromethane (DCM), 215
Dielectrics, 292
Diethyl phthalate, 142
Differential pulse voltammetry (DPV),
    213, 232

Dilution, 268, 269
Dimethoxy methyl phosphonate (DMMP),
    215
Direct
    electron transfer (DET), 223
    mixing of polymer and nanofillers, 157,
        159
Direction of motion, 285
Discrete
    evaporation of pock in high vacuum,
        199
    thermal evaporation, 193
Dispersion stabilizer, 244
Dithionite, 21, 60
DNT (2,4-dinitrotoluene), 215, 216
Domoic acid, 225
Dopamine (DA), 213–215, 225
Double-stranded, 232
Double-walled CNTs (DWNTs), 224
Drinking water treatment, 4, 6, 7, 15, 17,
    18, 20, 28–30, 32, 42, 49, 51, 53–58, 60,
    67, 74, 79, 82, 99, 107, 113, 114
Drug
    bioavailability, 266
    carrier system, 270
    carriers, 262, 268
    delivery, 103, 229, 231, 233, 242, 259,
        262, 265–267, 269
    system, 231, 265, 267, 269
    loading, 266
        micelles, 268
    resistance, 266
Dynamic light scattering (DLS), 267

E

E. coli, 226–229
Ease of integration, 217, 221
Ecological biodiversity, 3, 4, 8, 21, 23
Electric current, 282
Electrical
    cell impedance sensing biosensor, 232
    conductivity, 152, 160, 230, 281, 283,
        284, 286, 287, 289, 290, 292, 293
    devices, 160
    resistance, 291, 292
    resistivity, 291–293

Electricity, 5, 43, 56, 81, 121, 123, 283
Electrocatalysis, 221
Electrocatalytic effects, 225
Electrochemical
  bio-sensor (ECB), 222
  sensing, 213
Electrodialysis, 14, 74
Electrogenerated chemiluminescence
  (ECL), 135, 137, 138
Electromagnetic interference shielding,
  160
Electron
  acceptor, 143
  mass, 170
  repulsion energy, 170
  transfer, 219, 223
  transport layer (ETL), 122, 123
Electrons, 120, 122, 125, 127, 135,
  142–144, 169, 170, 204, 232, 284, 285,
  289–292
Electrospun fibers, 152
Electrostatic dissipation, 160
Empirical parameter, 174
Energy
  acceptor, 138
  engineering, 69, 102
Environmental
  engineering, 3–11, 13–16, 19–24,
    28–31, 33, 34, 37–43, 46, 48–51,
    54–56, 60, 62, 72–74, 79–81, 83,
    94–96, 98, 99, 102, 104, 107,
    111–115
  protection, 5, 6, 8, 9, 11, 12, 14, 16, 17,
    19, 21, 23, 27–30, 31, 33, 34, 36–38,
    40, 42, 45–51, 54, 55, 57, 59, 60,
    62–64, 72, 74, 76, 80–83, 94–96, 99,
    100, 107, 112, 115
  sustainability, 11, 14, 16–19, 28, 32, 33,
    43, 45, 47, 48, 54, 56, 58, 60, 64, 65,
    100, 109, 111, 112
Enzyme-linked immunosorbent (ELISA),
  225, 228, 229, 231
Epidermal growth factor (EGF), 231
  EGF receptors, 231
Epitaxy processes, 197
Epoxy-polyurethane (EPU), 225
Escherichia coli, 226, 227, 228

Ethylene glycol dimethacrylate (EGDMA),
  221
Ethylenediamine central group, 256
External force, 171
Extrusion molding, 160

F

Fabrication, 75, 129, 131, 135, 136, 144,
  151, 154, 157, 159, 161, 167, 208, 216,
  222, 223, 225, 242
  polymer matrix nanocomposite, 157
Faraday efficiency, 126
Fermi
  Dirac distribution, 287
  energy, 127, 287, 289
  level, 284
Field effect transistor (FET), 160, 209,
  222, 223, 225, 227, 230, 233
Fire extinguishing, 185–190, 199
  nanoaerosol, 187, 200
  systems, 166
Flame retardancy, 155
Fluorescein isothiocyanate (FITC), 136
Fluorescence lifetime imaging microscopy,
  136
Fluoride, 65, 223
Förster resonance energy transfer (FRET),
  136
Fringing field switching (FFS), 131
Fuel, 4, 27, 28, 30, 54, 81, 82, 94, 99, 106,
  113, 160, 180, 186

G

Gallium, 190–193, 195, 196
  arsenide (GaAs), 122, 126, 133, 292
  atoms, 195
  nanostructure, 195
Gas
  air mixture, 185, 187
  fire-extinguishing nanoaerosol gener-
    ator, 200
  sensors, 160, 207, 209, 210, 212, 214,
    216
Gaseous medium, 165, 168, 169, 176–181,
  187, 188, 198
Gas-phase sensor, 204

Gene
  delivery systems, 265
  therapy, 270
  transfer technologies, 270
Genosensor, 226
Geometrical model, 254
Glass carbon electrode (GCE), 213, 215,
  217, 220, 221, 224, 227, 232
Glassy carbon (GC), 219
Global
  climate change, 4, 22, 27–29, 38, 40, 46,
    50, 54, 55, 81, 82, 94, 111, 113
  environment, 47, 70, 81
  warming, 27, 38, 49, 50, 51, 65, 94, 96,
    111, 113
  water
    crisis, 4, 7, 24, 76
    shortage, 4, 5
Glucose oxidase (GOx), 223–225
Glycidyl methacrylate (GMA), 221
Gold, 102, 138, 139, 183, 190–192, 195,
  198, 225–228, 241, 242, 244–246,
  248–252, 254, 256–262, 272
  nanoparticles (AuNPs), 138, 139, 192,
    241, 242, 248, 250, 257, 259, 262,
    264, 272
  nanoplates, 250, 258
  particles, 260, 261
Grains, 293
Graphene, 135, 166, 207, 209, 217
  oxide, 207, 217
  quantum dots (GQDs), 125, 134, 136,
    139, 140, 142, 143
  polyethyleneimines (GQDPEIs), 134
Graphical model, 243
Gravity forces, 173
Green
  chemistry, 93, 95, 96, 98, 99, 101–103,
    106–109, 113–115
  engineering, 94, 96, 98, 100–103, 106,
    109, 113, 114
  function, 288
  innovation, 109
  nanomaterials, 93, 95, 96, 102, 107, 110,
    113
  nanotechnology, 85, 89, 93–111,
    113–115

synthesis, 107
technology, 106
Groundbreaking, 5, 29, 30, 31, 41, 45, 48,
  51, 56, 58, 71, 72, 80, 101, 104, 110

H

Hamiltonian (potential/kinetic energies),
  169
Head/neck squamous carcinoma cells, 231
Hectorite, 155
Heterogeneous materials, 290
Hexadecyltrimethylammonium bromide,
  132
Hexagonal NPs, 253
Higher stability, 217, 221
High-precision weapons, 293
Homogeneity, 179, 182, 183, 198
Homogeneous conductor, 283
Human
  civilization, 4–8, 10, 13, 14, 16–18, 21,
    24, 27–29, 31–44, 47–51, 53–58, 61,
    62, 64, 66, 68, 70, 72, 74, 79–82,
    93–96, 98–101, 106, 108, 110–114
  growth hormone, 265
  health, 86, 89, 105, 108, 136
  mankind, 3, 6, 8, 9, 14, 17–19, 21, 23,
    28, 39, 44, 46, 54, 56, 59, 79, 82, 98,
    103, 111, 113
  scientific endeavor, 3, 6, 14, 18, 21, 23,
    33, 43, 50, 53, 59, 97, 110, 113
Hydrochlorothiazide, 269
Hydrogen peroxide, 11, 12, 79, 80, 83,
  140, 212
Hydrogenation, 209
Hydrophilic lipophilic balance (HLB),
  266, 271
Hydrophilicity, 244, 267
Hydrophobic
  compounds, 265
  drug, 267
  interactions, 250, 258, 262
Hydrothermal process, 141, 143
Hydroxylamine hydrochloride, 210, 211
Hylocereus undatus, 138
Hypolipidemic drug, 267

I

Icosahedral gold nanoparticles, 256
Icosahedron
  particles, 254
  shapes, 256
ICs development, 86
Immunoglobulin G (IgG), 222
Impedance, 211, 226, 232
Imprinted conducting polymer (MICP), 223
In situ
  chelate flushing, 21, 60
  photoreductions, 143
  polymerization, 156, 157, 158
  soil flushing, 21, 60
  treatment, 21, 60
Indium, 190, 192, 213
  phosphide (InP), 130
Indomethacin, 267
Industrial wastewater treatment, 4, 5, 7,
  8, 10, 12, 13, 15, 17–19, 22, 28–30, 32,
  33, 36, 42, 51–58, 60, 67, 74, 79, 82, 99,
  107, 114
Injection molding, 156, 158, 160
Intercalation method, 157
Interdigitated electrodes (IDEs), 207, 208
Interfaces, 72, 121, 131, 284, 288, 290, 292
Interleukin-2, 265
Interleukin-6, 230
Internal electric field, 282
International Business Machines (IBM),
  87, 98, 104
Ion selective membrane, 217
Ionic
  liquid 1-butyl-3-methylimidazolium
    hexafluorophosphate, 232
  strength, 267, 268
Ion-selective electrodes (ISEs), 217
Iron
  based technologies, 21, 60
  deposition, 195
Isoelectric point, 139
Isotropic materials, 291

K

Kinetic energy, 170

L

Lamer mechanism, 248
Langmuir–Blodgett (LB) films, 211
Laser, 119, 144, 167, 209
L-cysteine, 141
Lead sulfide (PbS), 122, 125, 134, 228
Le-Chatelier's principle, 257
Leukemia, 229, 231
Ligand and metal charge transfer (LMCT),
  246, 261
Light emitting
  devices, 130
  diodes (LEDs), 119, 120, 129, 130, 144,
    160, 193
Liquid
  crystal displays (LCDs), 129, 130
  phase
    exfoliation (LPE) method, 132
  sensor, 204
Long-persistence phosphor, 123
Lorentz number, 286
Low dimensional systems, 281, 282
Low efficiency, 222, 292
Luminescent
  properties of zinc sulphide, 200
  solar concentrators (LSCs), 124
Lymphoblastoid–929, 138

M

Magnesium, 178, 181, 187, 198
Manganese sulfide, 197
Mangifera indica, 136
Material science, 71, 102, 141
Mathematical model/modeling, 165,
  167–169, 177, 188, 189, 192, 199, 293
Medicine, 69, 101, 102, 135, 166, 293
Melt
  compounding method, 159
  intercalation method, 156, 158
Membrane science, 4, 5, 7, 8, 10, 14, 15,
  22–24, 28–33, 36–38, 43, 48, 50, 55, 74,
  76, 81, 82, 113, 114
Mesodynamics, 168, 171, 173–176, 179,
  186, 197, 199
  mathematical model, 186

particles, 168, 171, 173, 174, 176, 179, 186, 197, 199
Mesoporous materials, 244
Mesoscopic titanium dioxide, 123
Metallic nanoparticles, 292
Metallo-supramolecular polymer (MSP), 210
Metal-matrix nanocomposites, 154
Methacrylic acid (MAA), 221
Methanol, 211, 215
Methidathion, 218, 219
Micellar, 241, 244, 259, 262, 266–269, 271
    assemblies, 262, 271
Micellization process, 244, 265
Microcomposites, 161
Microfiltration, 14, 74
Mineral fertilizers, 180
Moisture, 188, 203, 208, 233
Molar ratio, 253, 258
Molecular
    beam epitaxy, 166, 190–192, 199
    dynamics, 168, 170, 171, 173, 186, 197, 199, 218, 288
    weight, 158, 244, 256, 267
Molecularly imprinted polymer (MIP), 221, 223
Molybdenum disulfide, 216
Monte Carlo method, 288
Montmorillonite, 155
Motion of nanoparticles, 172
Mott
    approach, 287
    ratio, 287
Multidomain vertical alignment (MVA), 131
Multidrug resistant (MDR), 265, 267
Multilevel mathematical model, 165, 168, 176, 197
Multiwalled carbon nanotubes (MWCNTs), 207, 209, 210, 212–215, 218, 219, 221, 223, 225, 227, 228, 232, 233

N

Nafion composite film, 215
Nanoaerosol extinguishing generator, 186
Nanocells, 194

Nanoclusters, 198
Nanocoatings, 69, 83
Nanocomposites (NCs), 123, 130, 132, 134, 135, 139, 140, 151–157, 159–161, 167, 176, 192, 193, 198, 217, 220, 229
Nanocrystalline materials, 292
Nanocrystallization, 266
Nanoelectronics, 242
Nanoelements, 165, 166, 170, 174, 179–183, 194, 197–199
Nanofabrication, 89
Nanofillers, 157–159
Nanofilms, 166, 167, 170, 176, 190–200
Nanofiltration, 4, 6, 9, 10, 14, 15, 22–24, 27–33, 36–38, 50, 51, 55, 56, 67, 74, 75, 76, 81, 82, 104
Nanolayer coatings, 190, 194, 199
Nanomaterials, 35, 36, 49, 54, 65–67, 69, 70, 71, 83, 86, 89, 93–111, 113, 115, 140, 152, 157, 161, 166, 167, 192, 211, 271, 272, 281, 291–293
Nanometer-diameter cylinders, 152
Nanoobjects, 97, 165, 167, 169, 171, 175, 176, 179–181, 183, 184, 186, 192, 198
Nanoparticles, 67, 69, 97, 102, 103, 108, 120, 122, 130, 134, 135, 140, 143, 144, 152–154, 156–159, 165, 171–176, 178, 180–183, 189, 190, 192, 195, 197, 198, 200, 209–213, 221, 226, 229, 241–253, 256–259, 262, 265, 267–271, 284
Nanophase, 161
Nanoplatelets, 157, 158
Nanoprisms, 248
Nanoproduct, 89
Nanoscale systems, 284, 287
Nanoscience, 16, 27, 34, 35, 49, 50, 70, 71, 81, 94, 95, 97, 100, 101, 106–108, 110, 113, 241
Nanoscreen boundaries, 293
Nanosensor, 218
Nanostructured thermoelectrics, 284, 293
    materials, 284, 287
Nanostructures, 72, 124, 127, 140, 151, 165–168, 170–177, 179–190, 192–194, 196–200, 220, 248, 250, 253, 283, 287–289, 292, 293

Nanosystem, 169, 170, 171, 173, 175, 178, 179, 181, 182, 185, 188, 194, 198, 199, 284, 289, 291
Nanotechnology, 15, 16, 24, 27–40, 48–51, 53–57, 65–72, 81–83, 85, 86, 89, 93–98, 100–115, 151, 152, 160, 167, 168, 242, 281
Nanotest 600, 189
Nanotubes, 49, 67, 151, 152, 154, 194, 203, 204, 207, 229, 288
Nanowire (NW), 122, 126, 129, 131, 154, 194, 208, 285
Natural organic matter, 9, 31, 51
Near infrared (NIR), 121, 133, 137
Networked ZnO nanowires, 208
Nickel oxide (NiO), 125
Nimesulide drug, 269
Nitrides, 291
Nitrogen, 138, 178, 187, 198, 199, 216, 259
    doped carbon dots (N-CDs), 137–139
Nitrotoluene, 209
N-methylphenethylamine (NMPEA), 214
Non-toxic silicon quantum dots, 136
Non-traditional environmental engineering tools, 7, 23, 24, 37, 40, 49, 55, 114
Novel
    nanoscale structures, 133
    separation processes, 10, 22, 29, 30, 33, 38, 40, 42, 43, 48, 49, 54–56, 74–76, 83, 113, 114
Nuclear repulsion energy, 170

O

On/off ratio, 133, 214
One-dimensional, 151, 154, 204, 207, 209, 285, 286
Open circuit voltage, 123, 125
Optical, 35, 69, 103, 121, 124, 125, 127–130, 134–136, 139–141, 153, 155, 160, 186, 189, 193, 196, 211, 270
Organic light-emitting diodes (OLEDs), 129
Organophosphorus (OP), 219
    hydrolase (OPH), 219
Ostwald ripening, 248, 249, 258

Oxidative biological reactions, 213
Oxidizer, 186
Oxidizing agent, 244
Oxygen, 11, 12, 178, 187, 188, 194, 198, 199, 224
Ozone, 9, 11, 12, 42, 79, 83

P

Palladium, 195, 209
Parenteral system, 262
Pearson
    criterion, 183, 185
    statistics, 185
Peltier coefficient, 282, 283
pH, 37, 43, 126, 136, 139, 214, 217, 219, 220, 223, 225, 256–263, 267–269, 271
Phagocytic scavenger cells, 265
Pharmaceutical
    ingredients, 244, 262
    personal care products (PPCP), 13, 24
Pharmacological treatment, 215
Pharmokinetic properties, 270
Phonons, 284, 287–291
    component, 290, 291, 293
    scattering, 284, 293
    spectrum, 291
Photocatalyst, 119, 140–144
Photoconductors, 160
Photodetectors, 119, 120, 127, 130, 131, 133–135, 144, 193
Photoluminescent properties, 193
Photosensitizer, 125, 143
Photothermal therapy, 242
Photovoltaic (PV), 87, 89, 119–125, 144, 160, 190, 192, 199
    films, 190, 199
PH-responsive drug delivery vehicles, 259
Physiochemical
    properties, 266
    stability, 262
Pierson's hypothesis, 185
Pitard metal, 267
Planck's constant, 175
Platelet factor-4, 230
Platinum, 123, 195, 212

Pluronics, 244, 250, 252, 253, 255–259, 265–272
  copolymers, 255
  F-127, 251, 252, 256
  P105, 257
P-nitrophenol, 142, 219, 220
Pollution
  control, 29, 48, 54, 55, 57, 89, 111, 113
  free environment, 86
Poloxamers, 242–244, 256, 262, 265–267, 271
  comparison with poloxamines, 256
  micelles, 265
  poloxamer-407, 265, 266
Poloxamine, 241–243, 259, 262, 265, 267, 271
  poloxamine-1508, 265
  poloxamine-908, 265
Poly(3,4-ethylene dioxythiophene), 215, 216
Poly(brilliant cresyl blue), 223
Poly(diallyldimethylammonium chloride) (PDDA), 219, 220
Poly(neutral red), 223
Poly(vinyl chloride) (PVC), 217
Poly(vinyl pyrrolidone) (PVP), 132
Poly[3-(6-carboxyhexyl)thiophene–2,5-diyl)] (P3CT), 214
Polyaniline (PANI), 208, 215
Polyazine redox polymer, 223
Polycaprolactone (PCL), 269
Polycrystals, 284
Polyethylene oxide (PEO), 242–246, 250, 256, 257, 259, 265, 267–271
  blocks, 244, 259, 270
  corona, 257
Polyethylene terephthalate (PET), 216
Polyethyleneimine, 140, 228
Polymer
  matrix, 155–159, 161, 223
  nanocomposites, 154, 155, 157, 159–161
    applications, 160
    fabrication, 157
    preparation, 156
    processing, 155
    properties, 156, 157

Polymeric
  micelles, 271
  stabilizers, 266
Polymerization reactions, 157, 159
Polymethylmethacrylate (PMMA) aid transfer, 135
Polynanocomposites, 151, 155, 160
Polypropylene oxide (PPO)
  block, 257, 262, 265, 266
  segment, 266
Porphyrinato ligand, 208
Potassium
  carbonate, 178, 181, 187, 198
  nitrate, 186
Power conversion efficiency (PCE), 121–124
Processing techniques, 38, 151, 161
Propoxur, 218, 219
Prostate specific
  antigen (PSA), 230
  membrane antigen (PSMA), 230
Pseudocrown cavities, 257

Q

Quantum
  bits, 126, 128
  computers, 119, 126, 144
  dots (QDs), 119–136, 138–142, 144, 192, 193, 285, 286
    sensitized solar cells (QDSSCs), 124
  spins, 127
  information processing, 127
  mechanics, 168, 170, 178, 197, 199
  yield (QY), 123, 130, 137, 140
Quartz crystal microbalance (QCM), 211, 212
Qubits, 121, 126–129
  dynamics, 128
Quercetin (QN), 267, 268

R

Rabbits, 265
Raw materials, 75, 109, 166, 198
Reduced graphene oxide (RGO), 134, 135, 141, 142
Reduction process, 257, 259

Reference electrodes, 217
Reinforcing phase, 152, 153, 154
Relaxation time, 286, 290
Reliability, 177, 217, 220, 221
Renal excretion, 266
Reproducibility, 129, 215, 217, 219–221,
    225, 232
Reverse osmosis, 4, 6, 9, 10, 14, 22–24,
    32, 55, 56, 74, 76
Rhodamine B, 144
Risk evaluation processes, 68
Rotational molding, 160

S

Salmonella, 227, 228
Saponite, 155
Schrödinger equation, 169
Science and technology, 3–6, 8–10, 19, 22,
    28, 30, 33, 36, 40, 41, 43, 50, 53, 56, 57,
    66, 69–71, 95, 106, 111–114, 131, 293
Seebeck, 281– 283, 287, 292, 293
    coefficients, 281– 283, 287, 292, 293
    effect, 282
Seed growth method, 242, 248, 250
    soft template approach, 250
Seed-mediated growth processes, 248
Self-assembly, 72, 135, 170
Self-organization, 170
Semiconductors, 69, 86, 131, 196, 210,
    283, 287, 290, 292
Sensing, 69, 131, 135–137, 139, 140, 154,
    204, 208–212, 216, 217, 220, 221, 223,
    226, 229, 230, 242
Sensor development, 265
Shigella sonnei, 227
Short-circuit current density, 123
Side effects, 262
Silica optical fiber (SOF), 211, 212
Silicon
    dioxide, 218
    nanowire (SiNW)-based photodetector,
    134
    quantum dots (SiQDs), 136–138
Silver
    copper nanostructures, 182
    zinc nanoclusters, 181, 182

Single-walled carbon nanotubes
    (SWCNTs), 207–216, 218, 220, 223,
    224, 226–231, 233
Sodium
    borohydride, 98, 108, 139
    citrate, 98, 108
    dodecyl sulfate (SDS), 215
Software package, 165, 176, 198
Soil washing, 21, 60
Solar cells, 119–125, 130, 135, 144, 190
Sol-gel
    method, 142, 157, 159
    process, 156
    transitions, 262
Solid
    nanoparticles, 159
    phase sensor, 204
    state circuits, 86
Solubilization, 268, 269
Solution
    intercalation, 158
    mixing, 159
Solvent method, 159
Sonochemistry, 22
Sourirajan-Loeb synthetic membrane, 74
Space, 9, 16, 50, 51, 96, 128, 166, 175,
    210, 269, 293
Spherical or polyhedral morphology, 262
Square wave voltammetry, 215
Stabilizers, 266
Staphylococcus aureus, 226–229
Statistical analysis, 166
Steric repulsions, 265
Stimuli-responsive polymers, 262
Streptococcus pyogenes, 227
Structural
    defects, 291, 293
    formula, 243
Structured reducing agents, 245
Styrene, 155, 160
Successive ionic layer adsorption, and
    reaction (SILAR) method, 124, 125
Sulfonated poly(ether sulfone) (SPES), 213
Sulfur-doped graphene oxide quantum dots
    (S-GOQDs), 143
Supercapacitors, 135, 160
Superconductor devices, 160

Superlattices, 289
Surface plasmon resonance (SPR), 139, 143, 267
Surfactant stabilizer, 266
Sustainability, 4, 8, 9, 11, 14–19, 28, 29, 32, 33, 41, 43–48, 54–60, 64–66, 68, 69, 72, 74, 80–82, 86, 89, 93, 94, 96–100, 105, 107, 109–112, 115
Synperonics, 244

T

T904 monomer, 263
Telecommunications, 293
Temperature, 11, 122, 123, 132, 133, 137, 139, 154, 156–159, 167, 173, 181, 185, 186, 192, 195, 197, 199, 207, 208, 209, 211, 212, 216–218, 226, 244–246, 248, 256, 259, 261, 262, 265, 267, 269, 270, 281–283, 289–291
  coefficient, 291
Template
  directed patterning, 248
  synthesis, 156, 158
Tetragonal rutile-like structure, 141
Tetronics, 256, 257, 259, 262, 267, 268, 270–272
  T904, 259, 261, 267
Theoretical basis, 11, 281
Therapeutic reagents, 266
Thermal
  conductivity, 152, 153, 281, 283, 284, 286–293
  energy, 282, 283
  evaporation, 165, 166, 198, 199
    metals, 199
  printing, 265
Thermodynamic
  control, 259
  efficiency, 290
  parameter, 10, 198, 269
  stability, 256, 268
Thermoelectric, 193, 281–284, 286–288, 290–293
  coefficients, 283
  conversion, 281, 290
  converters, 193

efficiency, 293
energy conversion, 292
figure of merit, 281, 283, 284, 286, 287, 290, 293
nanomaterials, 281, 284, 288, 292, 293
properties, 281, 284, 293
refrigerators, 283
Thermoelectricity, 69, 281–283, 289
  nanoscale, 283
Thermo-emf coefficient, 284
Thermoforming, 160
Thermoresponsive
  behavior, 262, 265, 268
  nature, 241, 272
Thermostats, 170, 177
Thin-film-transistor (TFT), 214, 215
Thomson
  coefficients, 283
  effect, 283
Three-dimensional nanosystems, 285
Time efficiency factor, 232
Tissue detritus, 266
TNP (1,3,6-trinitropyrene), 143
Toluene, 211
Top-down approach, 157, 159
Transmission electron microscopy (TEM)
  images, 221, 252–254, 258, 260, 261
Transparent conductive coatings, 160
Treatment of cancer cells, 270
Tri-n-butyl phosphate (TBP)
  micelles, 245
  monomers, 245
Triangular NPs, 253
Triblock polymers, 245
Trojan horses, 271
Truncated triangular NPs, 253
Tunneling, 126, 127, 284, 292, 293
Two dimension nuclear overhauser effect
  spectroscopy, 267
Two-dimensional, 151, 155, 285
Two-phase reactions, 248
Tyr electrocatalytic characterization, 219

U

Ultrafiltration, 14, 74–76
Ultrasonication, 135, 213

Ultraviolet (UV), 11, 12, 79, 80, 120, 121, 125, 131–134, 208, 214, 215, 267, 269
light, 83
radiation, 11, 215
visible spectrophotometer, 267
Umbrella, 270
Urease, 265
Uric acid (UA), 213–215

V

Vacuum filtration method, 216
Visualization, 176, 177, 198
Volatile chemical compounds (VOC), 214, 125, 214

W

Waste gas emissions, 109
Water
filtration, 105
purification, 3–8, 10–12, 14–20, 22–24, 30, 32, 33, 36, 40, 42, 53–60, 62, 63, 65–68, 71, 74, 80–82, 100, 110, 111, 113–115, 152

quality deterioration, 65
treatment, 7, 11–15, 17, 18, 22, 30, 32, 33, 36, 38, 41, 55, 57, 58, 70, 76, 79, 80, 82, 113, 114
Wave
function, 127, 169, 170
spectrum, 287
Wear resistance, 155
Wiedemann-Franz law, 286, 289

X

X-ray
diffraction (XRD), 138, 140
photoelectron spectroscopy (XPS), 141
X-shape structure, 257

Z

Zinc
oxide network, 208
sulfide, 196, 197, 199, 200
Zwitter ionic surfactant, 248, 249